Probability Fundamentals

AN INTUITIVE GUIDE
TO UNDERSTANDING UNCERTAINTY

Jim Frost

Statistics By Jim Publishing
STATE COLLEGE, PENNSYLVANIA
U.S.A.

Published by: Statistics By Jim Publishing

State College, PA 16801

To contact the author, please email: jim@statisticsbyjim.com.

Visit the author's website at statisticsbyjim.com.

Ordering Information: Quantity sales. Special discounts are available on quantity purchases by educators. For details, contact the email address above.

Probability Fundamentals: An Intuitive Guide to Understanding Uncertainty / Jim Frost. —1st ed.

ISBN 979-8-9911935-9-7

Contents

INTRODUCTION

Accepting Uncertainty

My book teaches you how to understand and use probability in a practical way. You are not just going to memorize formulas. Instead, you'll learn what probabilities mean, know when to use them, and how to interpret the results correctly. By the end, you should feel comfortable working through probability problems and applying these ideas to real-world situations.

I do cover the underlying formulas and manual calculations throughout the book. Understanding how the calculations work helps remove the "black box" aspect from probability and gives you a deeper understanding of the concepts. It also prepares you for classes, homework, and exams where you might need to solve problems by hand before using software tools.

The material is roughly equivalent to an introductory college-level probability course. However, I intentionally keep the math simple. Most of what you'll need is basic arithmetic, addition, subtraction, multiplication, and some straightforward algebra. In a few places, I briefly touch on ideas from calculus, such as integration, but you won't need to perform those calculations. Software and a statistical table handle those details so you can focus on understanding.

I start with the foundations by using contingency tables to make probability concrete. This approach lets you see probabilities as proportions based on real data before moving into more abstract ideas. From there, I introduce how to define events and represent probabilities. Next, I cover the core probability rules and show how to apply them in realistic scenarios. You'll learn how to interpret and compare risks, work with probability tables and diagrams, and use tools like Bayes' Theorem to update probabilities.

Later chapters introduce counting methods, probability distributions, and models for discrete and continuous data, including the normal distribution. The final chapters connect probability to sampling, uncertainty, and the logic behind statistical inference.

All chapters contain practice problems. You can download the fully worked solutions for all problems for free from my website.

Free Resources for this Book

I provide free resources on my website for this book, including online statistical calculators and the practice problem solutions. We'll use the calculators throughout the book to perform calculations, visualize results, and give you extra practice so you can learn by doing. I also provide links to the free simulation software I use, Statistics101, and the scripts I wrote for some of the examples.

To obtain these files, go to the URL or scan the QR code:
https://statisticsbyjim.com/probability-fundamentals

CHAPTER 1

What is Probability?

Life is full of uncertainty. You can't know for sure if it will rain tomorrow, whether your favorite team will win, or if a package will arrive on time. These events all involve randomness, where you can't predict the outcomes with absolute certainty. Before performing any calculations, it helps to reflect on how we handle uncertainty in everyday life.

What are the chances of *that* occurring?! Have you ever asked yourself that after an unusual occurrence? You can use probabilities in many facets of your personal life. What are the chances of winning the lottery or being in a car accident? Are you more likely to be hit by lightning or win the lottery? Does wearing a seatbelt change the probability of being injured? How likely is it that you'll become pregnant?

People naturally form expectations about what *might* happen. "It's probably going to rain," or "There's a good chance the bus will be late." These statements are informal ways of expressing probability and show that you already use probabilistic thinking, even if not in a mathematical way.

Our everyday intuition about chance is useful but often unreliable. We tend to see patterns where none exist and treat independent events as if they're dependent. We overestimate rare events, underestimate common ones, and often mistake coincidence for cause. Our brains evolved to make quick judgments, not to calculate odds.

Humans struggle with probability. Our brains lean on shortcuts that work well in everyday life but fail badly when uncertainty gets complicated. These errors are natural, but they can lead to confident conclusions that are completely wrong. We are great at spotting patterns but terrible at judging how likely those patterns really are.

Psychologists Daniel Kahneman and Amos Tversky conducted pioneering research showing that humans consistently make systematic errors when reasoning under uncertainty. Working within the broader field of cognitive psychology and decision-making, they demonstrated that people rely on mental shortcuts, called heuristics, that often produce biased probability judgments.

Their work revealed that humans routinely misjudge randomness, overreact to vivid events, ignore base rates, and see patterns where none exist. These findings helped establish the modern scientific understanding that intuition alone is often unreliable for judging probabilities and making decisions.

One of the goals of this book is to show where intuition breaks down and why formal probability rules are essential. As you work through the chapters, you'll see examples where our built-in instincts mislead us and how the tools of probability provide a clearer, more reliable way to reason under uncertainty.

Humans aren't naturally good at judging probabilities.

That's why probability theory exists. Formalizing our thinking about uncertainty is crucial because humans struggle to judge probabilities accurately.

Probability theory offers a structured way to quantify uncertainty, moving us beyond intuition and guesswork. It translates vague feelings of "probably" and "maybe" into measurable quantities, providing a logical framework for decision-making. It doesn't remove randomness, but it helps you make sense of it and use it to make decisions.

Famous probability puzzles reveal how intuition often misleads. The Birthday Problem and Monty Hall Problem demonstrate surprising, counterintuitive results. The Two-Child Problem shows how the way we frame information can completely change the conclusions we draw. Understanding probability helps us interpret real-world events more accurately, such as understanding medical test results. It also helps us make sense of confusing scenarios, such as the False Positive Paradox, where highly accurate tests can yield mostly false positives under certain conditions!

These problems aren't just clever tricks. They reveal how our reasoning fails when we rely on instinct instead of math. This book covers all those probability problems and helps you make sense of them.

Probability theory gives us the tools to see through those errors. Whether you're judging the reliability of a medical test, deciding which investment carries more risk, or evaluating a scientific result, understanding probability helps you make better choices.

Everyday Examples of Probability

Probability plays a role in daily life, often in ways you might not notice. Consider these examples:

- **Weather forecasts:** A 70% chance of rain means it rains in 70 out of 100 similar weather situations.

- **Sports predictions:** When analysts say one team has a 60% chance of winning, they mean that outcome would happen most often if the game were played many times.
- **Medical testing:** A doctor might say there's a 5% chance of a false positive, meaning it rarely happens, but it can.
- **Traffic patterns:** Apps like Google Maps use probabilities to estimate travel times under changing conditions.
- **Lottery games:** You might hear the odds of winning are one in a million. That's a probability too.

Let's revisit one of those examples, sports predictions. Suppose a basketball team has a 60% chance of winning tonight's game. That doesn't mean they *will* win this exact matchup. It means that if we could replay this game many times under similar conditions, we'd expect the team to win roughly 60% of the time. It indicates they are slightly more likely to win than not.

Risks are the chances of adverse events, and modeling them is crucial for planning. Actuarial scientists and financial analysts need to understand the likelihood associated with risks to plan for them. Governments use probabilities to know how likely adverse events are to occur and to plan accordingly. How often do catastrophic floods or hurricanes happen in a particular area? What is the likelihood of flood water exceeding a particular level?

Manufacturers need to understand the probability of their products' failure over time to avoid unhappy customers and determine the length of their warranties. Have you ever had a warranty expire just before a product failed? That's no coincidence! Unfortunately for gamblers, casinos use probabilities to ensure they'll make profits. The house always wins in the long run!

Statistical hypothesis testing uses probabilities to help you evaluate hypotheses relevant to your study. P-values are a well-known type of likelihood, and they allow you to determine whether your results are statistically significant. Is the likelihood of contracting the flu lower if you are vaccinated? Probabilities are an integral part of experiments and statistical analyses.

Each example reflects uncertainty about whether an event will happen. Probability captures that uncertainty as a number. The higher the probability, the more likely the event. The lower it is, the less likely it is.

In everyday speech, people use words like *likelihood, chances, odds, percentages,* and *proportions*. Each describes the same basic idea of how frequently an outcome occurs in the long run.

Even though these terms sound different, they all reflect the same concept: probability describes the relative frequency of an event over repeated trials. You'll learn the precise mathematical relationships among them later, but for now, think of them as different ways of expressing the same idea.

The definition of probability is the likelihood that an event will occur. Probability theory analyzes the chances of events occurring. You can think of probabilities as being the following:

- The long-term proportion of times an event occurs during a random process.
- The propensity for a particular outcome to occur.

The most common ways to express probability are:

- **Percentages:** "There's a 60% chance of success."
- **Odds:** "The odds are three to two in favor."
- **Proportions:** "The probability is 0.6."

They all describe the same concept, but proportion is the most direct mathematical form. Percentages and odds are just different ways of expressing the same thing. You'll learn how to convert between them later in the book.

Probability as a Proportion

Probability is most often expressed as a proportion, a number between 0 and 1.

- A probability of **0** means the event cannot happen.
- A probability of **1** means the event is certain to occur.

Values between 0 and 1 denote uncertainty over whether the event will occur. As the probability increases, the event becomes more likely. The middle value of 0.5 signifies that the event is equally likely to happen or not. In a coin flip, the probability of heads occurring (0.5) equals the likelihood of it not occurring (tails, 0.5).

For example, a probability of 0.25 means the event occurs about one-fourth of the time. A probability of 0.75 means it occurs about three-fourths of the time. These numbers describe *how often*, not *when*, something will happen.

A higher probability means greater likelihood, but almost never a guarantee. Even when an event has a probability of 0.9, it can still fail one time out of ten.

You can apply probabilities to individuals, but it's often easier to understand them for groups. Saying "a card has a one-in-four chance of being a heart" makes more sense if you picture drawing a card from a shuffled deck. About one-fourth of all cards are hearts.

That group-based view helps interpret individual outcomes. When you draw a single card, it's either a heart or not. But over many draws, the proportion of hearts stabilizes near one-fourth. This group-level reasoning is at the heart of probability.

Probability helps us think clearly about uncertainty by using a single number that links individual outcomes to frequency patterns that emerge across many repetitions.

From Simple to Complex Problems

Probability can describe everything from flipping a coin to predicting the success of a surgery. What changes is the level of complexity. Simple situations involve only a few equally likely outcomes, whereas more complex real-world problems often involve many interacting factors.

However, a few key terms help describe any probability situation.

- An **experiment** is the process that produces an outcome.
- A **trial** is a single repetition of that experiment.
- An **outcome** is the result of one trial.
- An **event** is the outcome or set of outcomes you're interested in.

For example, rolling a die is an experiment. Each roll is a trial. The number that appears is the outcome, and rolling a six might be the event we care about.

All probabilities relate to a well-defined event of interest and the propensity for that outcome to occur relative to all possible outcomes.

A Simple Case

When the possible outcomes are equally likely, calculating probability is straightforward. If you roll one fair six-sided die, each side has a 1 in 6 chance of appearing. The probability of rolling a six is one favorable outcome out of six total possible outcomes:

$$P\,(6) = \frac{1}{6} = 0.167$$

That probability notation indicates that the probability of rolling a 6 is 0.167.

The simplest kind of probability involves counting how many ways the event of interest can happen out of all the equally likely ways it could happen.

A More Complex Case

Now imagine something more multifaceted. Suppose a hospital wants to estimate the probability that a patient will recover successfully after surgery. This situation is far more complex than a die roll.

Many factors can influence the outcome. Older patients might face higher risks than younger ones. Someone with heart disease, diabetes, or another chronic condition might have a lower chance of recovery than a healthy individual. The complexity grows further when you consider variables like the type of surgery, the surgeon's experience, the length of time under anesthesia, or even post-operative care quality.

With so many interacting factors, it isn't possible to list all potential outcomes or assume they're equally likely. Instead, doctors and researchers rely on data and models. They use historical data from past patients to build models that estimate the probability of success for new patients with specific characteristics.

These models describe the average success rate for groups of patients with similar traits.

The next section explores the different approaches we can use to describe probability: empirical, theoretical, and model-based.

Approaches to Probability

Now that we've seen how probability can describe both simple and complex situations, let's look at how we actually *obtain* probabilities.

There's no single way to do it. The right approach depends on the problem and the information available.

In this book, we'll explore three main approaches in the order shown below:

1. **Empirical probability**, based on observed data.
2. **Theoretical probability**, based on logical reasoning.
3. **Model-based probability**, built on statistical or mathematical models.

Each approach answers the same question: *how likely is this event?* But they rely on different kinds of information and methods.

Empirical Probability

Empirical probability comes directly from experience, experiments, and observation. You estimate probabilities by collecting data and calculating the proportion of times an event occurs.

Rather than being calculated from theory or logic, you observe it by repeating a process and recording how often a specific outcome occurs. This experimental probability is based on what actually occurred during the observations, and the results will likely vary with different groups, locations, or conditions. There's no simple theoretical expectation here, so the experimental value provides the best available estimate.

This type of probability answers the question: *"How often did this outcome happen when we actually tried it?"*

For example, a computer store might record customer purchases by gender and computer type over time. By tallying those outcomes, the store can estimate probabilities from the observed frequencies.

If roughly 60 percent of male customers buy a PC and 40 percent buy a Mac, those proportions represent the store's empirical probabilities based on real data. These estimates can change as more data are collected.

This approach is powerful because it reflects what actually happens rather than what's assumed to happen. However, it depends on having enough reliable data. When events are rare or information is limited, empirical probabilities can be unstable or misleading.

In Chapter 2, you'll learn how to use contingency tables to calculate various types of empirical probabilities.

Theoretical Probability

Theoretical probability is the probability of an event predicted through logical reasoning, a clear understanding of the situation, and a probability model. There is no experiment. You base these probabilities on a set of assumptions that define the likelihood of different outcomes. Instead of observing what happens, you use what you know about the process.

Of course, the accuracy depends on the theory's validity. If you use the wrong theory, you'll get inaccurate results!

This type of probability answers the question: *"What should happen in theory according to a probability model?"*

All purely theoretical probabilities rely on a theory defining how likely each outcome is, ranging from simple to complex. These models can come from logic, physical reasoning, or mathematical structure, but they do not rely on observed data.

The most basic theoretical model assumes that all outcomes are equally likely. At its most basic, theoretical probability is calculated using the classic formula:

$$Probability(Event) = \frac{Number\ of\ favorable\ outcomes}{Total\ number\ of\ outcomes}$$

This applies to familiar situations like flipping a fair coin, rolling a die, or drawing a card from a well-shuffled deck.

For these simplest of scenarios, one outcome doesn't affect the next outcome. Additionally, each outcome is mutually exclusive from the others. For example, the outcome of one coin toss does not affect the outcome of future coin flips, and you can only obtain a head or a tail for one toss, not both.

The numerator equals the number of ways an event can occur. We define what counts as an event based on our interests. For example, we can choose to consider heads in a coin toss or drawing a king from a deck of cards as events. If we define an event as rolling either a 1 or a 6 on a die, there are two ways the event can occur.

The denominator represents the number of possible outcomes. The subject matter defines this value. For example, coin tosses can have only two results, heads or tails. There are 52 cards in a standard deck of cards.

Let's start simple with a coin toss and define heads as the single outcome that counts as an event. There is only one way an event can occur, and there are two possible outcomes.

P(H) = 1/2 = 0.5.

I wrote that using standard notation, and it indicates that the likelihood of heads equals 0.5.

Now, let's calculate the probabilities for rolling a die. We'll find the likelihood of rolling a 6, a 1 or a 6, and rolling an even number. Notice how each example changes the number of outcomes that count as an event in the numerator. For a standard die, there are always six potential outcomes. Consequently, the denominator is always 6.

- P(6) = 1/6 = 0.167
- P(1 or 6) = 2/6 = 0.33
- P(Even) = 3/6 = 0.50

Finally, we'll calculate likelihoods for a randomized, full deck of cards. What are the chances of drawing any card with a heart (H), any king (K), and a king of hearts (KH)? In a full deck, there are 52 cards, as indicated in the denominator.

- P(H) = 13/52 = 0.25
- P(K) = 4/52 = 0.077
- P(KH) = 1/52 = 0.019

However, these chances only apply to the first draw from a full deck. Any card we remove affects the likelihood of the next card. Hence, drawing successive cards from a deck are not independent events like coin tosses and dice rolls.

More advanced theoretical models allow for outcomes with unequal likelihoods and dependent events. For example, a biased coin or non-uniform spinner can have defined, unequal probabilities for each outcome. And, as you draw multiple cards from a deck, the probability for the next card changes. As long as the model specifies those probabilities, analysts can use them to calculate theoretical probabilities.

Theoretical probability doesn't rely on data. These methods use logic to determine the likelihood of different outcomes. You understand the likelihood of each possible outcome and calculate the chances of favorable outcomes relative to all possible outcomes.

This approach works beautifully in structured settings like dice, coins, or cards, where the rules are simple and outcomes are clearly defined.

But most real-world problems don't offer such tidy conditions, which is why we often need data models.

Model-Based Probability

Model-based probability combines ideas from both observation and theory. It uses mathematical or statistical models to represent uncertainty, especially when the situation is too complex for either pure logic or empirical data alone.

Models can range from very simple to highly sophisticated.

At the simpler end, we use standard probability distributions that describe familiar patterns of randomness, such as the normal, binomial, and Poisson distributions. These distributions describe how outcomes are expected to behave under certain mathematical assumptions and allow analysts to calculate probabilities for specific ranges or values.

- A **binomial distribution** might represent the number of defective products in a batch when each item has an equal, independent chance of being defective.
- A **normal distribution** might describe natural variation in human height or measurement error around an average value.

These models help describe expected patterns even when we don't have full data.

You can also use a probability distribution to calculate theoretical probabilities. For example, if you toss a fair coin 10 times, the number of heads you expect to see follows a binomial distribution. For example, you can use the binomial distribution to find that the probability of getting exactly 6 heads in 10 coin tosses is:

$P(X = 6) = 0.205$, or 20.5%

This is a theoretical probability based on the binomial model. It tells you what should happen under those defined conditions. In Chapter 11, you'll learn how to use this and other distributions to find probabilities.

At the more complex end, models use many variables and data-driven relationships. A doctor might use a model that combines a patient's age, health conditions, and surgical type to estimate the probability of recovery. An insurance company might model accident probabilities for different driver profiles, road conditions, locations, and so on.

At the most complex level, logistic regression or other predictive models produce theoretical probabilities for outcomes based on a combination of predictor variables. Even though these models are estimated using data, the probabilities they output are theoretical because they reflect the model's theoretical expectations.

For instance, suppose researchers develop a logistic regression model to estimate the probability that a patient is readmitted to the hospital within 30 days. The model uses predictors such as age, length of stay, and number of prior admissions.

For a specific patient, the model outputs a predicted probability of 0.37. According to the model, there is a 37% probability that the hospital will readmit patients with similar characteristics. That value is not based on observed frequency alone but on what the model predicts under its assumptions and parameters.

Model-based probability connects the logic of theoretical reasoning with the evidence of real data. It's the bridge between structure and experience.

In the real world, randomness comes from countless factors, many of which we can't control. The weather depends on complex physical processes. The outcome of surgery depends on health, genetics, and chance events during recovery. A factory's defect rate depends on machinery, materials, and human behavior.

We can't capture every possible influence, but we can build models that describe how these random processes behave *on average*. Models act as simplified blueprints of uncertainty. They don't show every detail, but they focus on the features that matter most for predicting outcomes.

For example, meteorologists don't simulate every air molecule. Instead, they use equations describing temperature, humidity, and pressure to estimate the probability of rain. In the same way, doctors use data-based models to estimate recovery rates for patients with certain risk factors.

Probability models allow us to reason about unpredictable systems. They turn complex, uncertain realities into something we can analyze, test, and use for decision-making. Every model is an approximation. It is a structured guess about how the world behaves. With enough data and good assumptions, these models can describe uncertainty surprisingly well.

This book covers various probability distributions starting in Chapter 10. However, the more advanced regression models go beyond the book's scope. To learn more about those models, read my book, *Regression Analysis: An Intuitive Guide.*

Two Broader Views of Probability

You'll encounter two main philosophical perspectives on probability.

Frequentists/Objectivists

The frequentist view defines probability as a long-run relative frequency. How often does an event occur over many repetitions?

Objectivists numerically calculate likelihoods for objective conditions. Frequentist probability is the most common form you'll encounter, and it underlies the most common statistical hypothesis tests. In this branch, the likelihood of a random event defines the relative frequency of its occurrence in experiments if you were to repeat an experiment many times. In other words, probabilities are long-run frequencies of outcomes.

Frequentist methodologies provide guidance for applying mathematical probability theory to real-world situations. They offer distinct guidance on the design and execution of practical experiments and on evaluating competing hypotheses. Objectivists consider probabilities to be long-run proportions that you can calculate only by using repeated observations in experiments.

Bayesians/Subjectivists

The Bayesian view treats probability as a measure of belief that can change as new information becomes available.

Subjectivists incorporate beliefs into their probabilities. The most common form is Bayesian probabilities. This branch includes combining expert opinions with experimental data to produce likelihoods. Ideally, the expert opinions contain all known information about the subject matter. When combined with experimental data, the process produces a posterior probability distribution that defines the likelihood of a particular outcome.

Don't worry, we'll go over all that in much more detail in Chapter 8 after a thorough grounding in Bayesian approaches!

Subjectivists are more flexible about what they consider a probability. For example, they can use non-experimental data to calculate probabilities for a singular event, such as the outcome of an election.

Unsurprisingly, there are tradeoffs between these approaches. Objectivists do not rely on opinion, but their results can exclude relevant known information. On the other hand, subjectivists incorporate a degree of belief, but their analyses can include different types of information that affect the outcome.

Each methodology has its ardent supporters. Both perspectives aim to describe uncertainty; they simply take different paths to do it. We'll revisit these ideas later.

Common Probability Misconceptions

Even though people use probability ideas all the time, it's easy to fall into traps that distort what probability really means. Before moving on, let's clear up a few of the most common misunderstandings.

Not All Outcomes Are Equally Likely

Many beginners assume that if there are two possible outcomes, the probability must be 50/50. That isn't always true.

Imagine drawing a card from a standard deck. You could think of the outcomes as "red" or "black," which are equally likely because there are 26 of each. But if you change the categories to "ace" or "not ace," the probabilities shift dramatically: only 4 of the 52 cards are aces, so the probability of drawing one is $4/52 = 0.077$.

Having two outcomes doesn't automatically mean they're equally likely. The key is how many *ways* each outcome can happen, not how many labels we give them.

We'll explore this idea of how many ways an outcome can occur when we get to the Law of Total Probability in Chapter 7.

A High Probability Doesn't Guarantee the Outcome

A probability of 0.8 sounds almost certain, but it still means the event fails to occur once in five attempts. You might think, "If something has an 80% chance, it should happen." But that's not a certainty.

Think about weather forecasts again. If it rains 80% of the time when meteorologists predict an 80% chance, their forecasts are accurate even though it won't rain at all in some cases.

Probabilities Have Variability

Even when events are balanced in the long run, short runs can look very uneven. Flip a fair coin ten times, and you might get a run of heads in a row. That doesn't mean the coin is unfair. It just shows that randomness produces variation around the average.

This variability is a fundamental part of probability. For instance, if you flip a coin 10 times, you'd expect 5 heads because they occur half the time. However, that's an average outcome, and the actual results will vary around that mean. The binomial distribution you'll learn about later shows that there's only a 24.61% chance of getting exactly 5 heads. Hence, there's a 75.39% chance of getting something other than the expected 5 heads!

If you repeated the coin toss experiment thousands of times, the proportion of heads would hover near 0.5, but often not exactly at it for small samples. You'll see this with simulations in Chapter 10 when we get to the Law of Large Numbers. And, you'll learn how both the standard deviation and standard error describe it mathematically for individual and sample outcomes, respectively.

Probability Describes Likelihood, Not Cause

Another common mistake is to treat probability as an explanation for *why* something happens. Probability doesn't cause events. It only describes how likely they are. Saying there's a 25% chance of rain doesn't make the storm occur one out of four times; it simply summarizes the uncertainty based on known information.

Understanding this distinction helps prevent confusing description with causation.

Keeping Perspective

Probability helps you reason about uncertainty, but it never removes it. Events can be likely and still not occur, or unlikely and still happen. The goal isn't to predict every outcome perfectly but to understand the patterns and chances behind them.

Keep these misconceptions in mind as you move forward. They're the small traps that can twist your intuition, especially when problems become more complex.

Review and Next Steps

In this chapter, you learned that probability theory represents a structured way to describe uncertainty and make sense of randomness. We explored how probability appears in everyday life, how our intuition often misleads us, and why formal methods are essential for making sound decisions. You also saw that probabilities can come from three main sources: empirical data, logical reasoning, and models.

The next step is to connect those ideas to real observations. In the next chapter, we'll use observed data to estimate probabilities and see how they work in practice. You'll learn how to summarize information in contingency tables and use them to find different types of probabilities. These ideas turn probability from an abstract concept into something you can calculate and interpret directly from the world around you.

Chapter 1 Practice Problems

The fully worked solutions for all practice problems are available for free book on my website. Find the URL in the Introduction.

Question 1: A news report states that a new airline route arrives on time with a probability of 0.85. Explain what this probability means.

Question 2: In a survey of 20 students about their preferred study environment:

- 9 prefer studying at home
- 7 prefer studying in the library
- 4 prefer studying in a café

If one student is selected at random from the group:

A. What is the probability the student prefers studying at home?
B. What is the probability the student prefers the library?
C. What is the probability the student prefers either the library or a café?

Question 3: A student says that the probability of choosing a left-handed person at random must be 0.5 because people are either left-handed or right-handed.

Explain why this reasoning is incorrect.

Question 4: Identify whether each situation describes empirical probability, theoretical probability, or model-based probability.

A. A credit scoring system predicts the probability that a borrower will default using a statistical model based on income, credit history, and debt levels.
B. A quality control engineer estimates the probability that a product fails by analyzing thousands of past inspections.
C. A mathematician calculates the probability that a randomly chosen three-digit number is divisible by 5.

Question 5: A random number generator is designed to produce integers from 1 to 10 with equal probability. In the first 12 numbers produced, the number 7 appears four times.

A student claims this proves the generator is flawed.

Explain why this conclusion might be premature.

Question 6: Consider selecting a student at random from a class and recording their major.

A. What is the experiment?
B. What is a trial?
C. What is an outcome?
D. Give one example of an event.

CHAPTER 2

Empirical Probability from Contingency Tables

In this chapter, you'll start understanding and finding different types of probabilities. We'll start by using concrete, empirical data to ease into it. Contingency tables are a fantastic way to get your feet wet with various kinds of probabilities. They present observed data in an intuitive manner and allow you to calculate probabilities easily without getting too deep into the theory behind it all. Instead, you can just focus on the concepts.

In later chapters, we'll take the concepts you learn here and apply them to symbolic, theoretical probability calculations.

Contingency tables are a great way to classify outcomes and calculate different types of probabilities. These tables contain rows and columns that display bivariate frequencies of categorical data. Analysts also refer to contingency tables as crosstabulation (cross tabs), two-way tables, and frequency tables.

Statisticians use them for a variety of reasons. I love these tables because they both organize your data and allow you to answer a diverse set of questions. For our purposes, I focus on using them to calculate different types of probabilities, including joint, marginal, and conditional probabilities.

Contingency tables classify outcomes in rows and columns. The rows represent categories of the first variable, while the columns represent categories of the second variable. Table cells at the intersections of rows and columns indicate frequencies of both events coinciding for a specific combination of the two variables.

For example, the following table displays the results from a fictional computer usage survey amongst college students. Specifically, it describes the frequencies of the two variables: gender and preferred computer type. The counts of interior cells represent the number of PCs and Macs used by both genders. Additionally, the table contains sums for each row and column, as well as the grand total for all observations.

	PC	**Mac**	**Row Totals**
Male	66	40	106
Female	30	87	117
Column Totals	96	127	223

At first glance, it's easy to see how these tables both organize your data and paint a picture of the results. For example, 66 males use PCs while 87 females use Macs. Furthermore, there are 117 females, 106 males, 96 PC preferences, 127 Mac preferences, and a grand total of 223 observations in the study.

These tables are more flexible than they first appear because they allow you to answer a diverse set of probability questions. *What are the joint, marginal, and conditional probabilities of events occurring?*

As you work through the different types of probabilities, keep in mind that, in a general sense, all probabilities equal the following ratio:

$$Probability(Event) = \frac{Events}{Total\ Outcomes}$$

Events are a subset of all possible outcomes in which you are particularly interested. When using a contingency table to calculate different types of probabilities, it's just a matter of determining which table values go in the numerator and denominator. All the information you need is right there in the table!

Throughout this chapter, I'll first walk you through each type of probability and how to calculate it using a contingency table, allowing you to understand it intuitively. In the next chapter, I'll show you the formal probability notation and equations so you become familiar with them.

Joint Probabilities

Joint probabilities are the chances that events occur together. For example, what is the joint probability of Mac usage by a female?

Contingency tables really shine at highlighting joint probabilities because each cell shows how often the events occurred together. Those cell values are the joint events for the numerator. Each count inside the table corresponds to one joint event. For instance, the 87 in the Female–Mac cell represents the number of students who are both female and use Macs.

To calculate joint probabilities in a contingency table, use a joint cell value in the numerator. Then use the grand total for the number of outcomes in the denominator because we're interested in all possible outcomes.

For our example, the joint probability of females using Macs equals the value in that cell (87) divided by the grand total (223).

$$\frac{87}{223} = 0.390$$

Joint probabilities give us the foundation for more complex relationships. They show how two characteristics combine within the same individuals or cases. Later, when we study conditional probability more formally, we'll see that joint probabilities link directly to conditional probabilities through a simple relationship.

But for now, it's enough to see that each interior cell in a contingency table represents a joint event, and dividing by the total sample size converts those counts into joint probabilities.

Joint Probability Notation and Calculations

$P(A \cap B)$ is the notation for the joint probability of events "A" and "B" occurring together.

For our example, we determined that:
$P(\text{Female} \cap \text{Mac}) = 0.390$

The general process for calculating joint probabilities in a contingency table for a cell in row i, column j is the following:

$$P(Row\ i, Column\ j) = \frac{Count\ in\ cell\ (i,j)}{Grand\ Total}$$

1. The numerator equals the count of occurrences for the specific combination of events in which you're interested.
2. The denominator equals the grand total number of observations.

In the following table, the values in parentheses are the joint probabilities for the cells. Joint probabilities for an entire table always sum to 1.

	PC	Mac	Row Totals
Male	66 (0.296)	40 (0.179)	106
Female	30 (0.135)	87 (0.390)	117
Column Totals	96	127	223

Easy peasy, right?

Marginal Probabilities

Marginal probabilities are the likelihoods that a single event occurs with no regard to other events in the table. These probabilities do not depend on the condition of another outcome. This lack of dependency differs from joint probabilities and conditional probabilities. In our table, the single events are gender (male or female) and preferred computer type (PC or Mac).

In contingency tables, you use the marginal distributions to find marginal probabilities. A marginal distribution is a distribution of values for one variable that ignores a more extensive set of related variables in a dataset.

That definition sounds a bit convoluted, but the concept is simple. The idea is that when you have a larger set of related variables that you collected for a study, you might want to focus on one of them to answer a specific question.

These marginal probabilities summarize the overall distribution of each variable by itself. They tell us how common each category is when we ignore the other variable.

It's easiest to understand and find marginal distributions using a two-way contingency table. In fact, the term originates from these tables. To find them, look in the *margins* of a contingency table, hence the name. You can locate marginal distributions in the row and column totals.

I've circled the values in the margin that represent the dispersal of one variable without considering the other variable. There's a marginal distribution for computer type and another for gender.

	PC	Mac	Row Totals
Male	66	40	106
Female	30	87	117
Column Totals	96	127	223

Let's assess the distribution of computer types while disregarding gender. By looking at the margins in the table, we see the following preferences amongst computer types:

- PC: 96
- Mac: 127

Statisticians say you need to "marginalize out" the other variables to find a marginal distribution. In the previous example, we marginalized out gender to reveal the distribution of computer type usage.

When you have marginal distributions, you can calculate marginal probabilities.

Choose the individual event you're interested in and use the corresponding row or column total in the numerator. Then, use the grand total for the denominator.

For example, if you want to determine the probability of Mac usage and disregard gender, you simply take the column total for Mac (127) and divide it by the grand total (223). Or, if you want to determine the probability of a female respondent in the survey and not consider the type of computer, take the row total for Female (117) and divide by the grand total (223).

$$Probability\ (Mac) = \frac{127}{223} = 0.570$$

$$Probability\ (Female) = \frac{117}{223} = 0.524$$

These marginal probabilities summarize the overall distribution of each variable by itself. They tell us how common each category is when we ignore the other variable.

Where joint probabilities describe combinations of categories, marginal probabilities describe totals for one variable at a time.

Marginal Probability Notation and Calculations

P(A) denotes the probability of event A occurring on its own. For our example, we determined that:

P(Mac) = 0.570

The processes for calculating marginal probabilities in a contingency table for row i and column j are the following:

$$P(Row\ i) = \frac{Total\ Count\ in\ Row\ i}{Grand\ Total}$$

$$P(Column\ j) = \frac{Total\ Count\ in\ Column\ j}{Grand\ Total}$$

1. The numerator equals the row or column total for the individual event in which you're interested.
2. The denominator equals the grand total number of observations.

In the following table, the values in parentheses are marginal probabilities for each condition. The column marginal probabilities (PC and Mac) sum to 1. Similarly, the row marginal probabilities (Male and Female) also sum to 1.

	PC	Mac	Row Totals
Male	66	40	106 (0.475)
Female	30	87	117 (0.524)
Column Totals	96 (0.430)	127 (0.570)	223

Conditional Probabilities

Conditional probabilities are the likelihood that an event occurs given that another event has occurred. For example, given that a student is female, what is the probability she'll use a Mac?

These probabilities sound a bit more complicated, but they are easy to calculate using contingency tables. Let's answer the following conditional probability questions.

- What is the probability that a student will use a Mac, given that the customer is female?
- Given that the student uses a PC, what is the probability that the student is male?

Both are conditional probabilities because they provide a "given" event. Assuming that a particular event occurs, what is the probability of the other event occurring?

A conditional distribution is a distribution of values for one variable that exists when you specify the values of other variables. This type of distribution allows you to assess the dispersal of your variable of interest under specific conditions, hence the name.

Imagine that you want to assess the distribution of computer types used by only women. That's an example of a conditional distribution. We're conditioning computer types on the gender variable value of female.

The process of conditioning one variable on the value of another variable might sound complicated. However, it's simple to find a conditional distribution using a contingency table. Just look down a column or across a row.

	PC	Mac	Row Totals
Male	66	40	106
Female	30	87	117
Column Totals	96	127	223

I circle two examples of conditional distributions in the table. Let's stick with our original example of conditioning computer type on the gender value of female. By looking at the horizontal highlight in the table, we see that female survey respondents use the following:

- **PC**: 30
- **Mac**: 87

This process allows us to understand our data in the more specific context of only females.

We could also assess a different conditional distribution to understand a different context, such as gender conditioned on PC usage. That's the vertical example I highlight in the previous table.

When you have conditional distributions, you can calculate conditional probabilities. Fortunately, calculating conditional probabilities using contingency tables is straightforward. It's merely a matter of dividing a cell value by a row or column total.

As with a joint probability, we are interested in a particular combination of events that the table records in a cell. Use the cell value of interest in the numerator.

However, unlike joint and marginal probabilities, we do not use the grand total in the denominator. Instead, we're conditioning the probability on a particular outcome rather than the entire sample space. Consequently, we use the row or column total for the condition event (the "given" in the problem statement) in the denominator.

In this situation, we are no longer considering all 223 students. We are focusing on only the subgroup of 117 females, because that is our condition. Within that restricted group, we ask: what proportion uses Macs?

The logic here is straightforward: after we restrict our attention to females, the denominator of our probability becomes the total number of females rather than the entire sample. The denominator has changed, making it a *conditional* probability.

Thinking of it this way helps make conditional probability intuitive:

Conditional = probability under a restricted denominator.

Let's determine the probability that a student will use a Mac given that the student is female.

We need to use the female/Mac cell value (87) in the numerator and the female row total (117) in the denominator.

$$\frac{87}{117} = 0.744$$

Let's try another one. Given PC usage, what is the probability that the student is male?

We need to use the male/PC cell value (66) in the numerator and the PC column total in the denominator (96).

$$\frac{66}{96} = 0.688$$

Later, we'll formalize this idea symbolically, but for now, the key is understanding what happens conceptually. Conditional probabilities "zoom in" on the subset of data that meets the condition and then find the relative frequency of the event of interest within that group.

Conditional Probability Notation and Calculations

P(A | B) denotes the conditional probability of A occurring given that B has occurred. The second value indicates the "given" portion of the probability.

For our two examples of conditional probabilities, we determined the following:

- P(Mac | Female) = 0.744: The probability that a female student uses a Mac is 0.774.
- P(Male | PC) = 0.688: The probability that a PC user is male is 0.688.

Calculating a conditional probability involves using a joint probability in the numerator and a marginal probability in the denominator. The marginal probability is the given condition.

Use the more detailed steps in the previous sections for finding the appropriate joint and marginal probabilities. The process for calculating conditional probabilities using a contingency table is the following:

1. The numerator equals the count of occurrences for the specific combination of events in which you're interested. This count is in a cell.
2. The denominator equals the count of occurrences for the "given" portion of the question. This value can be either a row total or a column total that includes the cell in step 1.

Reverse Conditional Thinking

Conditional probability can feel straightforward at first, until we realize that reversing the condition changes the question entirely. Using the same table of students' gender and computer type usage helps make that distinction clear.

Earlier, we found the probability that a customer purchased a Mac given that the client is female:

P(Mac | Female) = 0.744

That question restricts the denominator to females and asks: "Among females, what proportion purchased Macs?"

But what if we turn the question around? What is the probability that a customer is female given that they purchased a Mac?

That's written as:

P(Female | Mac)

Now the condition is reversed. We are looking only at Mac users, not all students, and not only females. We are zooming in on a different group. There are 127 Mac users total, and 87 of them are female. So,

P(Female | Mac) = 87/127 = 0.685

Notice how these two conditional probabilities describe very different perspectives:

- P(Mac | Female) = 0.744 asks, "Among females, how many use Macs?"
- P(Female | Mac) = 0.685 asks, "Among Mac users, how many are female?"

The numbers aren't the same because the denominators, the groups we're focusing on, are different.

This distinction, though simple, is at the heart of many reasoning errors people make when interpreting data. We'll revisit this idea later when studying Bayes' Theorem, which provides a formal method for relating P(A | B) and P(B | A).

For now, it's enough to remember that reversing the condition focuses on a different subset and usually changes the probability.

Informal Check for Independence

One of the most important questions we can ask of a contingency table is whether the two variables are related. Do the values of one variable tend to correspond with the values of the other variable? For now, we will explore that idea informally by comparing conditional probabilities across groups rather than using formulas or hypothesis tests.

Here's the table of computer types by gender:

	PC	Mac	Row Totals
Male	66	40	106
Female	30	87	117
Column Totals	96	127	223

If gender and computer type were completely unrelated (that is, independent), then the proportion of Mac users should be roughly the same among males and females. Let's check.

For males:
P(Mac | Male) = 40/106 = 0.377

For females:
P(Mac | Female)= 87/117 = 0.744

Those proportions differ quite a bit, 37.7% versus 74.4%. That's a strong hint that gender and computer type are not independent in this dataset. Females are much more likely to purchase Macs than males.

To see what independence would look like, compare these values to the *overall* or marginal probability of using a Mac:

P(Mac) = 127/223 = 0.570

If the conditional probabilities P(Mac | Male) and P(Mac | Female) were both close to 0.570, we'd say there's no evidence of a relationship. Because they differ so much, it appears that the variables are associated. Type of computer usage depends on gender.

If conditional probabilities differ meaningfully across groups, the variables are probably related.

In Chapter 4, we'll define independence formally and show how these ideas connect to the symbolic probability rules. But for now, it's enough to recognize that differences in conditional probabilities suggest potential relationships between variables.

Review and Next Steps

In this chapter, we grounded probability in concrete, observed data.

Contingency tables are deceptively simple tools. They display frequency counts for pairs of categorical variables and summarize the multivariate relationship between several categorical variables.

Using frequency counts in contingency tables, you learned how to calculate and interpret joint, marginal, and conditional probabilities. These quantities show how events combine, how often each occurs overall, and how the likelihood of one event changes when another is known. You also saw that conditional probability is a probability under a restricted denominator (i.e., a specific subgroup). That's a simple but powerful way to think about how context changes what "chance" means.

Through examples, we began thinking about the relationships between variables. Comparing conditional proportions helped you informally assess independence, whether two characteristics appear related. We also introduced reverse conditional thinking, learning that P(A | B) and P(B | A) describe completely different perspectives, even though they involve the same two events. These ideas form the practical intuition behind more abstract probability principles that follow.

So far, our approach has been empirical: we estimated probabilities from observed data. In the next chapter, we shift from data to formal structure. You'll learn to describe events as sets within a sample space and to use set notation to express unions, intersections, and complements. From there, we'll develop the rules of probability (the additive, complement, and multiplication rules) that allow us to calculate probabilities.

Those rules, and the axioms of probability that support them, will turn the intuition you've built in this chapter into a formal mathematical framework.

In short, this chapter showed how probability lives in real data. The next several chapters show how the same ideas live in mathematics.

Chapter Summary

- **Empirical probabilities** come directly from observed data, such as counts in a contingency table. This chapter introduced probability through data and proportions.
- **Joint probabilities** describe the likelihood of two events occurring together (both A *and* B). P(A ∩ B) is the notation for the joint probability of events "A" and "B" occurring together. In a contingency table, Cell Total / Grand Total.

- **Marginal probabilities** describe the overall likelihood of a single event, ignoring the other variable. P(A) is the notation for the marginal probability of A. In a contingency table, Column or Row Total / Grand Total.
- **Conditional probabilities** describe the likelihood of one event *given that* another has occurred. In a contingency table, Cell Total / A Column or Row Total.
  - The symbol "|" in P(A | B) means "given that B has occurred."
  - Reversing the condition (e.g., P(A | B) vs. P(B | A)) changes the subgroup and usually changes the probability.
  - Comparing conditional probabilities across groups provides an informal way to assess independence. If those probabilities differ substantially, the variables are likely related.

Chapter 2 Practice Problems

The fully worked solutions for all practice problems are available for free on my website. Find the URL in the Introduction.

Question 1: Identify the type of each probability:

A. The probability that a student participates in intramural sports given that they commute.
B. The probability that a student commutes and participates in intramural sports.
C. The probability that a student lives on campus.

For the following questions, use this scenario and table. A survey asked 150 people about the streaming service they use most often and the device they usually use to watch it. Be sure to use proper notation in your answers!

	Mostly on TV	Mostly on Mobile	Total
Prefer Netflix	54	36	90
Prefer Other	24	36	60
Total	78	72	150

Question 2: Find the joint probability that a randomly selected person prefers Netflix and watches mostly on a TV.

Question 3: Find the marginal probability that a randomly selected person watches mostly on a mobile device.

Question 4: Find the conditional probability that a person watches mostly on a TV given that they prefer Netflix.

Question 5: Find the conditional probability that a person watches mostly on a TV given that they prefer another streaming service.

Question 6: Based on your answers to Questions 3 and 4, assess informally whether preferred streaming service and device used for watching appear to be independent.

Question 7: Determine which probability is larger:

- The probability that a randomly selected person prefers Netflix.
- The probability that a person prefers Netflix given that they watch mostly on a TV.

Calculate both probabilities and explain why they might differ.

CHAPTER 3

Events, Sets, and Probability Relationships

In the previous chapter, we worked with probabilities as proportions calculated from real data in contingency tables. When we have data, that approach is a direct way to measure how often events occur. It is practical and intuitive.

In this chapter, I introduce the concepts of sample space, sets, and events. The sample space is the set of all possible outcomes of a random process. An event is any subset of the sample space consisting of outcomes of interest, such as rolling a 6 on a die.

A set is a collection of distinct outcomes, and set notation is a standardized way to describe and organize these outcomes and their relationships. Understanding sets gives you a useful mental model for how probabilities behave. Set rules and notation also provide a shared language for describing events precisely and for understanding relationships between events. They allow us to describe and calculate different types of probabilities by providing clear definitions and relationships among outcomes.

After defining sets and events, we build on that foundation by revisiting marginal, joint, and conditional probabilities, and adding complement probabilities. These ideas follow naturally from the set operations because each type corresponds to a particular relationship between events. They apply in both empirical settings, where probabilities come from data, and in theoretical settings, where probabilities come from mathematical models.

Throughout the chapter, we'll express these ideas with simple symbols that help us work precisely and reveal relationships that are hard to see in words alone. By the end of this chapter, you will have a strong understanding of how probability works at a fundamental level. That structure prepares us for the next chapter, where we analyze more complex event relationships.

Sample Space

Probability always starts with a situation we can think of as a trial or experiment. A random process occurs, and it can produce different possible outcomes. To describe that situation clearly, we first need to list all possible outcomes before assessing their likelihood.

The sample space S is the set of all possible outcomes of a random experiment. Each individual outcome is an element of S. Any event we care about is a subset built from this set.

Think back to contingency tables. They displayed how often different observed outcomes occurred. Those outcomes come from a set of possible results for each trial. The sample space represents the full set of possible outcomes, expressed using set notation rather than a table.

Let's walk through a concrete example.

Suppose a small clinical study enrolls 10 patients. Imagine randomly selecting one patient from the study. We label the patients with ID numbers 1 through 10.

The sample space for this selection is:

$S = \{ 1, 2, 3, 4, 5, 6, 7, 8, 9, 10 \}$

Each number represents one patient in the study.

Sample spaces can take different forms:

- Finite sample spaces have a limited number of outcomes. Example: $S = \{ 1, 2, ..., 10 \}$ for our 10 patients.
- Countably infinite sample spaces have infinitely many outcomes that you can list in order. Example: $S = \{ 1, 2, 3, 4, ... \}$ for the number of customer calls a help desk receives in a day. While very high values are unlikely, there is no fixed upper limit.
- Uncountably infinite sample spaces have infinitely many outcomes that you cannot list one by one. Example: S is all possible values of a person's weight, or all real numbers between 0 and 1.

Using Inequalities in Sample Spaces

Some sample spaces contain so many possible outcomes that listing them individually is impractical. When the possible values form a range rather than individual points, we describe the sample space using inequalities. This keeps the notation compact while still defining all possible outcomes.

Here are several examples:

- **Blood pressure measurements**
  $\mathbf{S = \{ x : 80 \le x \le 200 \}}$

Plain language: "The systolic blood pressure for each patient can be any value between 80 and 200, inclusive."
This represents a continuous range of realistic values.

- **Daily high temperature**
  $S = \{ t : t > -50 \}$
  Plain language: "A day's high temperature can be any value above –50 degrees."
  There's no fixed upper limit, and temperatures vary along a continuous scale.
- **Time to complete a task**
  $S = \{ x : x \geq 0 \}$
  Plain language: "Completion time can be any non-negative real number."
  Because time cannot be negative, the sample space begins at zero and extends indefinitely.

Using inequalities is especially helpful when dealing with measurements or naturally continuous variables. Instead of listing outcomes individually (which can be impossible), we specify the full range of values the variable can take.

I will frequently use finite sample spaces in the examples because they are easy to visualize. But the same ideas apply to countable and uncountable sample spaces as well.

Naming Events

Events are subsets of the sample space S. An event is any collection of possible outcomes that share a specific property. We often designate events with capital letters such as A, B, or C. The choice of letter is arbitrary; what matters is the precise condition described for the event. For example:

- A = "the patient improved"
- B = "the patient had side effects"
- C = "the patient was in the control group"

Any subset of the sample space can be named as an event. If you can clearly describe a condition that outcomes either do or do not satisfy, you can define that condition as an event.

Suppose we define two events:

- Event A: patients who showed improvement
- Event B: patients who experienced side effects

If patients 2, 4, 6, and 7 improved, we can write:

A = { 2, 4, 6, 7 }

If patients 4, 5, and 10 had side effects, we can write:

B = { 4, 5, 10 }

Both A and B are subsets of S, with each event corresponding to a subset. When we write probabilities such as P(A) or P(B), we refer to the likelihood of these subsets

Up to this point, we defined events by listing the specific outcomes that meet a condition. For example, if A represents the event "the patient improved," we list the patients who improved because each patient is a separate outcome in the sample space. This works well when the condition is based on a discrete or categorical characteristic, such as improvement status, treatment assignment, or presence of side effects.

However, when the conditions involve numeric measurements rather than categories, it is more natural to describe the event using inequalities rather than listing individual values. Inequalities work better with continuous variables because they define the event with a range of values rather than distinct categories.

Regardless of the type of characteristic, the underlying idea is the same: an event is the set of all outcomes that satisfy the defining condition. The notation simply adapts to the type of information:

- For categorical or discrete conditions, we often show the event by listing the outcomes that meet it.
- For quantitative or continuous conditions, we describe the event with inequalities because the measurable values span a range.

Here are several examples of events defined by conditions involving continuous variables:

- **$A = \{ x : x < 120 \}$**
  Plain language: "All outcomes where the measured value is below 120."
- **$B = \{ x : 150 \leq x \leq 180 \}$**
  Plain language: "All outcomes where the value falls between 150 and 180."
- **$C = \{ t : t \leq 60 \}$**
  Plain language: "All temperatures at or below 60 degrees."

Even though the notation differs, both styles serve the same role in probability: defining subsets of the sample space as events based on conditions.

Set notation and event definition apply to both empirical and theoretical probability. Whether probabilities come from observed frequencies in a study or from a mathematical model of possible outcomes, the same set notation and event operations describe how outcomes combine and how events relate to one another.

After we define the sample space and label events as subsets of S, we can begin describing how these events connect. In the next section, I introduce basic operations on events and see how new events are built from existing ones.

Example: The Two-Child Problem or Why Carefully Defining the Sample Space Matters

Sometimes probability questions seem simple but produce surprising answers. In many cases, the difficulty arises because people do not carefully identify the correct sample space before reasoning about the event of interest. Always pay very close attention to the precise wording of probability problems!

Consider the following example of a well-known probability problem.

Suppose a family has two children. You are told that at least one of the children is a boy. What is the probability that both children are boys?

Many people initially think the probability is 1/2. The reasoning often goes like this: if one child is a boy, the other child could be either a boy or a girl with equal probability.

However, this reasoning overlooks the correct sample space.

If we assume boys and girls are equally likely and distinguish the children by birth order, the sample space for the possible outcomes of two children is:

S = { BB, BG, GB, GG }

Each element of S represents one possible outcome. The first letter represents the older child, and the second letter represents the younger child.

Now define the event:

A = "at least one child is a boy"

Because events are subsets of the sample space, we identify which outcomes in S satisfy this condition:

A = { BB, BG, GB }

The condition "at least one child is a boy" restricts the possible outcomes to the three equally likely elements in event A rather than four. The outcome GG is excluded because it does not contain a boy.

We are interested in the event:

B = "both children are boys"

B = { BB }

To find the probability that both children are boys given that at least one is a boy, we focus only on the outcomes that remain possible under event A. Among the three outcomes in A, only one corresponds to B.

Therefore,

$P(B \mid A) = 1/3$

The surprising result occurs because the correct sample space contains three possible outcomes, not two. The common intuition assumes there are only two equally likely possibilities, which would produce a 50/50 outcome. However, this reasoning ignores the distinction between the outcomes BG and GB, which are separate outcomes within event A. In the former, the boy is the older child, while in the latter, the boy is the younger child.

This example illustrates an important lesson. Probability calculations depend on correctly defining the sample space and identifying events as subsets of that space. If the possible outcomes are not specified carefully, intuition can easily lead to incorrect conclusions.

Event Operations

Now that we understand events as subsets of the sample space, we can describe how events relate to each other. Set operations help us form new events from existing ones by identifying how events overlap, interact, or exclude one another. Each new event is still a subset of the sample space, so we can assign it a probability.

Later, we will use these relationships to build the rules that govern how probability works.

Complement of an Event (Not A)

The complement of event A contains every outcome in the sample space S that is not in A. It represents the event that A does not occur. Together, a set and its complement must include the entire sample space. Logically, A and not-A must collectively include all outcomes. If that isn't true, there is a problem with how your outcomes are defined or categorized.

Notation: A^c

Recall our example where event A indicates improved patients:

- $S = \{ 1, 2, 3, 4, 5, 6, 7, 8, 9, 10 \}$
- $A = \{ 2, 4, 6, 7 \}$

Then:
$A^c = \{ 1, 3, 5, 8, 9, 10 \}$

These six patients have not improved.

In terms of set sizes, $A + A^c$ must equal the sample space total (N). Hence $A^c = N - A$.

For example, there are 10 total subjects in the sample space and 4 subjects in A. Hence, the number of patients in $A^c = 10 - 4 = 6$.

The complement changes the focus from the outcomes that satisfy a condition to those that do not. When A is small, A^c is large. When A is large, A^c is small. They are especially useful when it is easier to find the probability of something not happening than happening. As you'll see throughout this book, complements can serve as a helpful shortcut by simplifying calculations.

This book uses A^c, to denote the complement of event A, but there are several other standard notation styles. They include Ac (without the superscript) and A′ (A-prime).

Union of Events (A or B)

The union of events A and B includes every outcome that is in A, or in B, or in both. It represents the event that at least one of the two conditions occurs.

Notation: $A \cup B$

Example using the clinical study:

- A = { 2, 4, 6, 7 } (improved patients)
- B = { 4, 5, 10 } (patients with side effects)

Then:
$A \cup B = \{ 2, 4, 5, 6, 7, 10 \}$

These six patients have either improved, have side effects, or both.

The union often increases the size of a set by gathering all outcomes from both events. However, its size is not always equal to the sum of the sizes of the two sets because some elements can belong to both events. For example, there are 4 improved patients and 3 patients with side effects, but only 6 patients in the union because patient 4 belongs to both A and B. If we simply added the sizes of the two sets (4 + 3 = 7), patient 4 would be counted twice.

In the next chapter, we will see how the additive rule prevents this type of double counting when calculating probabilities for unions of events.

Intersection of Events (A and B)

The intersection of A and B includes only the outcomes that are in both events at the same time. It represents the event that both conditions occur.

Notation: $A \cap B$

Using the same example:

- $A = \{ 2, 4, 6, 7 \}$ (improved patients)
- $B = \{ 4, 5, 10 \}$ (patients with side effects)

Then:
$A \cap B = \{ 4 \}$

This single patient has both improved and has side effects.

The intersection often contains fewer outcomes because it includes only those that belong to both events. Its size can never exceed that of the smaller set because the overlap cannot contain more elements than the smallest group being compared.

De Morgan's Laws

Sometimes it is easier to calculate the probability that an event does not occur than to calculate the event directly. When events are combined using unions ("or") or intersections ("and"), it can be helpful to use complements and De Morgan's Laws to rewrite the expression in a simpler form.

According to De Morgan's Laws, the complement of a union equals the intersection of the complements:

$(A \cup B)^c = A^c \cap B^c$

This identity means that the outcomes that are not in the union of A and B are exactly the outcomes that are in neither A nor B.

Use this method when it is easier to identify the probability that none of the events occur than to calculate the probability that at least one occurs.

A useful way to remember De Morgan's Laws is to apply complements to the individual events and switch "or" to "and" (or vice versa). In probability terms, unions become intersections, and intersections become unions.

Similarly:

$(A \cap B)^c = A^c \cup B^c$

Here, the intersection ("and") becomes a union ("or") after applying the complements.

Empty Set

The empty set contains no outcomes.

Empty set symbol: $\emptyset$

If two events cannot happen together, their intersection is empty. For example:

- If event D = "patients over 200 years old," then D = $\emptyset$
- Its probability is always $P(\emptyset) = 0$

Empty sets represent impossible events. Their size is zero by definition.

Why These Operations Matter

These set relationships shape how to find probabilities, as you'll see in the next section. And they form the basis of the standard probability rules we'll cover in the next chapter:

- Unions →Additive Rule
- Complements → Complement Rule
- Intersections → Multiplication Rule

These rules allow you to calculate various probabilities directly from other probabilities without referring to the original data.

Types of Probabilities Involving Two Events

Now you should understand how events behave as sets. How they combine, overlap, and exclude one another. We can turn to the probabilities associated with them. These same set operations form the foundation for the different kinds of probabilities we work with in practice.

In this section, we'll build on these ideas by expressing unions, intersections, and restricted subsets in probability notation and learning how to interpret marginal, joint, and conditional probabilities symbolically.

Many real situations involve more than one event. Before we focus on each type of probability in detail, let's look at a quick overview.

Suppose A represents improved patients and B represents patients with side effects.

- **Marginal probability**: P(A) is the chance that a randomly selected patient improved and does not consider side effects at all.
- **Union probability**: $P(A \cup B)$ is the chance that a patient improved, *or* had side effects, *or* both.
- **Joint (Intersection) probability**: $P(A \cap B)$ is the chance that a patient improved *and* had side effects.
- **Conditional probability**: $P(A \mid B)$ is the chance that a patient improved *among those* with side effects.
- **Complement probability**: $P(A^c)$ is the chance a patient did *not* improve.

Previously, you calculated these types of probabilities using contingency tables and data. Now, we will express the same ideas symbolically using notation and formulas. This symbolic approach provides a more flexible framework that works with data, prior research, and theoretical reasoning.

Before we can calculate probabilities using this new framework, we need to think about the process we are using. We start by identifying the sample space, which contains all possible outcomes. Then we define an event as a set of outcomes that share a particular characteristic. After we know which outcomes belong to an event, we find its probability by calculating:

$$\frac{\textit{Number of event outcomes}}{\textit{Total number of outcomes in sample space}}$$

This method relies on an important assumption: each outcome in the sample space is equally likely to occur.

In our clinical trial examples, this means that each patient has the same chance of being randomly selected from the study. The events themselves (like "improved" or "had side effects") do not need to be equally likely. Some outcomes can be much more common than others, and that difference is reflected in the probabilities we calculate because those outcomes appear more often among the possible selections.

Set operations such as unions, intersections, and complements help us describe how events relate so we can determine the probability of combined or restricted conditions using this consistent approach.

Later in the book, we will study more advanced methods for situations where outcomes do not all have equal likelihoods, including when probability distributions model the likelihoods. But for now, this approach provides a clear, intuitive foundation.

Now that we have a framework for identifying events and calculating their probabilities, let's look at the different types of probabilities that involve two events and how they relate to each other.

Marginal Probability

Marginal probability is the chance that an event will happen without considering other variables. Statisticians write this as P(A), denoting the probability of event A. You can think of it as an unconditional probability. It tells you how likely something will happen on its own, independently of other variables.

For example, consider a bakery that sells bread and other products. Suppose you only care about how often customers buy bread, not whether they also buy coffee, pastries, or other products. The marginal probability of buying bread will give you this answer!

Notation: P(A)

Formula:
P(A) = Number of A events / Total number of outcomes

Answers the question:
Out of all cases, what fraction had a specific event occur?

Example using our clinical trial with 10 total patients:

- A = { 2, 4, 6, 7 } (4 improved patients)
- B = { 4, 5, 10 } (3 patients with side effects)
- P(A) = 4/10 = 0.40
- P(B) = 3/10 = 0.30

There is a 40% chance that a patient improved.

There is a 30% chance that a patient has side effects.

Marginal probabilities tell us how often an event occurs on its own, across the full sample space. A natural next step is to look at the opposite perspective. How often does the event *not* occur? That brings us to complement probabilities.

Complement Probability

The complement of an event is the set of outcomes in the sample space where the event does not occur. If A represents some condition of interest, then the complement of A contains every outcome that fails to meet that condition.

Notation: $P(A^c)$

Formula:
$P(A^c)$ = Number of outcomes where A does not occur / Total number of outcomes

Answers the question:
Out of all cases, what fraction had a specific event that did not occur?

Let A be the event "the patient improved." If 4 of the 10 patients improved, then the remaining 6 did not.

- $A = \{ 2, 4, 6, 7 \}$ (4 improved patients)
- $A^c = \{ 1, 3, 5, 8, 9, 10 \}$ (6 patients who did not improve)
- $P(A^c) = 6/10 = 0.60$

There is a 60% chance that a patient did not improve.

Complement probabilities are especially useful when events are easier to describe by identifying the outcomes that *don't* satisfy a condition. Instead of listing all outcomes in event A, it can sometimes be simpler to describe the outcomes that fall outside A and subtract from the full sample space.

This approach is particularly helpful when dealing with events such as "at least one occurrence" or other situations where describing what does **not** happen is easier than listing every outcome that does. The complement gives us a simple way to capture everything outside A, and it plays a key role in more advanced techniques later in the book.

One famous example appears in the birthday problem, where we want to know the probability that at least two people in a group share a birthday. Calculating this event directly is difficult because there are many ways a shared birthday can occur. Instead, it is much easier to find the probability that no one shares a birthday and then use the complement. We'll solve the birthday problem in Chapter 4!

Complement probabilities show how often an event does *not* occur. But many real questions involve combining events rather than opposing them. To handle situations where we want the probability of A or B happening, we turn to union probabilities.

Union Probability

The union probability of A and B represents the chance that at least one of the events occurs. It includes A alone, B alone, and any overlap between them where both A and B occur.

Notation: P(A ∪ B)

Formula:
P(A ∪ B) = Number of A ∪ B events / Total number of outcomes

Answers the question:
Out of all cases, what fraction had at least one of the events occur?

Example using our clinical trial with 10 total patients:

- A = { 2, 4, 6, 7 } (improved patients)
- B = { 4, 5, 10 } (patients with side effects)
- A ∪ B = { 2, 4, 5, 6, 7, 10 } (6 patients who improved, had side effects, or both)
- P(A ∪ B) = 6/10 = 0.60

There is a 60% chance the patient improved, had side effects, or both.

Union probabilities capture the chance that at least one of the events occurs. Sometimes we need the probability that two events happen together. That leads us to joint probabilities.

Joint Probability

Joint probability is the likelihood that two or more events will coincide. Knowing how to calculate them allows you to solve problems such as the following. What is the probability of:

- Getting two heads in two coin tosses?
- Consecutively drawing two aces from a deck of cards?
- The next customer being a woman who buys a Mac computer?

- A bike rental customer getting both a flat front tire and a flat rear tire?

Statisticians use the notation of P(A ∩ B) to indicate the joint probability of events "A" and "B" occurring together. For example, P(F ∩ Mac) denotes the likelihood of a female buying a Mac. Equivalent variants of this notation include P(A and B) and P(A, B).

The notation includes the symbol "∩," which signifies an intersection in set rules. This intersection specifies how two or more events, such as A and B, coincide. Consequently, joint probability is also known as the intersection of events.

The joint probability of A and B represents the chance that both events occur together. It focuses on the overlap between the two sets.

Notation: P(A ∩ B)

Formula:
P(A ∩ B) = Number of A ∩ B events / Total number of outcomes

Answers the question:
Out of all cases, what fraction had both events occur?

Example using our clinical trial with 10 total patients:

- A = { 2, 4, 6, 7 } (improved patients)
- B = { 4, 5, 10 } (patients with side effects)
- A ∩ B = { 4 } (1 patient who improved and had side effects)

Only patient 4 is in both sets.

So:
P(A ∩ B) = 1/10 = 0.10

There is a 10% chance that a patient both improved and had side effects.

Joint probabilities describe overlap, or how often two events occur together within the full sample space. Sometimes, however, we want to focus on only the cases where one event has already occurred and ask how often the other event happens within that subset. This shift in perspective leads to conditional probability.

Conditional Probability

A conditional probability is the likelihood of an event occurring given that another event has already happened. Conditional probabilities allow you to evaluate how prior information or known conditions affect the likelihood of an outcome. For example, what is the probability that A occurs given that B has occurred?

Unlike joint probability, which considers the full sample space, conditional probability restricts attention to a subset of outcomes where a given event has occurred. In other words, instead of asking how often A and B occur together overall, we ask how often A occurs within the cases where B has already happened.

Typically, the problem statement for conditional probability questions assumes that one event occurred or indicates that an observer witnesses it. The goal is to calculate the chances of the second event under the condition that the first event occurred.

This idea might sound complicated at first, but it makes sense that knowing an event occurred can affect the chances of another event.

For example, if someone asks you, "What is the likelihood that you're carrying an umbrella?" Wouldn't your first question be, "Is it raining?" Obviously, knowing whether it is raining affects the chances that you're carrying an umbrella.

P(A | B) denotes the conditional probability of event A occurring given that event B has occurred.

For example, imagine we're assessing the likelihood that someone owns a cat given the presence of an empty cardboard box on their floor. We'd use the following notation:

P(Cat | Open box on floor) = 0.8

This notation indicates that the probability of someone owning a cat, given the presence of an open box on their floor, is 0.8.

I know because I own cats and often have an empty box on my floor for them to enjoy!

For a more serious example of conditional probability, consider medical testing, such as COVID tests. In this context, evaluating conditional probabilities is critical. You need to know the likelihood of obtaining a positive test result when a person has COVID and the chances of a negative result when a person does not have COVID.

In studies that assess medical tests, researchers already know whether a volunteer has COVID (or whatever disease they're testing for). They then administer a COVID test and record the results. Consequently, we're evaluating the likelihood of a test result given a participant's known status.

The following conditional probability notation represents these two cases:

- P(Positive COVID test | Person has COVID)
- P(Negative COVID test | Person does not have COVID)

Tests can perform well for one, both, or neither of these conditions. They need high conditional probabilities in both cases to limit the risk of false negatives and false positives.

Notation and Formula

A conditional probability focuses on the probability of one event given that another event has already occurred.

Notation:
$P(A \mid B)$ = Probability that event A occurs given that event B has already occurred.

Formula:
$P(A \mid B) = P(A \cap B) / P(B)$

Answers the question:
Out of the cases where B happened, what fraction also had A?

In the conditional probability formula, the numerator of the ratio is the joint chance that A and B occur together. We use the joint probability in the numerator because we are interested in the subset of outcomes in which both events occur.

The denominator is the marginal probability that B occurs. We use that value in the denominator because the occurrence of event B defines the restricted sample space for the conditional probability. In other words, conditional probability assumes that event B has occurred, and the denominator accounts for that condition.

Every outcome where both A and B occur must also be an outcome where B occurs. Hence, the numerator is a subset of the denominator. This means the joint probability $P(A \cap B)$ can never be larger than $P(B)$ (that is, $P(A \cap B) \leq P(B)$). As a result, conditional probabilities are always between 0 and 1.

The formula gives us the ratio of the probability that both events occur relative to the probability that the given event occurs. Therefore, if the ratio equals 1, event A always occurs when event B occurs. Conversely, if the ratio equals 0, event A never occurs when B occurs. In most conditional probability examples, the ratio lies between 0 and 1, indicating that A sometimes occurs after B.

This structure mirrors the calculations we performed with contingency tables. When we calculated conditional probabilities there, we used the count in the joint cell for the two characteristics and divided it by the total for the relevant row or column. That row or column total served as the restricted denominator.

The joint cell represented the cases where both conditions occurred together, while the row or column total represented all cases where the given condition was present. The value in a joint cell can never be larger than the row or column total that contains it. At a minimum, the joint cell can be zero. These limits again produce values between 0 and 1.

The conditional probability formula expresses this same idea using probability notation instead of counts from a table. Instead of counts, we use the joint probability in the numerator and the marginal probability in the denominator.

Examples

Let's start with an example using the clinical trial sets from before:

$P(A \mid B)$ = the conditional probability that a patient improved given they had side effects.

- $A \cap B = \{ 4 \}$ (1 patient who improved and had side effects)
- $B = \{ 4, 5, 10 \}$ (3 patients with side effects)
- $P(A \cap B) = 1/10 = 0.1$
- $P(B) = 3/10 = 0.3$

So:
$P(A \mid B) = P(A \cap B) / P(B)$
$P(A \mid B) = 0.1 / 0.3 = 0.33$

Among patients with side effects, 33% improved.

Up to this point, our examples have calculated probabilities by working directly with sets and counting outcomes in the sample space. Now that we have introduced the conditional probability formula, we can also work with probabilities that are already expressed numerically.

In the next example, we will apply the formula using probabilities rather than sets. Then, in the next chapter, we will expand this approach further by introducing probability rules that allow us to calculate more probabilities directly from other probabilities.

Example: Carrying an Umbrella

Assume that a study in a rainy city assesses the chances related to rainy days and carrying umbrellas. The researchers were out observing the weather and people.

We want to see how the probability of carrying an umbrella changes depending on whether it is raining. That requires calculating the following two conditional probabilities:

- P(Umbrella | Rain): The likelihood that someone is carrying an umbrella given that it is raining.
- P(Umbrella | No Rain): The likelihood that someone is carrying an umbrella given that it is not raining.

To find our answers, we'll use the following two conditional probability formulas:

$$P(\text{Umbrella} \mid \text{Rain}) = \frac{P(\text{Umbrella} \cap \text{Rain})}{P(\text{Rain})}$$

$$P(\text{Umbrella} \mid \text{No Rain}) = \frac{P(\text{Umbrella} \cap \text{No Rain})}{P(\text{No Rain})}$$

The researchers find the following probabilities:

- P(Umbrella ∩ Rain): 0.20
- P(Umbrella ∩ No Rain): 0.40
- P(Rain): 0.25
- P(No Rain): 0.75

If you look at the joint likelihoods in the first two bullets, it appears that the chance of carrying an umbrella when it's *not* raining (0.40) is higher than when it is raining (0.20). That seems backward, and we'll come back to that.

For the correct answer, we need to calculate the conditional probability. Let's plug these numbers into the conditional probability formula!

$$P(\text{Umbrella} \mid \text{Rain}) = \frac{0.20}{0.25} = 0.80$$

$$P(\text{Umbrella} \mid \text{No Rain}) = \frac{0.40}{0.75} = 0.53$$

Based on the conditional probabilities, we see that people are more likely to carry an umbrella given that it's raining (0.80) compared to when it's not raining (0.53). That makes sense!

The joint likelihoods were misleading because they do not account for the fact that days with no rain are three times as likely as rainy days (0.75 vs. 0.25)! In this study, you're more likely to see people carrying umbrellas when there's no rain because there are many more days with no rain. The conditional probabilities consider that fact.

This example of conditional probabilities involves dependent events. We know that's the case because when you change the initial event (Rain), the chances of the second event (Umbrella) change. Consequently, the likelihood of the second event *depends* on the first event.

We'll come back to independent vs. dependent events in the next chapter!

Key Distinction: Joint vs. Conditional

Joint probabilities describe the likelihood of two events when we consider the entire sample space.

Conditional probabilities describe the likelihood of one event within a subset of the sample space where the other event has occurred.

If you forget which is which, you can reach incorrect conclusions, as you saw with the umbrella example.

Partitioning an Event into Two Parts

Sometimes it is helpful to break an event into two non-overlapping pieces by using joint probabilities with complements. For example, event A can be separated into two parts: the outcomes where A occurs together with B, and the outcomes where A occurs without B. These two subsets form a complete and non-overlapping breakdown of event A. Together they contain all outcomes where A occurs.

Consequently:
$P(A) = P(A \cap B) + P(A \cap B^c)$

This identity is often useful when information is given about the overlap between events and the cases where one event occurs without the other. It simply divides event A into two parts: when B occurs and when B does not occur.

Use this method when you know the probabilities for two of these quantities and need to find the third. Most commonly, it is used to find P(A) when you know the probabilities for the cases where A occurs with B and where A occurs without B.

We'll return to partitioning events in this manner when we create Venn and tree diagrams in Chapter 6.

Summary and Next Steps

In Chapter 2, you learned how to calculate probabilities from data using contingency tables. That approach is rooted in real observations and works well whenever data are available. In this chapter, we shifted from data-driven probabilities to a general symbolic framework applicable to any setting.

You learned how to describe outcomes using the sample space and how to represent events as sets. We used set notation to define the basic types of probability: marginal, complement, union, joint, and conditional. And you saw how each one reflects a different relationship between events. Set operations helped us understand how events combine, overlap, and relate to each other, forming the backbone of probability reasoning.

This structure leads directly to the core probability rules in the next chapter. Now that you understand the main types of probabilities and how they arise from set relationships, you are ready to learn the rules that connect them and allow us to calculate new probabilities from ones we already know.

Chapter Review Summary

- **Sample Space (S):** The set of all possible outcomes. Example: S = {1, 2, ..., 10} for ten study participants.
- **Events (A, B, ...):** Subsets of the sample space defined by conditions.
  Example: A = {patients who improved}.
- **Event Operations:**
  - **Union ($A \cup B$)**: "A or B": all outcomes in A, in B, or in both.
  - **Intersection ($A \cap B$)**: "A and B": outcomes in both A and B.
  - **Complement (A' or A^c)**: "not A": outcomes not in A.
  - **Empty Set ($\emptyset$)**: event with no possible outcomes.
- **Types of Probabilities:**
  - **Marginal Probability: P(A)**: probability that event A occurs.
  - **Complement Probability: $P(A'$ or $A^c)$**: probability that A does not occur.
  - **Union Probability: $P(A \cup B)$**: probability that A or B occurs.
  - **Joint Probability: $P(A \cap B)$**: probability that A and B occur together.
  - **Conditional Probability: $P(A \mid B)$**: probability of A given that B has occurred.
    - $P(A \mid B) = P(A \cap B) / P(B)$

Chapter 3 Practice Problems

Question 1: For each expression, identify the type of probability.

A. $P(A)$
B. $P(A \cap B)$
C. $P(A \mid B)$
D. $P(A \cup B)$
E. $P(B^c)$

Question 2: Suppose events A and B occur in the same sample space. Is the following statement always true, sometimes true, or never true?

$P(A \mid B) = P(B \mid A)$

Explain your reasoning.

Question 3: A school survey records whether students participate in two activities:

A = plays a sport
B = participates in the music program

The sample space consists of 12 students labeled 1–12.

$A = \{1, 2, 4, 5, 7, 9, 11\}$
$B = \{2, 3, 5, 6, 9, 10\}$

A. Find $A \cap B$.
B. Find $A \cup B$.
C. How many students participate in at least one of the two activities?
D. How many students participate in neither activity?

Question 4: A quality inspector checks 20 products coming off an assembly line.

Event A = product passes inspection

Suppose 16 of the products pass inspection.

A. Find P(A).
B. Find $P(A^c)$.
C. Explain in words what event A^c represents.

Question 5: A dataset contains 15 employees labeled 1–15.

Event A = employees who work remotely
Event B = employees in the marketing department

A = {2, 4, 5, 8, 11, 13}
B = {1, 4, 6, 8, 9, 12}

A. Identify A ∩ B.
B. Find P(A ∩ B).
C. Interpret this probability in words.

Question 6: Consider the following description and determine which is more likely.

Linda is 31 years old, single, outspoken, and very bright. She majored in philosophy. As a student, she was deeply concerned with issues of discrimination and social justice and participated in anti-nuclear demonstrations.

Which of the following statements is more likely?

A. Linda is a bank teller.
B. Linda is a bank teller and is active in the feminist movement.

Question 7: Suppose two events A and B occur in the same sample space.

A student calculates the conditional probability using the formula and obtains a value of **1.25**.

$P(A \mid B) = P(A \cap B) / P(B)$

The student concludes that event A is very likely to occur when B occurs.

Is this conclusion possible? Explain why or why not.

Question 8: A small online store tracks purchases for 40 customers.

Event A = customer bought headphones
Event B = customer bought a phone case

The store records:

- 10 customers bought both items
- 25 customers bought a phone case

Calculate the following:

A. $P(A \cap B)$
B. $P(B)$
C. $P(A \mid B)$

Interpret the result in words.

Question 9: A transportation survey records whether commuters drive to work and whether they live in the city center.

Event A = commuter drives to work
Event B = commuter lives in the city center

Researchers observe the following probabilities:

$P(A \cap B) = 0.15$
$P(A \cap B^c) = 0.35$
$P(B) = 0.20$

Answer the following:

A. Calculate P(A).
B. Calculate P(A | B).
C. Explain why looking only at P(A) might give a misleading impression about commuters who live in the city center.

Question 10: A family has two children. Assume that each child is equally likely to be a boy or a girl, and that each day of the week is equally likely as a day of birth.

You are told that at least one of the children is a boy who was born on a Tuesday.

What is the probability that both children are boys?

CHAPTER 4

Combining Probabilities: Rules and Relationships

So far, we have identified marginal, complement, joint, and conditional probabilities by counting outcomes in the sample space. This approach builds intuition about how events relate by determining which outcomes belonged to one or more events and dividing by the total number of possible outcomes. That approach helps build intuition for how events relate to each other.

In this chapter, we shift from counting outcomes to working with the probabilities themselves. Instead of listing or counting outcomes each time, we will use several core rules to combine existing probabilities, split them, or calculate new ones more efficiently. We did a little of that with the umbrella example and conditional probabilities in the previous chapter.

Now, you'll learn more methods for other types of probabilities and combinations of events using several standard rules. These tools allow us to combine probabilities from many contexts, whether from observed data, previous analyses, scientific research, or theoretical reasoning.

Probability does not always begin with new data. Often, we already know probabilities from other sources. Weather forecasters estimate the chance of rain tomorrow. Medical researchers publish survival probabilities for different treatments. Engineers have reliability estimates for components. And some situations have mathematical structures that allow us to reason about probability theoretically without any data at all, such as for rolling dice, flipping coins, and drawing cards from a deck.

To take advantage of these different sources of information, we need systematic methods for working with probabilities. These methods include the additive rule, the complement rule, and the multiplication rule. They allow us to combine known probabilities to find new ones and follow directly from how sets work. After you understand these rules, you can calculate probabilities in a wide range of situations, even when events become complex.

However, applying these rules correctly requires understanding how events relate to one another. In particular, we must consider whether events are independent, dependent, or mutually exclusive. These relationships affect how probabilities combine and determine which rules apply in a given situation.

Before we learn the rules, it helps to recall three basic properties that probabilities always follow. These properties will act as a foundation for all the rules you're about to learn:

1. A probability value must always be between 0 and 1, inclusive.
2. The probability of a combined event within a sample space equals the sum of the probabilities of its outcomes.
3. The probabilities of all possible outcomes in the sample space must add up to 1.

Everything we do with probability must follow these three principles. These properties are informal versions of the deeper foundations of probability, called the Kolmogorov axioms, which we examine later in the chapter.

The following rules let us work directly with probabilities rather than returning to the sample space and counting outcomes, giving us a faster and more flexible way to solve problems that involve combinations of events. We will refer to them throughout future chapters.

Additive Rule

The additive rule tells us how to find the union probability that A or B occurs: $P(A \cup B)$.

General case: $P(A \cup B) = P(A) + P(B) - P(A \cap B)$

From the previous chapter, we found the following probabilities for our clinical trial example using set notation with the outcome counts. Now, imagine we have only these probabilities to solve the problem:

- $P(A) = 0.40$
- $P(B) = 0.30$
- $P(A \cap B) = 0.10$

To find the union probability, plug the values into the additive rule:

$P(A \cup B) = P(A) + P(B) - P(A \cap B)$
$P(A \cup B) = 0.4 + 0.3 - 0.1 = 0.60$

That's the same union probability we found using the outcome counts. But why do we subtract out the joint probability $P(A \cap B)$?

If we just add P(A) and P(B), we double-count outcomes where both A and B occur. Subtracting the joint probability fixes that. Let's look at the sample space briefly.

- $A = \{ 2, 4, 6, 7 \}$
- $B = \{ 4, 5, 10 \}$
- $A \cap B = \{ 4 \}$
- $A \cup B = \{ 2, 4, 5, 6, 7, 10 \}$

Notice that A has 4 members and B has 3. However, the union $A \cup B$ has only six members. That's because patient 4 is in both sets, and you need to subtract them out once to get the correct number. Hence, you must subtract the intersection of $A \cap B$, which contains the overlap.

When events A and B are mutually exclusive, there is no double-counting, and the additive rule simplifies to the following:

$$P(A \cup B) = P(A) + P(B)$$

More about this simplification in the upcoming Mutual Exclusivity section!

Complement Rule

The complement rule focuses on what does not happen. Use it when you want the probability that event A fails to occur, given that you know the probability that A occurs.

$$P(A^c) = 1 - P(A)$$

This rule reflects the idea that A and not A cover the entire sample space. In other words, the probability that A or A^c occurs is 1. One must occur. Hence, if you know the chance of A happening, you automatically know the chance it does not happen.

Example:
A = "It rains tomorrow."
A^c = "It does not rain tomorrow."

If the weather service reports $P(A) = 0.30$:
$P(A^c) = 1 - 0.30 = 0.70$

There is a 70% chance it will not rain tomorrow.

Use the complement rule when the event involves many outcomes that are tedious to list and calculate individually. Finding the complement is often the easiest path when calculating the direct event requires adding up lots of small probabilities.

Multiplication Rule

The multiplication rule tells us how to find the joint probability of A and B happening together when we know:

- The marginal probability of B, $P(B)$, and
- The conditional probability of A given that B has occurred, $P(A \mid B)$.

General rule:
$P(A \cap B) = P(A \mid B) \times P(B)$

You can think of the multiplication rule as describing events that occur in sequence. First, event B occurs with probability $P(B)$. Then event A happens within that restricted situation with probability $P(A \mid B)$. Multiplying these probabilities gives the probability that both events occur together.

Here's how it works:

- $P(B)$ narrows the sample space to only the cases in which B occurs.
- $P(A \mid B)$ tells us how often A occurs within that smaller set.

Multiplying them gives the joint probability where both A and B occur.

This rule is one of the most useful shortcuts in probability because it allows us to find a joint probability using just a marginal and a conditional probability without the original data.

The general multiplication rule has an important symmetry that often goes unnoticed at first. When two events, A and B, can both occur, either one can serve as the conditioning event. The joint probability $P(A \cap B)$ can be written in two equivalent ways:

$$P(A \cap B) = P(A) \times P(B \mid A) = P(B) \times P(A \mid B)$$

Both forms describe the same overlap between A and B, just viewed from different starting events (A or B) in the sequence described earlier. Be sure to multiply each conditional probability by the marginal probability of its conditioning event.

The multiplication rule also extends to sequences involving more than two events. When several events occur in order, the joint probability equals the probability of the first event multiplied by the conditional probability of each subsequent event. For three events A, B, and C, the joint probability is the following:

$$P(A \cap B \cap C) = P(A) \times P(B \mid A) \times P(C \mid A \cap B)$$

Each step conditions on the outcomes that have already happened. The term $P(C \mid A \cap B)$ appears because, by the time we consider event C, both A and B have already occurred. Therefore, the probability for the third step must be conditioned on the intersection of those earlier events.

Simple Examples

From the previous section, we found the following probabilities for our clinical trial example using set notation with the outcome counts. Now, imagine we have only these probabilities to solve the problem:

- $P(A \mid B) = 0.33$ (improved patients given side effects)
- $P(B) = 0.30$ (patients with side effects)

To find the joint probability, apply the multiplication rule:

$$P(A \cap B) = P(A \mid B) \times P(B)$$
$$P(A \cap B) = 0.33 \times 0.3 = 0.10$$

This matches our earlier result using outcome counts in the sample space: 10% of patients improve and have side effects.

Let's try another example.

Suppose we need to calculate the likelihood of drawing two aces consecutively from a standard deck of 52 cards when we don't replace the cards. Initially, the deck contains four aces, so the likelihood of drawing an ace on the first draw is 4/52 or 1/13. If we draw an ace (event A1), only three aces and 51 cards remain in the deck. Consequently, the conditional probability of drawing another ace (event A2) is now 3/51.

Using the general multiplication rule, we can find the joint probability of drawing two aces in a row:

$$P(A1) = 4/52 = 1/13$$
$$P(A2 \mid A1) = 3/51$$

$$P(A1 \cap A2) = P(A2 \mid A1) \times P(A1)$$
$$= (1/13) \times (3/51) = 3/663 = 1/221$$

So, the joint probability of drawing two aces in a row is 1/221 or 0.0045.

Let's combine two of the previous rules to solve several confusing problems!

The next two examples combine the complement and multiplication rules in ways that also introduce an important idea: whether events are independent or dependent. In the first example, the probabilities remain constant because the events are independent. In the second, the probabilities change from step to step because earlier outcomes affect later ones.

Example: Repeated Independent Attempts

Suppose the probability of success on a single attempt is 0.01. The events are independent, meaning one trial does not affect another, and the probability of success remains constant across trials.

You repeat the attempt 100 times. What is the probability of at least one success?

At first glance, many people expect the probability to be very high because there are so many attempts. After all, you have a 1-in-100 chance of succeeding, and you attempt it 100 times.

However, simply adding the probabilities does not work here because success can occur on more than one attempt. For example, success could occur on both the 10th and 20th attempts. Adding 100 separate 0.01 probabilities would incorrectly count these joint occurrences multiple times and produce an inflated probability.

One approach would be to add the probabilities and then subtract out the overlaps between attempts, just as we subtract joint probabilities in the additive rule to avoid double-counting. However, that quickly becomes complicated because many different combinations of overlapping successes are possible.

Instead, it is much easier to use the complement rule to calculate the probability of no successes and then apply the multiplication rule.

First, apply the complement rule:
P(at least one success) = 1 – P(no successes)

Now find the probability of no successes. Each attempt has a 0.99 probability of failure. Because the attempts are independent, each failure probability remains 0.99 from trial to trial, allowing us to multiply the probabilities together:

P(no successes) = 0.99^{100}

Therefore,
P(at least one success) = 1 – 0.99^{100} = 0.634

So even after 100 attempts, the probability of at least one success is only about 63.4%, far from 100% as many people expect. Some sets of 100 attempts will contain no successes at all, while others will contain one or more successes.

This example shows how combining the complement rule with the multiplication rule allows us to solve problems involving repeated independent events. In the next example, we'll use a similar strategy, but with an important difference: the probabilities change at each step rather than remaining constant.

Example: Birthday Problem

The Birthday Problem asks: How many people are needed in a group for a 50% chance that at least two share a birthday? Go ahead and think about that for a moment. The answer surprises most people.

Many people guess 183 because that is half of all possible birthdays, which seems intuitive. Unfortunately, intuition fails to solve this problem as badly as it does the Two-Child Problem, though for different reasons. So, let's get straight to calculating probabilities for people sharing birthdays.

For these calculations, we'll make a few assumptions. First, we'll disregard leap year. That simplifies the math and doesn't change the results by much. We'll also assume that all birthdays have an equal probability of occurring and that different people's birthdays are independent. One person's birthday does not influence the next person's birthday.

At first glance, calculating this probability directly seems complicated. There are many ways people in a group can share a birthday. For example, the match could occur between the first and second person, the third and tenth person, or several people could share the same birthday. Accounting for all those possibilities individually would require adding together many different probabilities.

In the previous example with repeated events, each trial was independent and the probability remained constant from one attempt to the next. The birthday problem uses a similar strategy, but with an important difference: the probabilities change at each step. As we consider each additional person, we condition on the birthdays of people already in the group, which creates a sequence of dependent probabilities.

We'll use the same strategy that combines the complement rule and the multiplication rule. Rather than calculating the probability that at least two people share a birthday, we first calculate the complement: the probability that everyone in the group has a different birthday.

P(at least one match) = 1 – P(no match)

This situation is easier to analyze because we can think about the birthdays sequentially. The first person can have any birthday. The second person must have a birthday different from the first. The third person must differ from the first two, and so on.

We then apply the general multiplication rule, which allows us to combine a sequence of conditional probabilities. Each probability is conditional on the birthdays that have already appeared in the group, because the next birthday must differ from all previous birthdays.

After the first person's birthday is set, the second person must avoid that one specific day, leaving 364 acceptable days out of 365. If the first two birthdays are different, the third person must avoid both of those days, leaving 363 acceptable days. Each additional person must avoid all previously used birthdays, so the number of available days decreases by one each time.

As a result, we multiply a chain of conditional probabilities where the numerator decreases by one at each step. This sequence gives the joint probability that every person in the group has a different birthday. After finding that probability, we apply the complement rule to determine the probability that at least two people share a birthday.

Let's start with one person and then add people one at a time to illustrate how the calculations work.

For the first person, there are no birthdays already covered, which means that there is a 365/365 chance that there is not a shared birthday. That makes sense. We have just one person.

$$1 - \left(\frac{365}{365}\right) = 0$$

Now, let's add the second person. The first person covers one possible birthday, so the second person has a 364/365 chance of not sharing the same day. We need to multiply the probabilities of the first two people and subtract the result from 1.

$$1-\left(\frac{365}{365}\right)\times\left(\frac{364}{365}\right)=0.0027$$

For the third person, the previous two people cover two dates. Hence, the third person has a probability of 363/365 for not sharing a birthday.

$$1-\left(\frac{365}{365}\right)\times\left(\frac{364}{365}\right)\times\left(\frac{363}{365}\right)=0.0082$$

Now, you're seeing the pattern for calculating the probability for a given number of people. Here's the general form of the equation:

$$1-\left(\frac{365}{365}\right)*\left(\frac{364}{365}\right)*\left(\frac{363}{365}\right)*\cdots*\left(\frac{365-n+1}{365}\right)$$

I can calculate and graph the probabilities for any size group using Excel. This spreadsheet is available at my website on the resource page for this book.

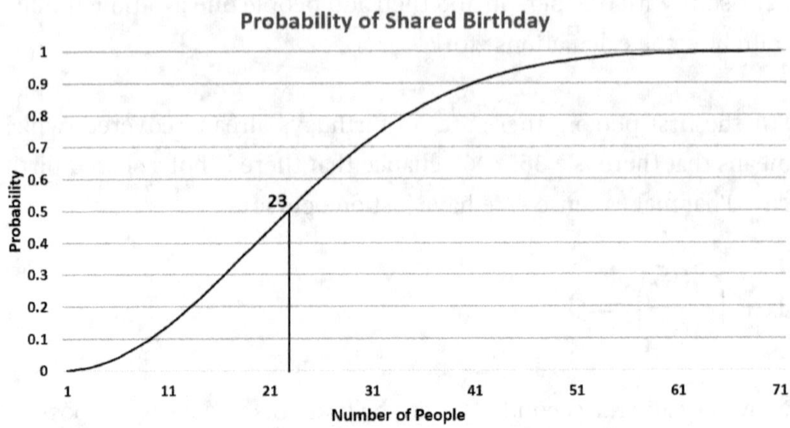

The answer is that you need a group of 23 people to have a 50.73% chance of at least 2 people sharing a birthday. Most people don't expect the group to be that small. Also, notice that a group of 57 has a probability of 0.99. It's virtually guaranteed!

Why is the group size so small? Most people think that 23 is surprising, and it hurts their brains a bit! Let's examine why it is so low.

Often, people think of their birthday and the probability of someone matching that specific date. However, the problem asks about any two people sharing a birthday. Hence, you must compare all possible pairs, which causes the number of comparisons to increase rapidly.

People	Pairs
1	0
2	1
3	3
4	6
5	10
6	15
7	21
8	28
9	36
10	45
11	55
12	66
13	78
14	91
15	105
16	120
17	136
18	153
19	171
20	190
21	210
22	231
23	253

The formula for the number of pairs for N people is: $(N \times (N-1))/2$. As you can see, the number snowballs to 253 for only 23 people!

People tend to think linearly, but the number of pairs increases exponentially. As the number of possible pairs grows, the number of opportunities for a match also increases.

For sharing a birthday, each pair has a 1/365 (0.0027) chance of matching. That's low for just one pair. However, as the number of pairs increases rapidly, so does the likelihood of a match. With 23 people, you need to compare 253 pairs. With that many comparisons, it becomes increasingly unlikely that none of the pairs share a birthday.

When there are 57 people, there are 1,596 pairs to compare, and it's virtually guaranteed with a 0.99 probability that at least one pair will match birthdays.

I love problems like this where intuition leads you astray, but math saves the day!

Now that we can combine probabilities using the core rules, we can examine how the relationship between events affects those rules. Two of the most important relationships are independence and mutual exclusivity.

You saw how independence and dependence affected how we solved the previous two problems. Let's start by examining the concept of independence more closely.

Independence

Independent events are those in which one event does not affect the next event. More specifically, the occurrence of one event does not affect the probability of the following event happening.

I've covered this loosely up to this point. Now we'll get into it more formally.

- **Independent events**: The occurrence of one event does not affect the probability of the other event. For example, when flipping a coin, getting "heads" does not change the likelihood of getting "heads" on the next coin flip. Or, when rolling a die, one outcome does not impact the next die roll.

- **Dependent events**: The occurrence of one event does affect the probability of the other event. For example, if you draw a King from a deck of cards and do not replace it, it causes the probability of getting another King on the next draw to decrease.

Conceptually, the idea behind independence is that information about a previous event doesn't change the probability of the subsequent event occurring.

Crucially, some events influence each other. Others do not. And you need to determine which case applies to calculate the correct probabilities. That's based on understanding how the system works.

Sometimes you can make that determination based on theory, such as knowing that coin tosses are independent events. The outcome of one flip does not affect the result of the next. If you have gotten tails four times in a row, the probability of tails on the next flip is still 50/50. Independent events have no "memory" of previous events.

Conversely, a deck of cards has a kind of a memory. When you draw a King from the deck, it reduces the number of Kings in the remaining cards and reduces the future probability of drawing another King.

Other times, you can use subject-area knowledge about how something works. For instance, if a patient has the flu (previous event), they have a greater probability of having a fever.

In many real-world settings, independence is assumed because it simplifies calculations. But convenience doesn't guarantee accuracy, so it's important to think critically about whether the events truly do not influence one another.

Now, let's move on to the probability perspective. We will use both conditional and joint probabilities to describe and assess independence.

Definition Based on Probabilities

Two events A and B are independent when learning that one occurred does not change the probability of the other. Conversely, when the events are dependent, they do change.

This idea is captured in two equivalent mathematical forms. These forms allow you to understand what independence looks like probabilistically and to assess independence when you have the necessary probabilities.

Conditional Probability

A and B are independent when conditioning on one event does not change the probability of the other:

- $P(A \mid B) = P(A)$
- $P(B \mid A) = P(B)$

This form highlights the core meaning of independence:

New information does not change the probability.

Ok, let's take a step back. When you assess conditional probabilities of independent events, the following is true:

$P(A \mid B) = P(A)$

But what does that mean?

The probability of A given that B occurred equals the probability of A without considering B at all. Whether B occurs or not has no effect on the chances of A. That's exactly what it means for events to be *independent*!

For independent events, knowing that an event has occurred does not change the probability of the remaining events.

Conversely, if conditioning on B changes the probability of A, the events are dependent. That shift in probability is the clearest sign of dependence: $P(A \mid B) \neq P(A)$.

In practice, you check for independence by seeing whether this equivalence exists.

Joint Probability

You can also assess independence using joint probabilities. The multiplication rule has two forms: the general form that applies to all events, and a specific form that only applies when events are independent. Earlier, you learned the general form.

In a later section, we'll cover the multiplication rule for independent events in more detail. For now, understand that this equation says the joint probability is simply the product of the individual probabilities for independent events because neither event affects the other.

For independent events, these two forms give the same result. That is:

- $P(A \cap B) = P(A) \times P(B \mid A)$ (General)
- $P(A \cap B) = P(A) \times P(B)$ (Independent)

This equality occurs only when conditioning on B does not change the probability of A.

In practice, you can check independence by comparing the general multiplication rule to the product of the marginal probabilities. When those values match, the events behave independently. When they differ, the events are dependent.

These conditional and joint probability method of assessing independence are mathematically equivalent. Both express the same idea from different angles. You can use whichever version is more convenient based on the information available.

Now that we've established two ways to identify independence, let's look at numerical examples.

Assessing Independence Using Existing Probabilities

Let's start with an example where the events are genuinely independent to see how that works out numerically.

Let:

A = Prefers coffee over tea.
B = Exercises in the morning.

These are unrelated personal habits. Liking coffee does not cause someone to exercise, and exercising does not affect beverage preferences. Because of that, we expect:

$P(B \mid A) = P(B)$

Let's use realistic probabilities based on common survey patterns. Suppose in a large community:

- $P(A) = 0.60$: 60% prefer coffee over tea
- $P(B) = 0.25$: 25% exercise in the morning
- $P(B \mid A) = 0.25$: among coffee drinkers, 25% exercise in the morning

Now compare $P(B \mid A)$ with $P(B)$:

- $P(B \mid A) = 0.25$
- $P(B) = 0.25$

The probabilities are identical. Learning that someone prefers coffee does not change the likelihood that they exercise in the morning. This equality is the hallmark of independent events.

Next, let's check independence using the multiplication rule. For this method, we compare the results of the specific rule for independent events and the general multiplication rule. If the two results match, the events are independent.

- **Independent case**
  $P(A \cap B) = P(A) \times P(B)$
  $= 0.60 \times 0.25$
  $= 0.15$

- **General Rule**
  $P(A \cap B) = P(A) \times P(B \mid A)$
  $= 0.60 \times 0.25$
  $= 0.15$

Both methods produce the same joint probability of 0.15.

Because these values match and the conditional probability didn't change, we conclude the events are independent. Knowing that someone prefers coffee gives you no information about whether they exercise in the morning.

These patterns indicate independence numerically:

- The conditional probability equals the marginal probability.
- The two forms of the multiplication rule give identical results.

Example of Dependent Events

Now, consider the following dependent events.

A = Has the flu.
B = Has a fever.

These events are related because the flu commonly causes a fever. If someone has the flu, the chance they have a fever increases, and you'd expect the following:

$P(B \mid A) > P(B)$

To see this numerically, let's use probabilities based on typical seasonal influenza data. Suppose in a community during flu season:

- $P(A) = 0.08$: 8% of people have the flu.
- $P(B) = 0.10$: 10% of people have a fever (from any cause).
- $P(B \mid A) = 0.70$: 70% of people with the flu develop a fever.

Now compare $P(B \mid A)$ with $P(B)$:

- $P(B \mid A) = 0.70$
- $P(B) = 0.10$

These probabilities are very different. Learning that someone has the flu raises the chance of fever from 10% to 70%. That is a strong shift and a clear sign of dependence.

Next, let's check independence using the two forms of the multiplication rule. If the results are equal, the events are independent. When they're not equal, they're dependent.

If A and B were independent, $P(A \cap B)$ would equal:

- $P(A) \times P(B)$ (Independent rule)
  $= 0.08 \times 0.10$
  $= 0.008$

But using the general form of the multiplication rule with the conditional probability, the actual joint probability is:

- $P(A \cap B) = P(A) \times P(B \mid A)$ (General Rule)
  $= 0.08 \times 0.70$
  $= 0.056$

These two numbers are nowhere close:

- Expected joint probability under independence: 0.008
- Actual joint probability: 0.056

This gap shows that the events are dependent. Having the flu sharply changes the chance of fever.

The Multiplication Rule Simplifies with Independent Events

Earlier, we learned the general multiplication rule:

$P(A \cap B) = P(A \mid B) \times P(B)$

This formula works for all cases. However, as discussed in the previous section, when A and B are independent, then:

$P(A \mid B) = P(A)$

This equivalence allows you to replace the conditional probability ($P(A \mid B)$) in the general rule with the marginal probability ($P(A)$) in a simplified multiplication rule for independent events:

$P(A \cap B) = P(A) \times P(B)$

In short, for independent events, you simply multiply their marginal probabilities and don't need to bother with conditional probabilities. The joint probability of A and B occurring equals the probability of A occurring multiplied by the probability of B occurring.

Example

Imagine you particularly like wearing tan pants with a blue shirt. However, in the morning, you're sleepy and grab your pants and shirt randomly from the closet. The pants are on one side of the closet while the shirts are on the other. These are independent events because grabbing a pair of pants doesn't affect the probabilities for shirts.

You have ten pairs of pants, and three are tan. Consequently, the probability of drawing a tan pair (event TP) is 0.3.

You have 16 shirts, and four are blue. Hence, the probability of grabbing a blue shirt (event BS) is 0.25.

These are independent events because selecting a pair of pants doesn't affect the likelihood of drawing a blue shirt and vice versa.

Using the multiplication rule for independent events:

$P(TP \cap BS) = P(TP) \times P(BS)$
$0.3 \times 0.25 = 0.075$

Or, the joint probability of randomly selecting a pair of tan pants and a blue shirt is 7.5%.

The likelihood of getting your preferred combination is low! You might want to drink some coffee so you choose more carefully!

Understanding independent events is crucial to calculating probabilities in statistics. When events are independent, you can easily calculate the likelihood of multiple occurrences by simply multiplying the individual probabilities because the occurrence of one event does not impact the chances of the next.

Although independence is straightforward in theory, applying it correctly requires careful attention. Even small differences in what we know can change the sample space and affect the probabilities.

Independence vs. Dependence: How Getting It Wrong Causes Trouble

Independence is one of the most useful ideas in probability, but it is also one of the most misunderstood. Determining whether events are independent or dependent is not just a technical step. It shapes every calculation you make afterward. When you classify a situation correctly, you can perform the calculations accurately.

When you get it wrong, the consequences can be serious. Problems arise in several ways. Sometimes people misunderstand what information they have and apply independence incorrectly.

- Treating independent events as if they were dependent can be costly, leading people to believe that past outcomes influence future ones when they do not, resulting in bad decisions and incorrect predictions.
- Treating dependent events as independent can lead people to underestimate risks, overestimate reliability, or misjudge the likelihood of rare events.

One of the most subtle sources of error is failing to recognize how the information provided in a problem changes the sample space. Even when events are independent, conditioning on incomplete or ambiguous information can lead to surprising results.

Because of this, one of the most important skills you can develop is the ability to recognize which situations involve independent events and which do not.

The next several examples illustrate how misunderstandings about independence can lead to errors in different ways. First, we examine what happens when applying independence without correctly identifying the sample space. Next, we consider the Gambler's Fallacy, where people incorrectly believe independent events must "balance out." Finally, we examine the opposite error in real-world cases, where analysts mistakenly treated dependent events as independent.

Example: Rolling a Double-Six on Two Dice

Suppose you roll two dice. You don't see the results, but an observer tells you that at least one die shows a six. What is the probability that both dice show a six?

The outcomes of rolling two dice are independent, and the probability that any single die shows a six is 1/6 (0.167). However, the answer to this problem is actually 1/11 (0.091).

This result seems surprising at first. If the dice are independent, why isn't the answer 1/6?

The key is understanding what information we have. The statement "at least one die shows a six" does not tell us which die shows the six. That uncertainty changes the sample space we must consider.

As with the Two-Child Problem, the precise wording determines the set of possible outcomes and can sometimes produce surprising results. Conditioning on this information subtly creates a new sample space that includes all outcomes with at least one six, not just those where a specific die is six.

Again, the specific wording matters. In the Double-Six Problem, as stated, we do not know which die shows a six. Similarly, in the Two-Child Problem, we do not know which child is a boy.

Case 1: At Least One Die Is a Six

First, list all possible outcomes where at least one die is a six. Using ordered pairs where the first number represents Die 1 and the second represents Die 2, the possible outcomes are:

	Die 2					
Die 1	1	2	3	4	5	6
1	(1,1)	(1,2)	(1,3)	(1,4)	(1,5)	(1,6)
2	(2,1)	(2,2)	(2,3)	(2,4)	(2,5)	(2,6)
3	(3,1)	(2,2)	(2,3)	(3,4)	(3,5)	(3,6)
4	(4,1)	(4,2)	(4,3)	(4,4)	(4,5)	(4,6)
5	(5,1)	(5,2)	(5,3)	(5,4)	(5,5)	(5,6)
6	(5,1)	(5,2)	(5,3)	(5,4)	(5,5)	(5,6)
6	(6,1)	(6,2)	(6,3)	(6,4)	(6,5)	(6,6)

Possible outcomes when rolling two dice. The highlighted row (Die 1 = 6) and column (Die 2 = 6) show outcomes where at least one die equals six. These 11 outcomes form the conditional sample space.

A = { (6,1), (6,2), (6,3), (6,4), (6,5), (6,6), (1,6), (2,6), (3,6), (4,6), (5,6) }

There are 11 possible outcomes consistent with the information that at least one die is a six.

Only one of these outcomes contains two sixes:

(6,6)

Therefore, P(both dice are six | at least one six) = 1/11.

Notice something interesting here. The outcomes with a six on the first die form one set:

D_1 = { (6,1), (6,2), (6,3), (6,4), (6,5), (6,6) }

The outcomes with a six on the second die form another set:

D_2 = { (1,6), (2,6), (3,6), (4,6), (5,6), (6,6) }

Each set contains six outcomes. However, the union of these two sets does not contain 12 outcomes because the pair (6,6) appears in both sets. Remember the caution to not double-count the overlap in unions! The overlap must be counted only once. As a result, the union contains 11 unique outcomes (not 12) to include in the denominator.

Additionally, both sets contain the same overlapping outcome: (6,6). This shared outcome represents the only case where both dice equal six, so the numerator contains only one favorable outcome, not 2.

Taken together, this overlap explains why the probability is 1/11 rather than 2/12 (or 1/6).

P(both dice are six | at least one die is a six) = 1/11.

Case 2: A Specific Die Is Known to Be Six

Now consider a slightly different question where independence plays a more intuitive role.

Suppose we know that Die 1 shows a six. What is the probability that both dice are six?

The relevant sample space now contains only outcomes where the first die is six:

B = { (6,1), (6,2), (6,3), (6,4), (6,5), (6,6) }

There are six possible outcomes, and one of them is (6,6). Therefore, P(both dice are six | Die 1 is six) = 1/6.

This result matches our intuition. After we know that the first die is a six, the probability that the second die is also a six is still 1/6. As expected, the Die 1 outcome does not affect the Die 2 outcome. Yes, they are truly independent!

The apparent conflict with independence arises because the two situations provide different information.

- When we know that a specific die is a six, the other die remains independent, and the probability is 1/6.
- When we only know that at least one die is a six, we do not know which die produced that outcome. This uncertainty changes the sample space and the probabilities within it.

This example illustrates an important lesson. Independence alone does not determine the probability. You must also consider what information the problem gives and how it restricts the sample space.

Gambler's Fallacy: A Misunderstanding of Independence

In the Double-Six example, the mistake comes from misunderstanding how given information affects the sample space. A different kind of error occurs when people misunderstand independence itself.

The gambler's fallacy is a cognitive bias that leads people to mistakenly believe that past outcomes influence the likelihood of independent, random events. It's also known as the "Monte Carlo Fallacy," named after a casino in Monaco, where it was famously observed in 1913. The wheel hit black 26 times in a row, and gamblers lost enormous sums betting on red because they believed a reversal was inevitable.

The fallacy assumes that independent, random events are "due" to balance out over time. However, for independent events, previous events don't influence the probability of future events.

Have you ever flipped a coin and thought it must be "heads" next because it's been "tails" several times in a row? Or have you been in a casino and believed a particular slot machine is due to pay out because it hasn't hit the jackpot in a while? If so, you've fallen victim to the gambler's fallacy.

The gambler's fallacy is not limited to gambling scenarios. It can also occur in situations that involve chance events, such as predicting the gender of newborn babies. Consider a family that has three boys and is expecting their fourth child. The parents might fall victim to the gambler's fallacy by assuming that they are "due" for a girl and that the probability of having a girl has increased because they already have three boys.

However, random processes determine a baby's gender, and the sex of previous children does not affect the probability of having a boy or a girl. The chances of having a boy or a girl are 50/50, regardless of the previous outcomes. Therefore, assuming the probability of having a girl has increased exemplifies this cognitive bias.

The gambler's fallacy can be dangerous in many situations, but it's especially problematic in gambling. Some people believe they can beat the odds by using it to their advantage. They might increase their bets or change their strategy based on the false belief that a particular outcome is "due" to happen. Unfortunately, this can lead to significant financial losses and even addiction.

Let's take an example to illustrate the gambler's fallacy. Imagine you're playing roulette, and the ball has landed on "red" for the last five spins. You might think it's more likely to land on "black" next, but each spin is independent, and the odds of "red" or "black" are always the same. Betting on "black" just because "red" has come up several times in a row is the gambler's fallacy in action.

But if the spins are independent:

P(Black) stays exactly the same every time. The wheel has no memory.

Independent events do not correct for streaks or “balance things out.”

The gambler's fallacy occurs because we look for patterns and meaning in everything we encounter, including the outcomes of random events. When we observe a series of similar events, such as several coin flips resulting in heads, our brains naturally assume that the next event will be tails to "balance it out."

This cognitive bias is driven by the incorrect assumption that random chance is fair and balanced rather than just arbitrary. People tend to believe that if an event occurs too frequently or not frequently enough, it must be due to some external factor or influence. Independent events are entirely random, and no external force balances their outcomes. Past outcomes don't influence future results.

To minimize this problem, be aware of circumstances where you feel something is "due" or will "balance itself out.” Carefully evaluate the situation and determine whether it is a random process involving independent events. If so, remind yourself that past events do not affect future outcomes. The next coin toss is always 50/50 regardless of the number of consecutive heads before.

The Danger of False Independence Assumptions

While the gambler's fallacy treats independent events as if they were dependent, the opposite mistake can be even more dangerous: treating dependent events as if they were independent. It might seem easy to tell the difference, but professionals who should've known better have made drastically incorrect forecasts by confusing independence and dependence.

Long-Term Capital Management (LTCM) was a prominent hedge fund in the late 1990s. It used sophisticated financial models and assumed market movements were independent from day to day. However, during the Russian financial crisis of 1998, market conditions showed that movements were not independent, as extreme events were more likely to be followed by more extreme events.

This incorrect assumption of independence led to a massive underestimation of risk, contributing significantly to LTCM's eventual collapse and bailout. LTCM was famous for using quantitative models for trading, yet they made such a fundamental error. Its collapse in 1998 was a significant event in financial history.

Similarly, the 2007-2008 housing market crash was partly due to a critical misjudgment in financial risk assessment. Financial institutions believed that mortgage defaults were independent events, underestimating the risk of widespread failures.

However, these defaults were actually interdependent, linked by broader economic factors such as housing prices and interest rates. Defaults were correlated because they were driven by shared economic conditions, meaning mortgage failures were not isolated events, but moved together.

When the market declined, a ripple effect of defaults occurred, revealing that the risks were much more interconnected than previously assumed. This oversight in recognizing the interdependencies among financial risks significantly contributed to the crash's severity, underscoring the need for more complex risk assessment models in finance.

Even seasoned professionals fall victim to these errors!

In the next section, we study mutual exclusivity, which is often confused with independence. But it describes a completely different relationship between events.

Mutual Exclusivity

Two events are mutually exclusive if they cannot happen at the same time. If one event occurs, the other is impossible.

Notation: $A \cap B = \emptyset$

The intersection contains no outcomes.

For example, let:

- T = Patient completed the treatment.
- D = Patient dropped out before treatment began.

A patient cannot both complete treatment and drop out beforehand. So these events are mutually exclusive:

$T \cap D = \emptyset$

There is no overlap because they describe contradictory situations.

When events are mutually exclusive:

- They never occur together.
- If one event occurs, the other has zero probability.
- They divide outcomes into separate, non-overlapping categories.

Other examples include:

- A customer makes a purchase or does not.
- A pet is a cat or a dog.
- A product is defective or not defective.

Mutual exclusivity is about structural impossibility, not just low frequency.

Additive Rule Simplifies with Mutually Exclusive Events

The general additive rule is:

$P(A \cup B) = P(A) + P(B) - P(A \cap B)$

But for mutually exclusive events:

$P(A \cap B) = 0$

Because the overlap is 0, the rule simplifies to:

$P(A \cup B) = P(A) + P(B)$

Example

Let:

- T = Patient completed treatment → $P(T) = 0.80$
- D = Patient dropped out before treatment → $P(D) = 0.10$

Because these two events cannot both occur, use the simplified rule:

- $P(T \cup D) = P(T) + P(D)$
- $P(T \cup D) = 0.80 + 0.10 = 0.90$

There is a 90% chance that a patient either completed treatment or dropped out beforehand. The remaining 10% might represent patients still in progress, lost to follow-up, etc.

Mutual Exclusivity vs. Independence

These concepts are often confused, but they are completely different.

Mutually exclusive refers to cases where one event prevents the other. Both cannot occur. The probability of the other event is 0 when the first event happens.

Independence refers to cases where one event has no effect on the other. Both can occur. The probability of the other event does not change when the first event happens.

In fact, when two events are mutually exclusive, they cannot be independent unless at least one event is impossible. Here's why.

If A and B are mutually exclusive, then $P(A \cap B) = 0$.

But if they were independent, then $P(A \cap B) = P(A) \times P(B)$.

The only way both can be true is if at least one event is impossible (i.e., has probability 0).

Why this difference matters

In medical research and many applied fields:

- Independence is about no influence.
- Mutual exclusivity is about no overlap.

Misinterpreting one as the other can lead to modeling errors and incorrect assumptions.

The Kolmogorov Axioms of Probability

Probability is built on a small set of fundamental rules, known as the Kolmogorov Axioms. These axioms ensure that all probabilities behave consistently and logically. Every probability rule we use comes from these ideas.

For any event A in the sample space S:

1. Probabilities are never negative: $P(A) \geq 0$.
2. The entire sample space has probability 1: $P(S) = 1$.
3. If two events are mutually exclusive, the probability of either occurring is the sum of their probabilities. If A and B cannot both occur, then: $P(A \text{ or } B) = P(A) + P(B)$.

These simple principles guarantee that probabilities make sense. They give us the complement rule, the additive rule, and many other results used throughout this book. You don't need to memorize the axioms as separate facts, just understand that they are the foundation supporting the rules we already use.

Summary and Next Steps

In this chapter, you learned the three core probability rules: the additive rule, the complement rule, and the multiplication rule. These rules let us combine, reverse, and break apart probabilities in flexible ways.

You also saw how the relationship between events changes how these rules behave. For the multiplication rule, independent events allow you to multiply only the marginal probabilities, while dependent events require conditional probabilities. Mutually exclusive events simplify the additive rule. However, mutually exclusive events cannot be independent unless one has a probability of zero.

With these rules in hand, you now have the full toolkit for working with probabilities in symbolic form. You can follow how events combine and translate almost any event relationship into a mathematical expression and calculate its probability.

In the next chapter, we will shift to understanding comparisons using measures that build on conditional probability: odds, odds ratios, and relative risk. You'll learn how to express probabilities in different forms and compare risks across groups.

Formula Quick Reference

- **Additive Rule (general case) for union probabilities:**
  $P(A \cup B) = P(A) + P(B) - P(A \cap B)$
  Use when events overlap and you must avoid double-counting.
- **Additive Rule (mutually exclusive case):**
  $P(A \cup B) = P(A) + P(B)$
  Use only when the events cannot both occur ($A \cap B = \emptyset$).
- **Complement Rule:**
  $P(\text{not } A) = 1 - P(A)$
  Useful when the outcomes you do not want are easier to identify or count.
- **Multiplication Rule (general case) for joint probabilities:**
  $P(A \cap B) = P(A \mid B) \times P(B)$
  $P(A \cap B) = P(B \mid A) \times P(A)$
  Use when events are dependent.
- **Multiplication Rule (independent events):**
  $P(A \cap B) = P(A) \times P(B)$
  Use only when the events do not influence each other.
- **Independence (definitions):** Independence means one event does not affect the probability of the other.
  - $P(A \mid B) = P(A)$
  - $P(B \mid A) = P(B)$

    - $P(A \cap B) = P(A) \times P(B) = P(A \mid B) \times P(B)$
- **Mutual Exclusivity:** Events cannot occur together. Mutually exclusive events with non-zero probabilities cannot be independent.
    - $A \cap B = \emptyset$
- **Key Distinctions:**
    - **Independence**: events do not affect each other.
    - **Mutual Exclusivity**: events cannot occur together.
- **Axioms of Probability:**
1. Probabilities range from 0 to 1.
2. The probability of the entire sample space is 1.
3. For mutually exclusive events, $P(A \cup B) = P(A) + P(B)$.

Chapter 4 Practice Problems

Question 1: A university survey found the following probabilities among students:

- $P(A) = 0.55$ prefer online classes
- $P(B) = 0.40$ prefer morning classes
- $P(A \cap B) = 0.25$ prefer both online and morning classes

What is the probability that a randomly selected student prefers online classes or morning classes?

Question 2: A person flips a fair coin twice.

What is the probability of getting heads on both flips?

Question 3: A jar contains 5 red marbles and 7 blue marbles. Two marbles are drawn without replacement.

What is the probability that both marbles drawn are red?

Question 4: A small technology company tracks whether its employees complete a voluntary training program.

- 70% of employees complete the training.
- Among those who complete the training, 80% report improved productivity.
- Among employees who report improved productivity, 90% say they would recommend the training to others.

What is the joint probability that a randomly selected employee completes the training, reports improved productivity, and recommends the training to others? That is, find P(training ∩ productivity ∩ recommend).

Question 5: A study examines two events:

A = A person owns a bicycle
B = A person drinks coffee daily

Researchers find:

- P(A) = 0.35
- P(B) = 0.60
- P(A ∩ B) = 0.21

Based on these probabilities, are the events independent or dependent? Explain briefly.

Question 6: Two events A and B have the following probabilities:

- P(A) = 0.30
- P(B) = 0.40
- P(A ∩ B) = 0

Are these events independent, mutually exclusive, both, or neither? Explain.

Question 7: A cybersecurity system scans incoming emails for phishing attempts. Each email has a 2% probability of being a phishing email, and the events are independent.

If the system scans 50 emails, what is the probability that at least one phishing email is detected?

Question 8: A bookstore is assembling a display of 4 featured books from a shipment that contains 6 hardcover books and 10 paperback books. The books are selected without replacement.

What is the probability that all 4 selected books are hardcover?

CHAPTER 5

Understanding Risk Measures

What's the risk? People frequently discuss risk, but they don't always clearly understand it. It is your exposure to danger or adverse outcomes. Statistically, we define risk as the probability of a negative outcome occurring, and there are several ways to calculate it.

Imagine deciding whether to take a medication to reduce the chances of an illness. Ideally, this decision involves understanding different types of risk. These include knowing your absolute risk for the medical condition and the relative risk reduction if you take the medication. How many people need to take the drug to prevent one case?

Understanding and calculating these statistics can help you make informed decisions in your daily life. Risk assessment and management play a crucial role in countless fields, including the following:

- **Healthcare**: Comparing patient outcomes between groups.
- **Finance**: Weighing potential gains against the likelihood of loss.

- **Public Safety**: Evaluating policies that affect the chances of harmful events.
- **Engineering**: Assessing how design choices change the probability of failure.

In each case, the key question is not "What is the probability?" but "How much more or less likely is the outcome in one group compared to another?"

In earlier chapters, you learned how to calculate probabilities and conditional probabilities. However, in many real-world problems, a single probability is not enough. Decisions often depend on comparing the likelihood of an outcome under different conditions.

In Chapter 2, you used contingency tables to find probabilities within groups. You also saw how changing the conditioning, from rows to columns, changes the probability you calculate. These ideas form the foundation for understanding how outcomes differ between groups.

In Chapter 4, you compared marginal and conditional probabilities to assess independence. When $P(A \mid B) = P(A)$, the events are independent. Conversely, differences between marginal and conditional probabilities indicate an association. Here, we build on that idea by measuring the magnitude of those differences in applied settings.

Rather than focusing on a single probability, this chapter builds directly on those earlier ideas by introducing standard tools for comparing outcomes across groups: risk, odds, odds ratios, relative risk, risk reduction, and the number needed to treat. These measures quantify and compare the occurrence of adverse outcomes under different conditions, each having its own strengths and weaknesses.

Absolute Risk and Odds

Before comparing probabilities, you must know two basic terms: risk and odds. These absolute measures find likelihoods within a group but don't compare groups. Later, we'll build on them using conditional forms of both for comparing groups.

Risk

Risk is simply the probability that an event occurs:

Risk = P(event)

Probabilities are intuitive because they range from 0 to 1 and directly answer the question, "What is the chance this event will happen?"

Absolute risk (AR) is a straightforward measure that reflects the likelihood of an event occurring within a specific population over a defined period. It's expressed as a simple probability, focusing on a single group without comparing it to others. For instance, if you're looking at the risk of developing a medical condition within a group, AR gives us the direct probability.

The absolute risk formula is the following:

$$Absolute\ Risk = \frac{Number\ of\ Events\ in\ Group}{Total\ Number\ in\ Group}$$

For example, suppose we have a group of 100 people who don't take the medication, and 12 of them get sick during the year-long study. The absolute risk for this group is the following:

$$Absolute\ Risk = \frac{12}{100} = 0.12$$

The calculated absolute risk is 12%. This result indicates that individuals in the untreated group have a 12% probability of developing the condition over the course of a year.

Odds

Odds provide an alternative way to express likelihood. In statistics, odds have a precise definition:

$$Odds = \frac{Probability\ Event\ Occurs\ (p)}{Probability\ Event\ Does\ Not\ Occur\ (1-p)}$$

Odds relate to a binary outcome where the outcome either occurs or does not occur. For example, study subjects were either infected or not infected. A person graduates or does not graduate from college. You win a game, or you lose.

Odds describe how often an event occurs relative to how often it does not occur.

For example, imagine playing a die-rolling game where a six is very good. Your odds of rolling a six are the following:

$$Odds\ of\ a\ 6 = \frac{p\ (\frac{1}{6} = 0.167)}{1-p\ (\frac{5}{6}\ or\ 1-0.167 = 0.833)} = 0.20$$

Your odds of rolling a six are 0.20 or 1 in 5. Because the number of die outcomes is a constant six, you can replace 1/6 and 5/6 in the formula with 1 and 5 to derive the same answer (1/5 = 0.20).

The odds are 1 to 5, meaning one favorable outcome for every five unfavorable outcomes.

Odds greater than 1 indicate that the event occurs more often than not. Odds less than 1 indicate that the event occurs less often than it does not occur.

For example, imagine you're playing a game. If your odds of winning are 2 (or 2 wins to 1 loss), that indicates you are twice as likely to win as to lose. On the other hand, if your odds of winning are 0.5 (or 1 win to 2 losses), you're half as likely to win as to lose.

Tip: When odds are expressed as a decimal, you can convert them to a "1 in X" format by taking the reciprocal. This tells you how many non-events occur for each event. For example, odds of 0.25 correspond to 1 / 0.25 = 4, meaning about 1 event for every 4 non-events (often stated as 1 to 4 odds).

People often use odds and probabilities interchangeably in everyday language, but in statistics, they are distinct quantities with different interpretations. Because odds are measured on a scale that extends from zero to infinity, they are unbounded above and behave differently from probabilities, which are restricted to values between 0 and 1.

Converting Between Probabilities and Odds

Odds and probabilities describe the same information but express it on different scales. Here's how you can convert between them easily.

To convert a probability to odds, divide the probability of the event by the probability that it does not occur:

$$\text{odds} = p / (1 - p)$$

For example, if an event has a probability of 0.20, the odds are:
$0.20 / (1 - 0.20) = 0.20 / 0.80 = 0.25$

These odds mean that one favorable outcome occurs for every four unfavorable outcomes.

To convert odds back to a probability, reverse the process:

$$p = \text{odds} / (1 + \text{odds})$$

For example, if the odds are 0.25, the probability is:

0.25 / (1 + 0.25) = 0.25 / 1.25 = 0.20

Warning: Be careful when interpreting odds, particularly when expressed in decimal form. It's easy to confuse them with probabilities.

For example, the *probability* of rolling a six on a die is 0.167, indicating it happens about 16.7% of the time. However, the *odds* of rolling a six are 0.2, meaning it happens 1 time for every 5 times it doesn't happen. You don't want to misinterpret the 0.2 odds as a percentage of all cases!

History

Historically, odds came before probability. In early gambling contexts, odds described payoffs. For example, three to one against an outcome meant that if you staked one coin and won, you would receive three in return.

When mathematicians began studying games of chance formally, they asked what odds would make a bet fair. That question led to the concept of probability as the long-run proportion of times an outcome occurs. Fair odds correspond to a probability p such that:

odds = p / (1 – p)

Today, probabilities are easier for most people to interpret, while odds offer several mathematical conveniences. Odds combine multiplicatively, which means that independent effects multiply on the odds scale and become additive when expressed as log-odds, making odds especially useful for statistical modeling. While this additive property isn't a key point for this introductory book, we'll briefly revisit it when assessing regression output in the upcoming odds ratio section.

Further, some study designs impose constraints that prevent probabilities and risks from being estimated directly. In those settings, odds-based measures remain usable and informative. Hence, both odds and probabilities remain useful in modern statistics.

Next, we'll look at the conditional forms of odds and risks that you can use to compare groups.

Conditional Measures

So far, we have worked with absolute risk and absolute odds. These measures describe the likelihood of an event *within a single group*. They answer questions such as:

- What is the probability of illness in this group?
- What are the odds of failure in this population?

However, many real-world questions involve comparisons between groups. For example:

- Is illness more likely among exposed individuals than unexposed ones?
- Are treatment outcomes better than control outcomes?

To answer these questions, we must move from absolute measures to conditional measures. Conditioning allows us to restrict attention to a specific group and then compare results across groups.

When comparing groups, we begin by finding conditional probabilities, such as the probability of disease given exposure relative to the probability of disease given no exposure.

In essence, we compare the following to determine whether exposure and non-exposure are associated with different probabilities of the event occurring.

P(Event | Exposure) and P(Event | No Exposure)

For instance, if the probability of the event occurring is higher in the exposure group than in the non-exposure group, it indicates that the exposure group has a higher event rate

Conditional Odds

Just as probabilities can be conditional, odds can be conditional as well. Conditional odds describe the odds of an event occurring *given* a particular condition, such as exposure, treatment, or group membership.

Conditional odds answer questions like:

- What are the odds of infection among exposed individuals?
- What are the odds of recovery given treatment?

Formally, conditional odds are derived from conditional probabilities. For example, the odds of an event given exposure can be written as:

odds(Event | Exposure) = P(Event | Exposure) / P(Non Event | Exposure)

Notice that it compares the probability of the event occurring to it not occurring, making it an odds. However, it adds the condition of exposure. So, you're assessing the odds of the event, but only in the exposure group.

Conditional Odds Formula

The table shows a standard format for risk in a two-way contingency table.

	Events	Non-Events	Conditional Odds
Exposed	A	B	A / B
Unexposed	C	D	C / D

Each letter represents a count of events or non-events in the exposed and unexposed groups. A row represents a group defined by exposure status, exposed or unexposed. In an experiment, the exposed group receives the treatment, while the unexposed group serves as the control.

In this layout, you condition on the table's rows. However, for conditional odds, you don't use row totals for the given conditions, unlike conditional probabilities. Instead, you divide events by non-events within a group.

When calculating odds from a table, you can enter either the raw counts or probabilities, depending on which the table supplies.

Example: Calculating Conditional Odds from a Contingency Table

Let's work through conditional odds calculations using real-world data. In this case, we'll use data from a flu vaccination study (Beran et al., 2009). This study was a randomized controlled trial, the gold standard for identifying causal relationships.

Suppose we have the following table with raw count data:

Treatment	Flu infections	Non-infections
Vaccine	49	5054
Placebo	74	2475

First, calculate the conditional odds for each group.

For the exposed group who receive the vaccination, the odds of the event are:

A / B = 49 / 5054 = 0.0097

For the unexposed group who didn't receive the vaccine, the odds of the event are:

C / D = 74 / 2475 = 0.0299

Each value describes how often the event occurs relative to non-occurrence within that group.

These conditional odds now place the two groups on a comparable footing. After conditioning on group membership, we can compare the odds directly. In this case, the odds of flu infection in the exposed (vaccinated) group are 0.0097, while they are 0.0299 in the unvaccinated group. Subjects in the vaccinated group have lower odds of flu infection.

That comparison leads naturally to the next concept, the odds ratio, which compares conditional odds across groups.

What is an Odds Ratio (OR)?

An odds ratio (OR) is the ratio of the odds of an event in one group to the odds of the event in another group. It quantifies the association between a condition (such as exposure to a risk factor) and an outcome. Analysts frequently use odds ratios to identify risk factors by assessing the relationship between exposure and a medical outcome. For example, is there an association between exposure to a chemical and a disease?

When you have a grouping variable, an odds ratio answers the question, are the odds of an event occurring higher or lower in one condition or another? It calculates the odds of an outcome occurring in one context relative to a baseline or control condition. For example, your grouping variable can be a subject's exposure to a risk factor (yes or no) to see how that relates to disease status.

For example, you can calculate the odds of infection among exposed and unexposed individuals and compare them using odds ratios. The odds ratio formula shows how to calculate it for conditions A and B.

$$Odds\ Ratio = \frac{Odds\ of\ an\ Event\ (Condition\ A)}{Odds\ of\ an\ Event\ (Condition\ B)}$$

The denominator (condition B) in the odds ratio formula is the baseline or control group. Consequently, the OR tells you how much higher or lower the odds are for the numerator events (condition A) relative to the denominator events.

If you have a treatment and control group, the treatment will be in the numerator while the control group is in the denominator of the formula. This odds ratio arrangement indicates how your treatment group fares compared to the control group.

For example, the flu study assesses infections in a vaccine group compared to a control group. Infections are the events for the binary outcome. By calculating the following OR, analysts can assess the odds of infections in the treatment group relative to the control group.

$$\frac{Odds\ of\ Infections\ (Treatment\ Group)}{Odds\ of\ Infections\ (Control\ Group)}$$

The interpretation of this odds ratio is that when the treatment is effective, the odds of infection in the treatment group will be lower than in the control group, yielding an OR less than 1.

Interpretations

In the odds ratio formula, the value of 1 is critical during interpretation because it indicates that the odds are equal in both conditions. Consequently, analysts always compare their OR results to 1 when interpreting the results. As the OR moves away from one in either direction, the association between the condition and outcome becomes stronger.

Odds Ratio = 1: The ratio equals one when the numerator and denominator are equal. This equivalence occurs when the odds of the event occurring in one condition equal those in the other condition. There is no association between condition and event occurrence.

Odds Ratio > 1: The numerator is greater than the denominator. Hence, the event's odds are higher for the group/condition in the numerator. This is often a risk factor.

Odds Ratio < 1: The numerator is less than the denominator. Hence, the odds of the outcome occurring is lower for the group/condition in the numerator. This can be a protective factor.

In the flu vaccination experiment, the researchers hope that the OR is less than one because that indicates the treatment group has lower odds of becoming infected than the control group.

Caution: ORs are a type of correlation and do not necessarily represent causal relationships unless they are from a randomized controlled trial!

Example Calculations

The following equation expands the earlier odds ratio formula for calculating an OR with two conditions (A and B). Again, it's the ratio of two odds. Hence, the numerator and denominator are also ratios.

$$\frac{Odds\ of\ Event\ (Condition\ A)}{Odds\ of\ Event\ (Condition\ B)} = \frac{\#\ Events\ (A) / \#\ Non\ Events\ (A)}{\#\ Events\ (B) / \#\ Non\ Events\ (B)}$$

In the flu vaccination study, we assessed the relationship between treatment and the odds of being infected. Our two conditions were the treatment (condition A) and the control group (condition B). On the right-hand side, we enter the numbers of infections (events) and non-infections (non-events) from our sample for both groups.

Returning to our flu vaccine study and its 2 × 2 contingency table, recall that:

Vaccine group odds in the numerator (treatment group) are:
49 / 5054 = 0.0097

Unvaccinated group odds in the denominator (control group) are:
74 / 2475 = 0.0299

Hence, the odds ratio is 0.0097 / 0.0299 = 0.3244

The odds of a flu infection (the event) in the vaccinated group are 0.3244 times as high as in the unvaccinated group, roughly one-third the odds. Equivalently, the unvaccinated group has about 1 / 0.3244 = 3.08 times the odds of infection.

Because this is a randomized controlled trial, differences in infection odds can be interpreted causally rather than as simple associations. Exposure decreases the odds of the event occurring. The vaccine is a protective factor against infections.

Example: Case-Control Study

Let's use data from an actual study to calculate an odds ratio for a different type of study design, a case-control study. In these studies, researchers select individuals based on whether they have the outcome of interest (cases) or not (controls), and then look backward to assess prior exposure to potential risk or protective factors.

Because the study is designed to identify such factors, the researchers construct the odds ratio by comparing the odds of exposure among cases versus controls rather than the odds of infection. As you'll see, that goal affects how we set up the OR.

The North Carolina Division of Public Health needed to identify risk factors associated with an *E. Coli* outbreak. We'll calculate the OR for one risk factor of visiting a petting zoo using their data, but they assessed multiple possibilities in their study.

In this study, the outcome of interest is *E. coli* infection. However, case-control studies focus on finding risk and protective factors rather than assessing treatments. Consequently, we need to change how we construct the OR. Rather than conditioning on rows (exposure), as in the flu vaccine study, we condition on the columns representing outcome status (sick versus not sick), and compare how frequently prior exposure occurs within each group. More about this difference in the next section.

The odds ratio is constructed from the two conditional odds using the two columns in the following table.

	Got Sick (Cases)	**Did Not Get Sick (Controls)**
Visited Petting Zoo	36	64
Did Not Visit	9	123

Because it is a case-control study, we condition on the columns, unlike the flu study. We need to compare odds(Exposure | Cases) to odds(Exposure | Controls) to determine whether the odds of exposure are higher among cases (i.e., those who got sick) than among controls who didn't get sick.

By plugging these numbers into the odds ratio formula, we can assess the association between visiting a petting zoo and *E. coli* infection. In case-control studies, put all infected cases in the numerator and all uninfected controls in the denominator. The resulting odds ratio compares the odds of prior exposure among cases to those among controls.

$$\frac{\#\,Visited\ Zoo\ (Cases) / \#\,Did\ Not\ Visit\ (Cases)}{\#\,Visted\ Zoo\ (Controls) / \#\,Did\ Not\ Visit\ (Controls)} = \frac{36/9}{64/123} = 7.7$$

The odds ratio indicates that those infected with E. coli (cases) have 7.7 times the odds of visiting the petting zoo than those who were not infected (controls). That's a big red flag for the petting zoo being the *E. coli* source!

This study also assessed whether awareness of the risk of disease from contact with livestock was a protective factor. For this factor, the study calculated the odds ratio:

$$\frac{Odds\ of\ Disease\ Risk\ Awareness\ Among\ the\ Sick}{Odds\ of\ Disease\ Risk\ Awareness\ Among\ the\ Healthy} = 0.1$$

The OR value of 0.1 indicates that those who became infected with *E. coli* had only one-tenth the odds of being aware of the disease risk from contacting livestock as those who were not infected.

Knowledge is power! Presumably, those who were aware of the risk took precautions.

Different Arrangements

You might have noticed differences in how we arranged the odds ratios for the randomized flu vaccine experiment and the case-control study. These differences arise because the two study designs ask different questions and, thereby, require conditioning on different variables.

Think back to the contingency tables and how switching from rows to columns reversed the conditional probabilities. We're doing the same here, but with conditional odds rather than conditional probabilities.

The condition determines how the odds are constructed using the table. The two arrangements reverse the direction of conditioning in the odds, from odds(Infection | Exposure) to odds(Exposure | Infection). In the table, that means switching from rows to columns.

For the randomized flu experiment, the researchers randomly assigned subjects to either the treatment or control group to determine the vaccine's effect. Consequently, we grouped (i.e., conditioned) individuals by treatment status, which appears in the table rows.

Within each experimental group, we calculate the odds of infection by dividing the number of infected individuals by the number of uninfected individuals. Placing the treatment group in the numerator and the control group in the denominator produces an odds ratio that compares the odds of disease given the two treatment conditions: odds(Flu Infection | Treatment) and odds(Flu Infection | Control).

In the case-control study, the researchers wanted to identify potential risk or protective factors using a retrospective, observational design. Case-control studies group individuals by outcome status, which appears in the table columns. One group consists of cases (those who became sick), and the other has the controls (those who did not).

Within each outcome group, we then calculate the odds of prior exposure by dividing the number of exposed individuals by the number who were not exposed. Placing the cases in the numerator and controls in the denominator produces an odds ratio that compares the odds of exposure history given the two outcomes: odds(Exposure | Cases) and odds(Exposure | Controls).

Continuous Variables

So far, we've looked at using odds ratios to compare binary outcomes between groups. However, using particular modeling methods, you can find odds ratios for continuous variables. These models use odds ratios because they fit the model using log odds. While those advanced methods go beyond the scope of this introductory book, let's look at ORs for several continuous variables. I will show you how to interpret them, because you're likely to encounter them even if you don't fit the models yourself.

When you perform binary logistic regression using the logit transformation, you can obtain ORs for continuous variables. Unlike the groups in the previous examples, a continuous variable can increase or decrease in value. Fortunately, the interpretation of an odds ratio for a continuous variable is similar and still centers around the value of one:

- **Greater than 1**: As the continuous variable increases, the odds of the event increase.
- **Less than 1**: As the variable increases, the odds of the event decrease.
- **Equals 1**: As the variable increases, the odds of the event do not change.

In my book *Regression Analysis: An Intuitive Guide*, I performed binary logistic regression and obtained ORs for two continuous independent variables. Let's interpret those odds ratios!

In that analysis, I assess whether measures of conservativeness and establishmentarianism predict membership in the Freedom Caucus within the U.S. House of Representatives in 2014.

Here's how you interpret these scores:

- **Conservativeness**: Higher scores represent more conservative viewpoints.
- **Establishmentarianism**: Higher scores represent viewpoints that favor the political establishment.

For this book, I'll focus on interpreting the ORs for this binary logistic model.

```
Odds Ratios for Continuous Predictors

                   Unit of
                    Change  Odds Ratio          95% CI
Conservativeness       0.1      2.6659  (1.7407, 4.0829)
Establishmentness      0.1      0.7251  (0.6050, 0.8691)
```

The odds ratio for conservativeness indicates that for every 0.1 increase (the unit of change) in the conservativeness score, the odds that a House member belongs to the Freedom Caucus increase by 2.67 times, holding the other predictor constant.

Conversely, the odds ratio for establishmentness indicates that for every 0.1 increase in the establishmentarianism score, the odds that a House member belongs to the Freedom Caucus is multiplied by 0.73, representing a reduction. Equivalently, you can say the odds decrease by about 27%.

Taking both results together, House members who are more conservative and less favorable towards the establishment make up the Freedom Caucus.

Next, we'll switch gears and assess another ratio, relative risk, but it uses probabilities rather than odds.

Relative Risk (RR)

Relative risk (RR) is a ratio of the absolute risks for two groups. Recall that absolute risk is the probability that an event will occur. This ratio tells us how the chances in one group relate to those in another. It helps you understand how exposure to a particular factor, such as a treatment or lifestyle change, affects the probability of the outcome.

Relative risk is closely related to odds ratios in that both compare outcomes across two groups and use a ratio to summarize the strength and direction of an effect. In fact, when outcomes are rare, relative risk and odds ratios often produce similar numerical results. The key difference is that relative risk compares probabilities directly, while odds ratios compare odds, which behave differently as outcomes become more common. Because of this distinction, relative risk is often easier to interpret when absolute risks can be estimated. However, odds ratios are necessary for some study designs and statistical modeling methods.

The relative risk formula includes the exposure or treatment group in the numerator and the unexposed or control group in the denominator, as shown in the following. This comparison shows how much a factor alters the probability of the outcome.

$$RR = \frac{AR\ of\ Exposed\ or\ Treated\ Group}{AR\ of\ Unexposed\ or\ Control\ Group}$$

As with odds ratios, exposure can be a protective factor that reduces the likelihood of an adverse outcome (such as a vaccine or an education program) or a risk factor that increases the likelihood of an adverse outcome (such as a toxin or a harmful environment). In this manner, the risk ratio determines whether exposure is a protective or risk factor.

Relative Risk Formula

Now, let's expand the relative risk calculation to show the formula in more detail. We return to the following table, which shows the standard two-way contingency table format for risk.

	Events	**Non-Events**	**Absolute Risk**
Exposed	A	B	A / (A + B)
Unexposed	C	D	C / (C + D)

Again, each letter represents a count of events or non-events in the exposed and unexposed groups. Contrast these calculations with the odds calculations. Here, we find conditional probabilities using row totals in the denominators (e.g., A + B) rather than non-events, as we do for odds (e.g., B).

The RR is the ratio of those two event probabilities. The relative risk formula uses the letter notation from the table:

$$\frac{A/(A+B)}{C/(C+D)}$$

This relative risk calculation emphasizes how it is a ratio of two ARs. To calculate RR, you first need to calculate the absolute risks (ARs) for both groups. Then, divide one by the other.

How to Interpret Relative Risk

The interpretation of relative risk follows the same rules as odds ratios. The interpretation of relative risk follows the same rules as odds ratios. The value of 1 again serves as an important benchmark, indicating that the exposed and unexposed groups have equal absolute risks. As the ratio moves away from one in either direction, the relationship between exposure and the outcome strengthens.

Relative Risk = 1: There is no association between exposure and the outcome.

Relative Risk > 1: The risk is greater in the exposed group than in the unexposed group. Hence, exposure is a risk factor.

Relative Risk < 1: The risk is lower for the exposed group than for the unexposed group. Exposure is a protective factor.

While the interpretation process is similar for RRs and ORs, the values represent different properties. I cover that in the upcoming Odds Ratio vs. Relative Risk section.

Relative risk is a measure of the association between exposure status and the outcome. As you have undoubtedly heard, correlation outside randomized experiments does not necessarily imply a causal relationship! Use proper experimental designs to uncover causation.

Example: Calculating Relative Risk from a Contingency Table

Let's return to the flu vaccine study. Previously, we calculated the odds ratio, and now we'll find the relative risk.

Treatment	Flu infections	Non-infections
Vaccine	49	5054
Placebo	74	2475

Now, let's plug these numbers into the relative risk formula:

$$\frac{A/(A+B)}{C/(C+D)} = \frac{49/(49+5054)}{74/(74+2475)} = \frac{0.0096}{0.0290} = 0.3310$$

The risk ratio is 0.3310, indicating that the vaccine is a protective factor. The vaccinated are about 1/3 as likely to catch the flu as the unvaccinated.

Note that RR is different than vaccine effectiveness, although they're related because effectiveness is calculated using: 1 – RR. In this case, the vaccine's effectiveness is 1 – 0.331 = 0.669.

This RR indicates that the probability of contracting the flu in the treatment group is one-third that of the control group, suggesting that the vaccine effectively reduces the likelihood of flu infections. Lower RRs indicate greater effectiveness or protection.

In this case, the OR and RR results are nearly identical. However, that's not always the case, as you'll see in the upcoming comparative section.

Odds Ratios vs. Relative Risk

You have seen how both odds ratios and relative risk compare the likelihood of an event across groups. Because both measures are ratios and both use the value 1 as a reference point, they can appear to convey the same information. However, they compare different underlying quantities and can behave quite differently in practice.

Understanding how these measures relate to one another and why both exist is essential for correct interpretation.

As you saw, relative risk compares conditional probabilities:

Relative Risk = P(Event | Exposed) / P(Event | Unexposed)

It answers a direct and intuitive question: how many times more (or less) likely is the event in one group than in another?

In contrast, odds ratios compare conditional odds rather than probabilities:

Odds Ratio = Odds(Event | Exposed) / Odds(Event | Unexposed)

Instead of comparing probabilities directly, odds ratios compare how often an event occurs relative to non-occurrence within each group.

Both measures quantify association, but on different scales. Relative risk aligns with how people naturally think about chance, while odds ratios require thinking in terms of ratios of occurrences to non-occurrences and are sensitive to the event's frequency.

When Odds Ratios and Relative Risk Are Similar

When the event you are studying is rare, odds and probabilities are numerically close. In these situations, odds ratios and relative risk produce very similar values.

For example, when probabilities are only a few percent, doubling the probability also approximately doubles the odds. In such cases, interpreting an odds ratio as a relative risk introduces little distortion.

Odds ratios are often used in low-prevalence settings without serious interpretive problems, even though the measures are not formally the same.

Suppose a disease is rare.

- In an exposed group, 2 out of 1,000 people develop the disease (risk = 0.002).
- In an unexposed group, 1 out of 1,000 people develop the disease (risk = 0.001).

The relative risk is:

RR = 0.002 / 0.001 = 2.0

Now calculate the odds.

- Odds in the exposed group: 0.002 / 0.998 = 0.002004.
- Odds in the unexposed group: 0.001 / 0.999 = 0.001001.

The odds ratio is:

OR = 0.002004 / 0.001001 = 2.002

In this case, the odds ratio and relative risk are essentially identical because the event is rare.

In the flu vaccine study, we saw that the OR and RR were nearly identical: 0.3244 vs. 0.3310. That's because the infection rate was low.

When Odds Ratios Diverge from Relative Risk

As event probabilities increase, odds and probabilities separate. Odds grow nonlinearly, increasing faster than probabilities as outcomes become more common.

Because of this nonlinearity, odds ratios can appear substantially larger than relative risk when outcomes are common. An odds ratio can suggest a strong association even when the probability change is modest.

Suppose the event is common.

- In an exposed group, 60 out of 100 people experience the event (risk = 0.60).
- In an unexposed group, 30 out of 100 people experience the event (risk = 0.30).

The relative risk is:

RR = 0.60 / 0.30 = 2.0

The exposed group is twice as likely to experience the event.

Now calculate the odds.

- Odds in the exposed group: 0.60 / 0.40 = 1.50.
- Odds in the unexposed group: 0.30 / 0.70 = 0.43.

The odds ratio is:

OR = 1.50 / 0.43 = 3.5

Here, the odds ratio suggests a much stronger association than the relative risk. The probability doubled, but the odds more than tripled.

This divergence occurs because odds increase nonlinearly as probabilities become larger.

This behavior explains why you should not interpret odds ratios as probabilities or as relative risks.

Why Analysts Use Odds Ratios Despite Their Interpretation Challenges

As you saw, relative risk is calculated using absolute risks in each group. Those risks require meaningful denominators that reflect the total number of individuals at risk. When that information is available, relative risk provides an intuitive comparison of probabilities.

However, many real-world study designs do not allow analysts to estimate absolute risks reliably.

Odds ratios exist precisely for situations like these because they do not have this requirement. Odds ratios depend on only relative odds within the observed data, not on the event's population prevalence. This property makes them usable even when population prevalence is unknown or distorted by the study design. That is one of the core reasons odds ratios exist in applied work, despite being less intuitive than relative risk.

For example, in case-control studies, researchers begin by selecting individuals based on outcome status, such as whether they are sick or not, and then look backward to assess prior exposure. Because the sample is constructed around outcomes, the proportions of cases and controls reflect the study design rather than the population prevalence.

Consequently, absolute and relative risks cannot be estimated from case-control data. Odds ratios remain valid because they compare relative odds within the sampled groups, and that comparison does not depend on how common the outcome is in the population.

This tradeoff explains why odds ratios are so common in retrospective studies and epidemiology. They sacrifice some interpretive simplicity in exchange for applicability when probabilities cannot be estimated. This property makes odds ratios the standard measure in case-control studies and many retrospective designs.

Another reason odds ratios appear so frequently is their role in statistical modeling. Many commonly used models for binary outcomes work on the log odds scale, expressing effects as changes in odds rather than probabilities. When these models include predictors, whether categorical or continuous, their coefficients naturally translate into odds ratios. That is why models, such as logistic regression, report results in terms of odds ratios.

Common Misinterpretations of Odds and Odds Ratios

As discussed in the previous sections, odds and odds ratios are less intuitive than relative risk because they compare odds rather than probabilities. This difference makes odds ratios harder to interpret and sensitive to the event's frequency.

That lack of intuition leads to a set of recurring interpretation errors.

Common pitfalls include the following:

- **Treating odds as probabilities:** Odds are not probabilities. Odds of 1.5 do not mean a 150% chance. They mean the event occurs 1.5 times as often as it does not. Odds must be converted to a probability (e.g., odds of 1.5 = a probability of 0.6).
- **Treating odds ratios as probabilities:** An odds ratio does not describe any single group and cannot be converted into a probability without knowing the baseline risk. An odds ratio indicates how strongly odds differ between groups, not how common the event is overall.

- **Interpreting an odds ratio as relative risk:** Odds ratios and relative risk are similar only when events are rare. When events are common, odds ratios exaggerate the effect size relative to changes in probability.

So far, we have focused on measures that compare likelihood across groups using ratios, such as odds ratios and relative risk. These statistics describe the strength and direction of an association, but they do not directly show how much the probability changes in absolute terms.

To answer that question, we now turn to measures that emphasize absolute differences rather than ratios.

Absolute Risk Reduction (ARR)

Absolute risk reduction (ARR) is the difference between the ARs for the control and treatment groups. It's another method for understanding the effect of a treatment or condition. For this calculation, you're again working with ARs for two groups, but instead of dividing them, you use subtraction.

The absolute risk reduction formula is the following:

$$ARR = AR\ Control - AR\ Treatment$$

Using the values from the previous flu vaccine study, the ARR calculation is:

ARR = 0.0290 - 0.0096 = 0.0194.

This absolute risk reduction shows that treatment reduces the chance of getting the flu by 1.94%. In other words, if 100 people got the annual vaccination, it would prevent nearly two cases in that year.

Relative Risk Reduction (RRR)

Relative risk reduction (RRR) compares the probability of an event occurring in two groups by showing the proportionate reduction in probability.

The relative risk reduction formula is the following:

$$RRR = \frac{(AR\ Control\ Group - AR\ Treatment\ Group)}{AR\ Control\ Group}$$

Alternatively, use the following formula with RR to obtain the same results:

$$RRR = 1 - RR$$

Let's return to the flu vaccine example. The calculated relative risk reduction is the following:

RRR = (0.0290 - 0.0096) / 0.0290 = 0.6690

Or, using the RR we found previously:

RRR = 1 – 0.3310 = 0.6690

This result indicates the medication reduced the disease's relative probability in the treatment group by 66.9% compared to the untreated group. In the context of vaccines, doctors refer to this value as vaccine effectiveness.

RRR measures can range from 0 – 100%, with higher values indicating greater reduction. 100% represents the complete elimination of risk. RRR is particularly useful when you want to compare the impact of a treatment or intervention across populations.

Number Needed to Treat (NNT)

The number needed to treat (NNT) is the number of people who need to undergo a specific treatment to prevent one additional adverse outcome.

The formula for the number needed to treat is the inverse of the absolute risk reduction:

$$NNT = \frac{1}{ARR}$$

For example, our previously calculated ARR of 1.94% for the treatment group corresponds to the following number needed to treat calculation:

$$NNT = \frac{1}{0.0194} = 51.5$$

Analysts usually round NNT up to the next whole person in applied settings because you cannot treat a partial person. This NNT result means you need to vaccinate 52 people to prevent one flu infection.

A lower number needed to treat is generally preferable because fewer people need the treatment to prevent an adverse outcome. As NNT increases, weighing the treatment benefits against side effects becomes more important.

Tips for Using the Types of Risk

Given this range of statistics, which ones should you use? The answer depends on both the risks themselves and what you want to communicate. Different measures emphasize different aspects of risk, and no single statistic tells the whole story.

In the following discussion, we'll consider treatments applied to higher- and lower-risk populations.

Using Relative Measures (RR & RRR)

Relative measures such as relative risk (RR) and relative risk reduction (RRR) compare probabilities using ratios. Because they are ratios, they often remain stable across populations with different baseline risks.

For example, suppose a treatment reduces the chances of an illness from 20% to 10% in a higher-risk population and from 4% to 2% in a lower-risk population. The RRR is 50% in both cases. The treatment halves the probability, irrespective of the starting level.

This property facilitates comparisons between different populations.

However, relative measures can be less effective for communicating the absolute magnitude of the effects because they mask the baseline risk level. Consider that a low baseline AR can inflate both RR and RRR, making them sound more dramatic than warranted.

For instance, if the treatment has an RRR of 50%, that sounds impressive. But if the absolute risk reduction is only 2% (from 4% to 2%), you might reconsider its effectiveness.

For this reason, relative measures should almost always be reported alongside absolute risks.

Using Odds Ratios

Odds ratios compare conditional odds rather than probabilities. As discussed earlier, odds ratios and relative risk are similar when events are rare but can diverge substantially when events are common.

Odds ratios are especially useful when absolute risks cannot be estimated, and their use is often dictated by the experimental design and analysis method. For example, in case-control studies, researchers fix the number of cases and controls by design. Because the sample does not reflect population prevalence, absolute risks and relative risk cannot be calculated. Odds ratios remain valid in these settings because they do not depend on prevalence.

For this reason, researchers frequently use odds ratios in retrospective studies and commonly report them in applied research, even though they can be less intuitive than relative risk.

However, when absolute risks can be estimated, measures such as relative risk and absolute risk are usually more intuitive because they compare probabilities directly. In randomized experiments, random assignment ensures comparable groups and allows absolute risks to be estimated, making relative risk and absolute risk natural alternatives to odds ratios.

When odds ratios are reported, it is important to avoid interpreting them as probabilities or as relative risk.

Using Absolute Measures (AR, ARR, & NNT)

Absolute risk measures tend to provide more concrete information about the impact of a treatment in a specific population because they describe the actual probability of an outcome occurring within that group. Measures such as absolute risk reduction (ARR) and number needed to treat (NNT) build directly on absolute risk and therefore convey the practical impact more clearly.

Returning to the earlier example, the higher-risk population experienced a reduction from 20% to 10%, producing an absolute risk reduction of 10%. In contrast, the lower-risk population experienced a reduction from 4% to 2%, producing an absolute risk reduction of only 2%.

Although the relative risk reductions were identical, the absolute effects differed substantially.

These differences have practical consequences. In the previous example, the number needed to treat in the higher-risk group is 10, whereas in the lower-risk group it is 50. This information can influence whether a treatment is considered worthwhile in different risk-level groups, especially when side effects, cost, or feasibility are concerns. But you wouldn't know that from the RRs!

Absolute measures are often the most effective way to communicate real-world impact to non-experts. However, be aware they might be applicable only to a specific population (e.g., high- or low-risk).

Putting the Measures Together

You gain deeper insight into managing adverse outcomes by understanding and applying these measures to describe the relative and absolute risks, their reductions, and the number needed to treat (NNT). Understanding these statistics can help you make decisions.

Each type of measure answers a different question:

Relative risk and relative risk reduction describe proportional change and often apply across populations with different baseline risk levels.

Odds ratios describe differences in odds and are frequently dictated by study design and modeling methods rather than interpretability.

Absolute risk, absolute risk reduction, and NNT describe practical impact but depend directly on the baseline risk of the population being studied.

Using these measures together provides a more complete and honest picture of the risks and benefits.

Review and Next Steps

In this chapter, you learned several ways to compare probabilities across groups. Rather than focusing on single probabilities, you explored measures that describe how outcomes differ between exposed and unexposed groups, treatments and controls, or cases and noncases. Different study designs condition on different variables, which determine how odds ratios are constructed and interpreted.

You saw that different measures answer different questions. Odds ratios compare conditional odds and remain usable in study designs where probabilities cannot be estimated.

Relative risk compares probabilities directly and is often easier to interpret when absolute risks are available. They can apply across populations with differing baseline risks.

Absolute risk reduction and the number needed to treat translate statistical comparisons into practical impact, but might apply only to specific populations.

Understanding what a measure represents helps prevent common interpretation errors and supports better decision-making.

In the next chapter, we will introduce visual tools that help make probability relationships easier to see.

Formula Quick Reference

Risk Measure	Definition	Formula
Absolute Risk (AR)	Probability that an event occurs in a specific group.	AR = events / total in group
Odds	Ratio of the probability an event occurs to the probability it does not occur.	Odds = p / (1 – p)
Odds Ratio (OR)	Ratio of the odds of an event (or exposure) between two groups.	OR = $odds_1$ / $odds_2$
Relative Risk (RR)	Ratio of the probabilities of an event between two groups.	RR = AR_1 / AR_2
Absolute Risk Reduction (ARR)	Difference in absolute risk between two groups.	ARR = AR_control – AR_treatment
Relative Risk Reduction (RRR)	Proportional reduction in risk relative to the control group.	RRR = ARR / AR_control
Number Needed to Treat (NNT)	Number of individuals who must receive treatment to prevent one additional adverse event.	NNT = 1 / ARR

Chapter 5 Practice Problems

Question 1: A manufacturing defect occurs in 8% of products.

A. What are the odds of a defect?
B. Interpret the odds in plain language.
C. If the odds of a defect were 0.25 instead, what would the probability be?

Question 2: A sports team has odds of winning equal to 2.

A. What does this mean in terms of wins and losses?
B. Is the team more likely to win or lose?
C. What is the probability of winning?

Question 3: A study examines whether exposure to a chemical is associated with a respiratory illness.

	Illness	No Illness
Exposed	40	160
Not Exposed	20	180

Answer the following:

A. Calculate the absolute risk for each group.
B. Calculate the relative risk.
C. Calculate the odds for each group.
D. Calculate the odds ratio.
E. Interpret the relative risk and odds ratio in context.

Question 4: A study reports that a treatment has:

- Relative risk = 0.75
- Odds ratio = 0.70

Answer the following:

A. What does the relative risk tell you?
B. What does the odds ratio tell you?
C. Are these values consistent with each other?

Problem 5: Researchers conduct a case-control study of a disease and prior exposure to a pollutant. They compare:

- the odds of exposure among cases
- the odds of exposure among controls

Answer the following:

A. Why can't researchers calculate relative risk in this study?
B. What does the odds ratio represent in this context?

Problem 6: Two studies examine the effect of the same exposure on a disease. Both studies report an odds ratio of 2.0.

Study 1: In the unexposed group, 1 out of 100 individuals develop the disease.

Study 2: In the unexposed group, 40 out of 100 individuals develop the disease.

In both studies, the exposed group has an odds ratio of 2.0 relative to the unexposed group.

A. For each study, calculate the probability of disease in the exposed group.
B. For each study, calculate the relative risk.
C. Compare the relative risks between the two studies.
D. Explain why the same odds ratio can correspond to different relative risks.

CHAPTER 6

Visualizing Probability: Venn & Tree Diagrams

Probability rules can feel abstract until you see how events fit together. Pictures help. Visual tools make it easier to understand how events overlap, how they combine, and how conditional probabilities unfold step by step. In this chapter, we use two powerful diagrams, Venn and tree diagrams, to bring these relationships to life.

Venn diagrams show how events relate within a sample space, making it easy to see unions, intersections, complements, and whether events are mutually exclusive.

Tree diagrams tell a different story. They are ideal for situations that unfold in stages and when one event affects the next. They clarify sequences of conditional probabilities and effectively represent dependent events by showing how one outcome affects the next. These ideas play an important role in the Bayesian methods we cover in the next several chapters.

Some probability puzzles become much easier when viewed visually. We'll use a tree diagram to solve the Monty Hall Problem, a puzzle that confuses almost everyone until they see the structure. A tree diagram makes the answer clear and prepares you for the reverse-probability reasoning that appears in the next chapter.

By the end of this chapter, you'll be able to use Venn diagrams and tree diagrams to interpret and solve probability problems more clearly. Additionally, working through the examples in this chapter gives you practice at using the probability rules you've learned. These visual tools will make the relationships between events easier to understand and will set the stage for the Law of Total Probability and Bayes' Theorem, which we will develop in the next chapter.

Now, let's look more closely at Venn diagrams, beginning with what they represent and how to interpret them.

Venn Diagrams

Venn diagrams visually represent relationships between concepts. They use circles to display similarities and differences between sets of ideas, traits, or items. Intersections indicate that the groups have common elements. Non-overlapping areas represent traits that are unique to one set. Venn diagrams are also known as logic diagrams and set diagrams.

The real strength of Venn diagrams comes from their ability to represent interesting relationships in reports and presentations. The alternative is to present lists of attributes and then explain the relationships, which can be tedious. Venn diagrams graphically organize the same information in an easy-to-digest format. Because these charts are so efficient, educators have used them extensively since the mid-20th century as tools to present complex data.

John Venn popularized these diagrams in the 1880s to teach set theory, although Leonard Euler created them in the 1700s. Educators now use them in numerous contexts, including statistics, probability, logic, business, education, and computer science, among others. Here, we'll focus on using them to represent probabilities.

To interpret Venn probability diagrams, start by examining the circles. Each circle represents an event, and the rectangle that contains them represents the sample space. Areas where circles overlap indicate that both events can occur together. The area outside all circles represents outcomes where none of the events occur.

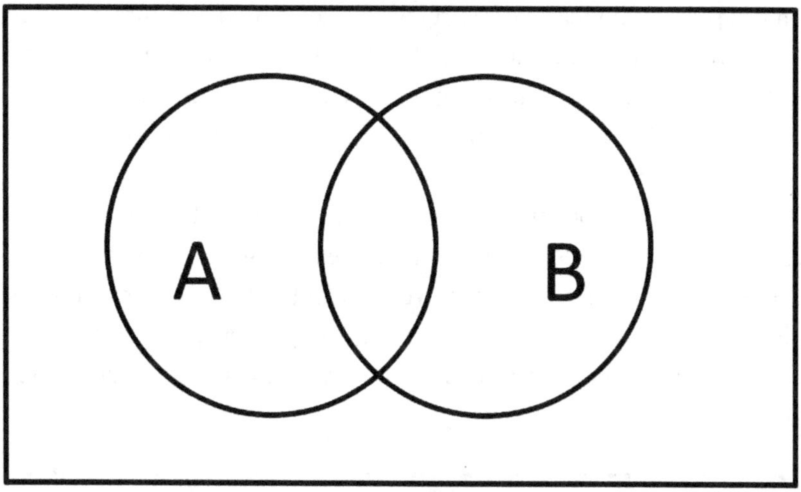

Conversely, circles that don't overlap represent mutually exclusive events. They can't occur together.

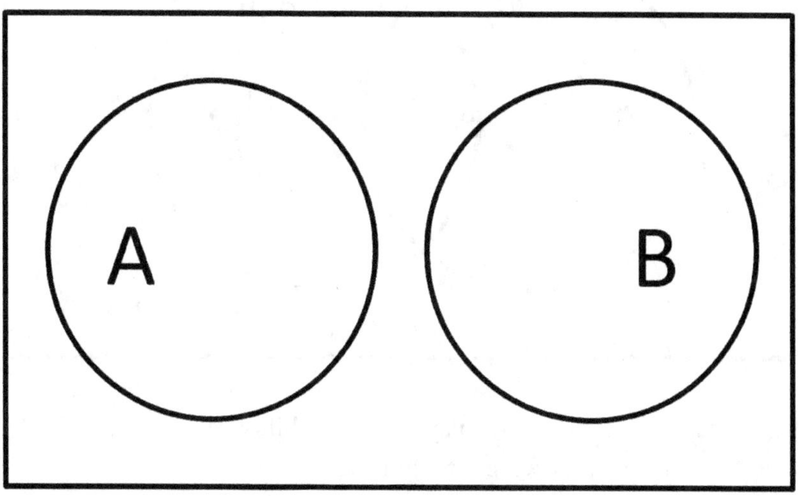

Typically, circles in Venn diagrams don't reflect probabilities to scale.

To connect these diagrams to the probability ideas you've already learned, remember that each region in a Venn diagram represents a portion of the sample space. The circles represent events; their overlaps indicate joint outcomes; and the areas outside them indicate complements.

In Chapter 4, you learned the rules that govern how events combine. However, relating probabilities for events occurring separately, together, or not at all can be complex. Fortunately, Venn diagrams turn those numerical relationships into a clear visual representation.

The following probability Venn diagram shows the typical information they provide. The rectangle represents the sample space, while the two circles represent events A and B. This diagram indicates that the two events can occur simultaneously because the circles overlap. However, when A and B are mutually exclusive, the circles won't overlap.

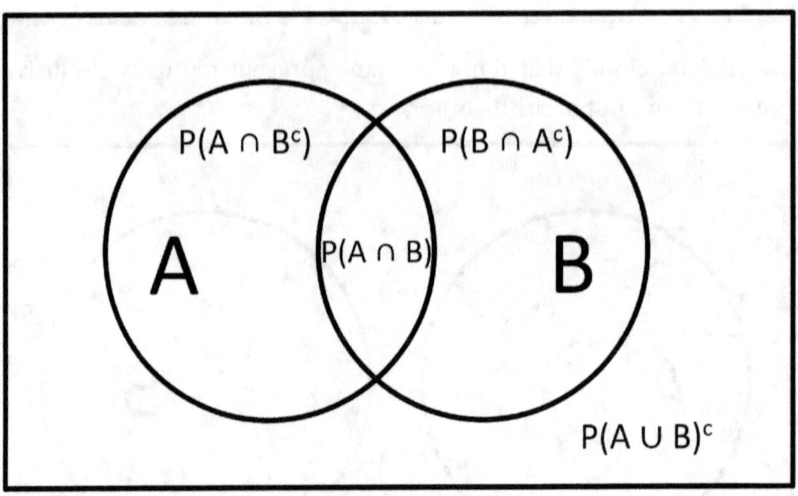

These diagrams typically display four probability values, shown in the diagram with the following notation:

- $P(A \cap B^c)$: Joint probability of A occurring and B not occurring.
- $P(B \cap A^c)$: Joint probability of B occurring and A not occurring.
- $P(A \cap B)$: Joint probability of A and B both occurring.
- $P(A \cup B)^c$: Union probability of neither A nor B occurring.

These four regions are mutually exclusive, meaning that no outcome can fall into more than one of them. Together, they represent all possible outcomes in the sample space, so the probabilities across all four regions must sum to 1.

When to Use This Method

Consider using a Venn diagram when a problem focuses on how two events relate within the same sample space, particularly when you're given marginal probabilities. They are especially helpful when you need to understand how the sample space splits into various regions. Whenever the task is to understand how events overlap, separate, or exclude each other, and you aren't dealing with stages or conditional updates, a Venn diagram is often the clearest visual tool.

A Venn diagram gives you a clear visual sense of how the events relate. However, you can't determine whether two events are independent by visually assessing a Venn diagram. To assess independence, use the numeric methods in Chapter 4.

Let's work through an example!

Bike Rental Example

For this example, imagine we run a mountain bike rental shop in an area with rough terrain. Flat tires are a problem, and we want to understand the probability of the front, back, both, or neither tire going flat during a rental. After collecting data, the owners display the probabilities in a Venn diagram.

The following example also serves as a refresher for using the various types of probabilities and rules you've learned throughout this book! I'll cover how you calculate each probability that the Venn diagram displays.

The bike has two tires, and either one, both, or neither can fail. Let:

- R = rear tire goes flat
- F = front tire goes flat

Suppose the shop has good maintenance records and estimates:

- P(R) = 0.10
- P(F) = 0.15

Joint Probability of Two Flat Tires

The shop also believes that front and rear flats happen independently. One flat does not make the other more or less likely. Because the events are independent, we can use the independent multiplication rule to find the joint probability:

$P(R \cap F) = P(R) \times P(F) = 0.10 \times 0.15 = 0.015$

That overlap represents the region where both circles intersect. It's the chance a rider ends the trip with two flat tires.

Note: If the events had been dependent, we would have needed to know the conditional probability, $P(R) \times P(F \mid R)$, and use the general multiplication rule to obtain this joint probability value.

Tip: We need this joint probability value to calculate the other probabilities in the remaining steps. So, if you're given a Venn Diagram problem to fill in, start by calculating the joint probability!

Probability of an Individual Tire but Not Both Tires

Next, we find the probabilities for the parts of each circle that do not overlap. When two events can occur together, the probability of only one of them happening must exclude the overlap.

For example, to find the probability that only a rear flat occurs, you start with the total probability of a rear flat P(R). That total includes two types of outcomes:

1. Outcomes where the rear tire goes flat **but** the front is fine.
2. Outcomes where **both** tires go flat.

When you want the probability that *only* one event occurs, complements become especially useful. Earlier in the book, you learned that complements carve out everything not in an event. Here, that same idea helps isolate the part of a set that excludes the overlap and express the probability of only one tire going flat.

The region for "rear flat only" is denoted $R \cap F^c$, meaning the rear tire goes flat while the front tire does not. To calculate this probability, start with the total probability of a rear flat, P(R), and then partition it into the following non-overlapping parts.

1. Rear flat with no front flat: $P(R \cap F^c)$.
2. Rear flat and front flat together: $P(R \cap F)$.

The second group includes outcomes where both events occur, so we must remove that overlap to find the probability of a rear-only flat.

Recall from Chapter 3 that we can partition an event into non-overlapping parts. Using that identity, the following is true for our bike example.

$P(R) = P(R \cap F) + P(R \cap F^c)$

We were given P(R), and then calculated $P(R \cap F)$ in the previous step. Now, we can use this identity and those values to find $P(R \cap F^c)$. Using algebra, we can solve for it with the following:

$P(R \cap F^c) = P(R) - P(R \cap F)$

It's simple subtraction at this point!

This same logic applies to the front-only region $F \cap R^c$.

Rear-only probability (rear flat and no front flat)

$P(R \cap F^c) = P(R) - P(R \cap F)$
$= 0.10 - 0.015$
$= 0.085$

Front-only probability (front flat and no rear flat)

$P(F \cap R^c) = P(F) - P(R \cap F)$
$= 0.15 - 0.015$
$= 0.135$

Together, the three regions inside the circles represent every way a rider can get at least one flat tire:

- Rear only: 0.085
- Both: 0.015
- Front only: 0.135

No Flat Tires

Now consider riders who finish their trip with no flat tires. This outcome is the complement of getting at least one flat, so we can express it as $P(R \cup F)^c$. Earlier, we used complements to isolate parts of a set. Here, the same idea lets us describe the entire region outside both circles in the Venn diagram.

First, calculate the union probability of getting at least one flat using the additive rule:

$P(R \cup F) = P(R) + P(F) - P(R \cap F)$
$= 0.10 + 0.15 - 0.015$
$= 0.235$

So, the probability of getting no flats is the complement of that event:

$P(R \cup F)^c = 1 - P(R \cup F)$
$= 1 - 0.235$
$= 0.765$

This large outside region indicates that most rides end without any flat tires. Let's put all the probability values into our Venn Diagram! At a glance, the Venn diagram helps you understand the probabilities of various combinations of flat tires occurring.

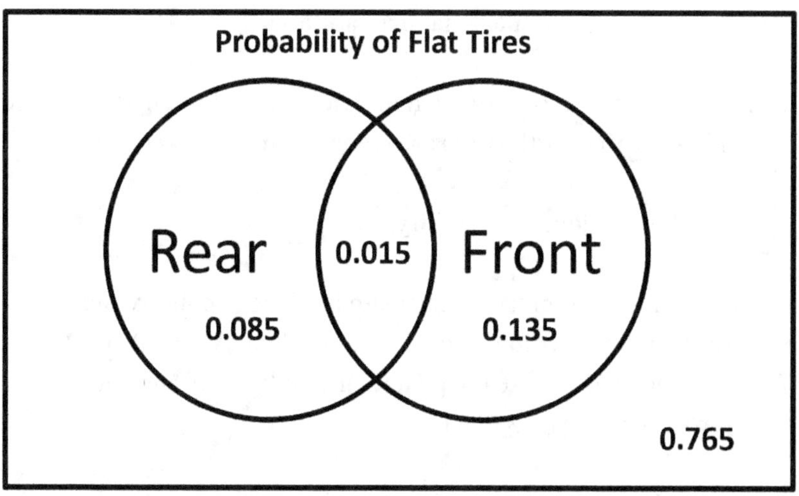

Each region communicates something different:

- **0.085** → only the rear tire failed.
- **0.015** → both tires failed.
- **0.135** → only the front tire failed.
- **0.765** →neither tire failed.

These values represent *all* possible outcomes, so every part of the rectangle is filled by exactly one of these four numbers. That's our entire sample space. So, the probabilities must sum to 1 (100%).

General Steps for Filling in a Venn Diagram

When working with a Venn diagram, it helps to follow a consistent process. Although problems can vary, the following steps will guide you through most situations. We used this procedure for the bike rental example.

In many problems, you are given the marginal probabilities for two events along with either the joint probability or enough information to find it. After you know the marginal and joint probabilities, you can fill in the remaining regions of the Venn diagram.

1. Start with the given marginal probabilities.

Identify the probabilities for each individual event, such as P(A) and P(B). These represent the total area of each circle.

2. Determine the joint probability.

Find $P(A \cap B)$ if it is given or calculate it. If the events are independent, use the multiplication rule for independent events: $P(A) \times P(B)$. If they are dependent, use the general multiplication rule with conditional probability: $P(A) \times P(B \mid A)$.

Place the joint probability $P(A \cap B)$ in the overlapping region of the circles. Use this value to calculate the remaining regions.

3. Partition each event to find the "only" regions.

Subtract the overlap from each marginal probability to find the parts of each circle that do not overlap:

$P(A \cap B^c) = P(A) - P(A \cap B)$
$P(B \cap A^c) = P(B) - P(A \cap B)$

4. Find the probability of neither event.

Use the complement of the union:
$P(A \cup B)^c = 1 - [P(A) + P(B) - P(A \cap B)]$

5. Check that all regions sum to 1.

Add the four regions to confirm they represent the entire sample space.

Venn diagrams show how events relate within a single sample space. But many problems involve sequences of events where probabilities change along the way. In those situations, we need a different tool, a tree diagram.

Tree Diagrams

Tree diagrams are perfect for problems that confuse people when described only in words. Particularly when they involve multiple steps where conditional probabilities apply along the way. They help you organize information and keep track of conditional relationships.

Later, we will use a tree diagram to solve the Monty Hall Problem, a puzzle that confuses almost everyone. But visualizing its structure really helps.

When to Use This Method

Consider creating a tree diagram whenever a problem gives you a marginal probability for one event and then provides conditional probabilities for what happens afterward. That structure often means you'll need to work out a probability, such as P(B), that isn't given directly.

Trees are also ideal when a scenario involves several possible sequences of events, and you need the probability of one specific combination out of many. More generally, if a problem describes stages, choices, or updated probabilities ("after A occurs, B happens with probability..."), or if the outcome depends on the path taken, a tree diagram is usually the clearest way to organize the information and calculate the correct probability.

For example, you might track customer service results through different support methods, evaluate defect rates from different factories, or understand different customer order patterns.

In short, tree diagrams give you a visual way to understand how probabilities unfold when events happen in stages or when one event depends on another. They clearly show the structure of a problem, step by step, and prepare you for more advanced ideas, including Bayes' Theorem.

General Probability Tree Setup

We'll start by covering how a probability tree works and how to create and interpret it in general. Then, we'll move on to three examples.

Each level of branches represents a stage in the process. Each split in a tree diagram shows all possible outcomes of the event at that stage. Trees adapt to whatever possibilities the problem requires. If an event has k outcomes, the branch splits into k paths.

The following probability tree shows events A and A^c at the first split and B and B^c at the second split, with the terminal nodes displaying the joint probabilities using the multiplication rule.

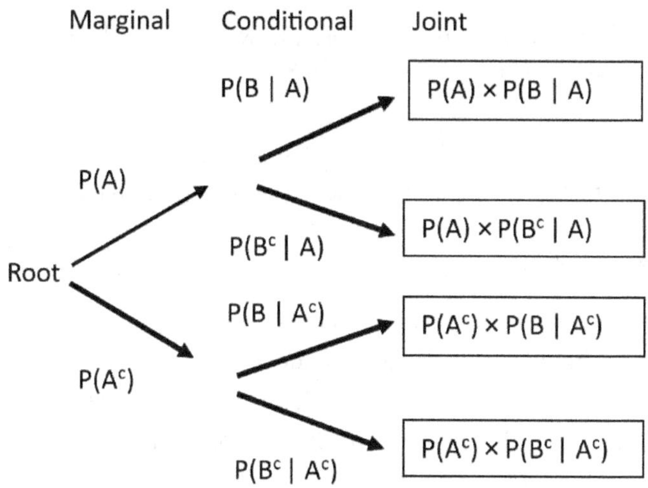

A probability tree always begins with a single starting point, often called the *root*. This point represents the full sample space before any events occur. The first split shows the possible outcomes of the first event in the full sample space. These first branches display marginal probabilities because no prior events have occurred to condition on.

The next split shows the possible outcomes of the second stage, but now each set of branches is conditional on the path taken so far. Hence, the second and later splits display conditional probabilities given the previous branches in a path. At each step, you are effectively zooming in on only the outcomes that remain possible after the earlier events. As you move right in the tree, the nodes represent increasingly restricted sample spaces.

Keep branching until every possible sequence of outcomes is represented. The endpoints of the tree represent complete, final outcomes and are called terminal nodes. The number of terminal nodes is the product of the number of branches at each stage. The probability of each terminal node outcome is a joint probability that you find by using the general multiplication rule. In other words, multiply all the probabilities along a path to find the terminal node probability.

Collectively, the tree must depict all possible ways of producing all possible outcomes. Specifically:

- Each terminal node has a unique path leading up to it.
- The set of all terminal nodes must cover all possible paths.

Common Uses for Probability Trees

Probability tree diagrams serve two main purposes. They allow you to determine the probability of a specific sequence of outcomes represented by a single path. They also help identify multiple paths that lead to the same event, allowing you to find the total probability for that event by combining those paths.

To accomplish these two things, tree diagrams follow a simple rule of thumb that you'll use repeatedly: multiply across a path and add down the terminal branches.

First, use the multiplication rule for the probabilities along a path to find the joint probability of that complete sequence of events. This process gives you the joint probability of a specific sequence of events. In the example tree, notice that the joint probabilities in the terminal node are just the product of the marginal and conditional probabilities along the path to the terminal node.

Second, add the probabilities of all paths that end in the same type of outcome. Because each complete path in the tree represents a mutually exclusive sequence of events, we can use the simplified additive rule and sum the probabilities of all paths that end in the same outcome.

This idea builds on a concept we first saw in Chapter 4, where we partitioned an event into non-overlapping pieces using complements. It also appeared in Venn diagrams. In a probability tree, this partitioning becomes even more concrete: the paths that lead to a given event divide it into mutually exclusive pieces that together include all possible ways that event can occur.

Because of this structure, adding the probabilities of those paths gives the total probability for the event. This same idea extends to more complex trees with many branches and forms the basis of the Law of Total Probability.

You can use this method to recover a marginal probability that is not provided. For example, in the previous tree, you can sum the probabilities in the first and third terminal nodes from the top to find P(B). You can identify which terminal nodes correspond to event B by looking at the second-stage branch labels. Any path where the second branch includes B (such as $P(B \mid A)$ or $P(B \mid A^c)$) represents an outcome where B occurs.

The terminal nodes represent all the mutually exclusive and exhaustive outcomes in the scenario.

Tip: Branch Probabilities Must Sum to 1
At every split in a probability tree, including the terminal nodes, the probabilities of all branches must sum to 1 because they represent all possible outcomes at that stage. If one probability is missing, you can find it by subtracting the other branch probabilities from 1.

Let's bring these to life using several examples.

Example: Coffee and Pastry Purchases at a Café

Tree diagrams are helpful when a situation involves two decisions that can occur together. In this example, a café tracks whether a customer buys coffee, a pastry, both, or neither. The diagram helps us visualize the connections between these choices and calculate the probability of each final outcome.

Suppose the café has reliable data showing:

- The probability a customer orders coffee: P(C) = 0.70.
- The probability a customer doesn't order coffee: P(C^c) = 0.30.
- The probability a customer buys a pastry: P(P) = 0.40.
- The probability a customer doesn't buy a pastry: P(P^c) = 0.60.

We begin the tree at a single point that represents a randomly selected customer. The first split shows whether the customer buys coffee. One branch represents coffee, and the other represents no coffee. The probabilities at this split must sum to 1 because the two outcomes cover every possibility at that stage.

From each of these branches, we split again for the pastry decision. The pastry probabilities also add to one at this split.

Here's the probability tree for the café purchases based on the given probabilities.

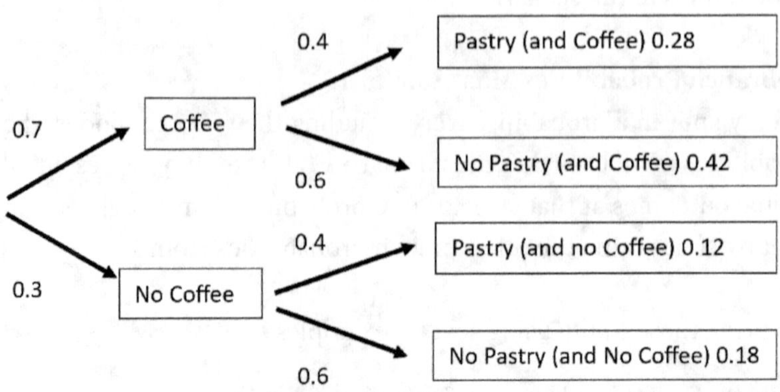

Tree diagrams make it easy to identify independent or dependent events. Just look at the branch probabilities for each level after the first.

If the probabilities for the second split stay the same no matter which first-level branch you follow, the events are independent. Conversely, when those second-level probabilities change depending on the earlier outcome, the events are dependent.

By looking at this portion of the tree, you can see that the coffee and pastry events are independent because the conditional probabilities of pastry (0.40) and no pastry (0.60) are the same under both the coffee and no coffee branches.

Reminder: Determine independence by assessing whether conditional probabilities are constant between groups of branches.

A tree diagram makes the problem's structure clear. To find the probability of a complete scenario, multiply the probabilities along the path. For example, the joint probability that a customer buys both coffee and a pastry is:

$P(\text{Coffee} \cap \text{Pastry}) = 0.70 \times 0.40 = 0.28$

You follow the same procedure for every path:

- **Coffee and pastry:** $0.70 \times 0.40 = 0.28$
- **Coffee and no pastry:** $0.70 \times 0.60 = 0.42$
- **No coffee and pastry:** $0.30 \times 0.40 = 0.12$
- **No coffee and no pastry:** $0.30 \times 0.60 = 0.18$

These four outcomes cover all possibilities and are mutually exclusive. They also add to 1, which confirms the diagram is internally consistent. The tree shows how you obtain the joint probabilities by multiplying the probabilities of each step.

Tree diagrams also make it easy to find marginal probabilities by adding across final branches. Notice that we were not given the marginal probability of purchasing a pastry, P(Pastry). However, we can recover that probability using this tree. That probability is the sum of the two pastry endpoints in the tree, which are the first and third endpoints from the top:

P(Pastry) = 0.28 + 0.12 = 0.40

Example: Customer Support Method and Satisfaction

Suppose a company offers two customer support options: phone support and email support. A customer has a 0.60 probability of choosing phone support and a 0.40 probability of choosing email. Satisfaction rates differ by support method, so this is a natural setting for a probability tree.

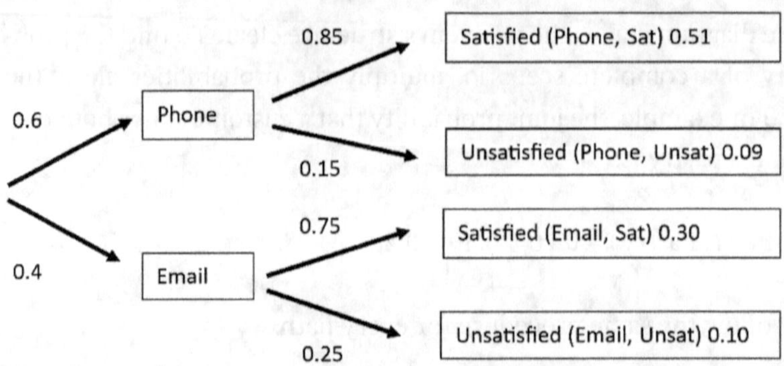

The tree diagram clearly shows these dependencies. On the phone branch, 85% of customers report satisfaction, compared with only 75% on the email branch. Because the conditional probabilities differ across branches, 0.85 vs. 0.75, we know satisfaction depends on the support method. If the events were independent, the satisfaction probabilities would be the same on both branches.

The tree makes it easy to see which combinations are most and least common. Again, just multiply along each path to find the endpoint probability. The largest endpoint probability is "Phone + Satisfied," reflecting that most customers use phone support and most of them report positive experiences. The smallest endpoint probability is "Phone + Not satisfied," indicating that it is relatively rare for a customer to both choose phone support and walk away unhappy.

You can also recover marginal probabilities by adding the relevant paths. For example, to find the overall probability that a randomly selected customer is satisfied, add the joint probabilities from both satisfaction endpoints:

- $P(\text{Satisfied} \cap \text{Phone}) = 0.60 \times 0.85 = 0.51$
- $P(\text{Satisfied} \cap \text{Email}) = 0.40 \times 0.75 = 0.30$

So the marginal probability of satisfaction is:

$P(\text{Satisfied}) = 0.51 + 0.30 = 0.81$

This means that, overall, about 81% of customers report being satisfied, regardless of which support method they used.

The tree displays the sequence, dependencies, and path outcomes in a single, compact diagram. Next, let's apply the same visual logic to a puzzle where intuition often fails.

Example: Monty Hall Problem

Who would've thought an old TV game show could inspire a statistical problem that has tripped up mathematicians and statisticians with PhDs? The Monty Hall problem has confused people for decades. In the game show *Let's Make a Deal*, Monty Hall asks you to guess which closed door a prize is behind. The answer is so puzzling that people often refuse to accept it!

The Monty Hall problem's baffling solution reminds me of optical illusions, where you find it hard to disbelieve your eyes. I think of this puzzle as a statistical illusion. Like an optical illusion, even after knowing the truth, it's hard to see and accept the correct answer, and the illusion can seem more accurate than the actual answer. This statistical illusion occurs because your intuition relies on a false assumption.

Here's how the game works. Monty randomly hides a prize behind one of three doors, while the other two have a goat (no prize).

Then he follows this order of events:

1. Monty Hall asks you to choose one of three doors. You state out loud which door you pick, but you don't open it.
2. Monty opens one of the other two doors that has a goat.
3. At this moment, there are two closed doors, one of which you picked. The prize is behind one of the closed doors, but you don't know which one.
4. Monty asks you, "Do you want to switch doors?"

Most people assume that both doors are equally likely to have the prize. It appears the door you chose has a 50/50 chance. Most stick with their initial choice because there is no perceived reason to change.

Time to shatter this illusion with the truth! If you switch doors, you double your probability of winning!

What!?

When Marilyn vos Savant was asked this question in her Parade magazine column, she correctly answered that switching doors gives you a 66% chance of winning. Her answer was so unbelievable that she received thousands of incredulous letters from readers, many of them with Ph.D.s! Paul Erdős, a noted mathematician, was swayed only after observing a computer simulation.

Fortunately, probability trees are a great way to solve the Monty Hall Problem because they can handle conditional probabilities and dependent events. The rules look simple: you pick one of three doors, Monty reveals a goat behind a different door, and then you choose to stay or switch. But the structure underneath is all conditional. Each stage depends on what happened earlier, and the tree tracks that.

We begin with a random event that occurs before the contestants do anything: Monty places the car. The first split in the tree has three branches because the car is equally likely to be behind any of the three doors.

For the tree, we label the contestant's choice as Door 1. That door choice does not change the math because the doors are symmetric. Nothing about the problem gives Door 1 any special status over Door 2 or Door 3. We're simply labeling the contestant's choice as Door 1. If you relabeled the doors, the situation would play out exactly the same way. By fixing the label, we simplify the tree without altering any probabilities.

At this point, each initial branch has a 1/3 probability of correctly choosing the door with the car.

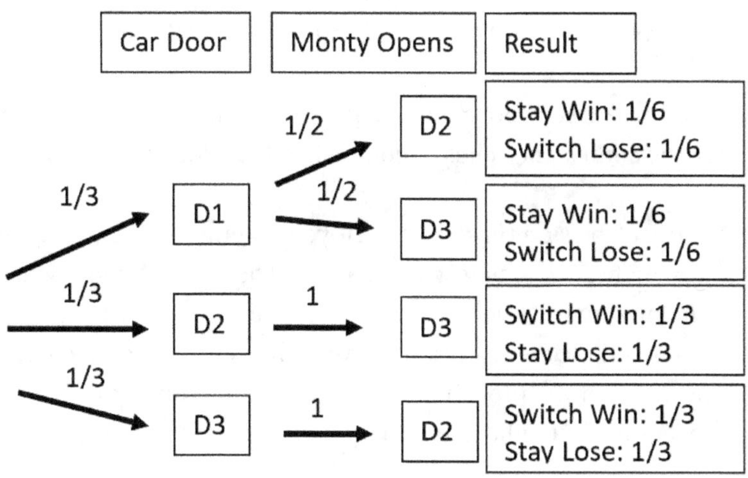

The second split shows Monty's action, and this is where the conditional thinking kicks in. Monty never opens the door you chose, and he never opens the door hiding the car. His choice *depends* on those two conditions. Due to this dependency, we'd expect the conditional probabilities in the second split to differ between the branches.

When the car is behind your chosen door, both the other doors contain goats. Monty can open either one. We show that with a 1/2 – 1/2 split on the Door 1 path. But when the car sits behind Door 2 or Door 3, Monty has no flexibility. One of the doors is your choice, and the other hides the car, so he must open the single remaining goat door. Each of those paths has a conditional probability of 1. The tree diagram emphasizes these forced choices by making those branches proceed in only one direction.

This second level is the heart of the problem. The conditional probabilities change because Monty's options change depending on the hidden car location and the door the contestant chooses. You can see immediately that some branches are rare (like "Car behind Door 1 and Monty opens Door 2," which has probability 1/6) while others are more common (like "Car behind Door 2 and Monty opens Door 3," which has probability 1/3). The tree captures these differences.

Each path in the tree represents one possible combination of where the car is, the contestant's chosen door, and which door Monty opens.

Next, the final labels on those paths tell you whether staying wins or loses and whether switching wins or loses. These labels show what happens if you stay with Door 1 and what happens if you switch to the only unopened door. As always, the probabilities for those outcomes are the products of the probabilities for the full path up to that point: the car location (first split) and Monty's constrained action (second split).

Now, to see which strategy is best, you must recover the overall (marginal) probabilities for winning by staying or switching. Simply use the same method that we used for the previous trees. Add the probabilities on the branches that lead to each outcome of interest in the final nodes.

On this tree, I've labeled the outcomes of interest as Stay Win and Switch Win.

- P(Stay Win) = 1/6 + 1/6 = 1/3
- P(Switch Win) = 1/3 + 1/3 = 2/3

As you can see, switching doubles the win probability compared to staying!

The tree ties everything together. It shows that Monty's behavior depends on information you do not see in the game, and his constrained choices shift the probabilities toward switching. Nothing mysterious here, just conditional probability organized visually.

The solution to the Monty Hall problem seems weird because our mental assumptions for solving it do not match the actual process. Subconsciously, we based our mental assumptions on independent, random events. However, Monty knows the prize location and uses this knowledge to affect the outcomes non-randomly. The results make sense after you understand how Monty uses his knowledge to pick a door.

I have written about the Monty Hall Problem on my website. And that blog post has received more comments than any other. Readers have filled the comment section with doubts that switching doubles your chances of winning, even after reading the explanation for why it is true. Most doubters have interesting reasons for why it must be 50/50. Many of these comments are from students, and I suspect the Dunning-Kruger Effect is occurring.

So, as you read this, I wouldn't be surprised if you find it hard to accept. Just know that both probability theory and computer simulations that play the game millions of times agree that switching doubles your chances. And remind yourself that Monty isn't acting randomly. He's selectively removing non-prize doors from the game.

Review and Next Steps

In this chapter, you learned how visual tools can make probability relationships easier to understand. Venn diagrams showed how events overlap, separate, or exclude each other within a sample space. By labeling regions and applying probability rules, you saw how to calculate joint probabilities, complements, and the likelihood of outcomes involving one event but not the other.

Tree diagrams offer a different type of insight. They lay out sequential or dependent situations step by step, showing exactly how probabilities change after each event occurs. You saw how to use the multiplication rule to follow a single path and the additive rule to recover marginal probabilities from the endpoints. These diagrams made the conditional structure of a problem visible in a way that Venn diagrams cannot.

We ended by applying these ideas to the Monty Hall Problem. That example reveals how misleading intuition can be and why visual organization of conditional events helps uncover the true probability structure. Problems that seem baffling when described in words become much clearer when represented on a tree.

These ideas set the stage for the next chapter. Tree diagrams naturally lead to two of the most important tools in probability: the Law of Total Probability and Bayes' Theorem.

Chapter Summary

- Venn diagrams help visualize unions, intersections, complements, and mutually exclusive events.
- Venn diagrams can represent dependent events but cannot show conditional structure.
- Tree diagrams represent sequential and conditional processes step by step.
- Multiply down a path to find joint probabilities.
- Add across final branches to recover marginal probabilities.
- Tree diagrams help identify independence (conditional probabilities stay the same across branches).

These tools prepare you for the Law of Total Probability and Bayes' Theorem in the next chapter.

Chapter 6 Practice Problems

Question 1: A streaming service tracks whether subscribers watch documentaries and whether they watch science fiction. Let:

- D = subscriber watches documentaries.
- S = subscriber watches science fiction.

Suppose the service estimates:

- P(D) = 0.40
- P(S) = 0.25

Assume these events are independent. Fill in the four regions of the Venn diagram:

- $P(D \cap S)$
- $P(D \cap S^c)$
- $P(S \cap D^c)$
- $P(D \cup S)^c$

Question 2: A fitness center tracks whether members attend yoga classes and whether they attend strength training classes. Let:

- Y = member attends yoga.
- S = member attends strength training.

Suppose the center estimates:

- P(Y) = 0.50
- P(T) = 0.40
- $P(Y \cup S) = 0.70$

Fill in the four regions of the Venn diagram:

- $P(Y \cap S)$
- $P(Y \cap S^c)$
- $P(S \cap Y^c)$
- $P(Y \cup S)^c$

Question 3: A bookstore tracks whether a customer buys a fiction book or a nonfiction book. It also tracks whether the customer buys a bookmark during the same visit. Let:

- F = customer buys fiction.
- N = customer buys nonfiction.
- B = customer buys a bookmark.

Suppose the probabilities are:

- $P(F) = 0.55$.
- $P(N) = 0.45$.
- $P(B \mid F) = 0.30$.
- $P(B^c \mid F) = 0.70$.
- $P(B \mid N) = 0.20$.
- $P(B^c \mid N) = 0.80$.

Draw a probability tree for this situation. Then answer the following questions:

- **A.** Are buying a bookmark and book type independent or dependent?
- **B.** What is the probability that a customer buys a nonfiction book and a bookmark?
- **C.** What is the overall probability that a customer buys a bookmark?

Question 4: A small company is selecting two employees at random from a team of five to attend a specialized training program. The team consists of:

- 3 engineers.
- 2 analysts.

The two employees are selected one after the other, without replacement. Draw a probability tree for this situation. Then answer the following questions:

A. What is the probability that both selected employees are engineers?
B. What is the probability that both are analysts?
C. And the probability of one engineer and one analyst?

CHAPTER 7

Bayesian Methods: Total Probability & Bayes' Theorem

Many real-world events can happen in more than one way. A product might be defective because it was manufactured at one of several factories. A customer might cancel a subscription for several possible reasons. A medical test can yield both true-positive and false-positive results. When an outcome has multiple underlying cases, we need a way to combine the contributions from each case.

The Law of Total Probability gives us that tool. It tells us how to calculate the overall probability of an event by breaking it into separate, mutually exclusive pieces and adding up their probabilities. You've already seen how tree diagrams represent these paths visually. In this chapter, we'll link that visual idea to a formal equation that captures the same structure.

In other scenarios, problems run in the opposite direction from the usual situation where causes precede outcomes. In real-world settings, we frequently observe the outcome first, and then want to know which underlying case is most likely.

For example, after observing a failure, engineers need to determine the most likely explanation. After a positive test result, doctors want to know the probability that the patient has the disease. These situations require reversing the conditional probability, and Bayes' Theorem provides the method to do exactly that. It combines what we already know about the possible cases with new evidence we observe.

In this chapter, you will learn how the Law of Total Probability decomposes complicated events, how Bayes' Theorem reverses conditional probabilities, and how both ideas work together in practical applications such as diagnostic testing, quality control, decision making, and evaluating evidence.

Understanding and using probability trees is crucial for these methods. Be sure you're completely familiar with using these diagrams in Chapter 6 before proceeding.

Law of Total Probability (LOTP)

Some events can only happen in one way. Others can occur through several different paths. When an outcome has multiple possible sources, the overall probability depends on the contribution from each source.

For example, a defect can come from any factory in a company's supply chain, and each factory can produce defects at different rates. The overall, company-wide defect rate depends on each factory's contribution to the total.

The Law of Total Probability (LOTP) finds the overall probability of an outcome by dividing the problem into mutually exclusive cases, assessing the likelihood of the outcome in each case, and combining those contributions across all cases.

Initially, that might sound complicated. But it becomes clear when we use a tree diagram because the branches naturally separate the distinct cases that lead to the same final outcome.

In the previous chapter, we used probability trees to calculate probabilities by multiplying along paths and adding across them. In doing so, we calculated overall (marginal) probabilities without naming the method. The Law of Total Probability formalizes that process and turns it into a general rule for combining probabilities across multiple cases that lead to the same outcome.

To understand the Law of Total Probability, start with the idea that an event can happen in more than one way. Different cases or groups can lead to the same outcome. These cases can include distinct groups, causes, or sources.

Here are a few simple examples of this idea:

- **Customer Service:** A customer might end up satisfied because they used online chat, called by phone, or sent an email. Each option is a different sequence of events, and only one can be true for any given customer.
- **Medical Test Results:** A positive test can occur in two mutually exclusive ways. The patient *has* the disease and receives a true positive, or the patient *does not* have the disease and receives a false positive. These two paths cannot happen together and cover all possible ways to get a positive result.
- **Factory Defects:** A defective part might come from Factory 1, Factory 2, or Factory 3. Each part follows one unique production path, and any of those paths can result in a defect.

LOTP Formula

The Law of Total Probability is the algebraic expression of this idea:

1. Break the event into its mutually exclusive cases.
2. Weight each case by its relative frequency in the population (marginal probability).
3. Determine how strongly each case predicts the outcome (conditional probability).
4. Multiply the relative frequency by the conditional probability to get each case's contribution.
5. Sum all cases that lead to your outcome of interest.

At first glance, the Law of Total Probability formula can be intimidating. However, connecting it to what you've already learned using probability trees makes it intuitive. So, let's start with a simple probability tree and use that to transition to the equation.

Instead of labeling the tree with probability values, I'll use notation that we'll link to the formula.

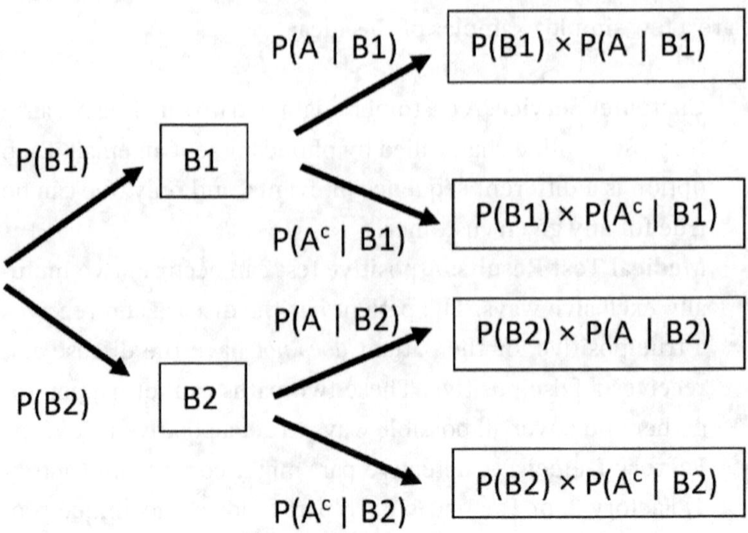

Stage one of the diagram starts with the marginal probabilities for cases B1 and B2 (P(B1), P(B2)). The first split divides the sample space into mutually exclusive cases, such as customer service method, disease status, or factory number. Each observation belongs to exactly one case. The initial probabilities represent the relative frequency of each case in the overall population. They are the starting proportions before considering any additional information.

Then stage 2 shows the conditional probabilities of event A and its complement (A^c) occurring given that either case B1 or B2 has already occurred. The second split indicates which event occurs next for each stage 1 case, such as customer satisfaction, a positive test result, or producing a defective part

In this setup, the B's represent the cases that define the different paths, while the A's represent the events that occur within each case. That's why we condition on B first and then consider A. Because the cases are mutually exclusive, we can calculate their individual contributions and then sum them to find the overall probability of the outcome of interest.

You can continue adding stages in this way if a problem involves additional steps or events. Each new stage represents what happens next, given everything that has already occurred along that path.

Next, we'll use the standard probability tree procedure of multiplying down branches and adding across them:

1. The joint probability at each final node equals the product of the probabilities along its path due to the multiplication rule. I've put the notation for those probabilities in the final node.
2. Then you sum the paths ending in your outcome of interest using the additive rule.

To find the overall, or marginal, probability of A, add the probabilities of all final nodes that end with A. To find the paths that end in A, look at the conditional probabilities in the terminal nodes. When A appears to the left of the bar in the conditional probability, that branch ends in outcome A. In this tree, they are at the top and the third-from-the-top terminal nodes.

In this example, the two relevant paths are:

- $P(B1) \times P(A \mid B1)$
- $P(B2) \times P(A \mid B2)$

So, we just add those probabilities:
$P(A) = P(B1) \times P(A \mid B1) + P(B2) \times P(A \mid B2)$

Guess what? That's the formula for the Law of Total Probability!

You can think of it as a weighted average across the different cases, where you're averaging the probability of outcome A. The weights are the marginal probabilities, P(B1) and P(B2). These weights reflect how often each case occurs in the overall population.

Paths that begin with higher-probability cases have a greater influence on the overall probability. If B1 is far more common than B2, then the term $P(B1) \times P(A \mid B1)$ plays a larger role in determining the final probability of A. In summary, outcomes associated with common situations influence the overall probability more than those associated with rare situations.

This formula matches the two-case tree. However, many problems involve more than two cases. A probability tree can split into three, five, or however many cases are required. The Law of Total Probability works the same way, no matter how many branches there are.

In general, if the cases are labeled B1, B2, B3, and so on, and they form a complete set of mutually exclusive possibilities, then the overall probability of A is the sum of the probabilities for each case that leads to A. The general equation for the Law of Total Probability is the following:

$$P(A) = P(B1) \times P(A \mid B1) + P(B2) \times P(A \mid B2) + \ldots + P(Bk) \times P(A \mid Bk)$$

Here, k is the total number of branches or cases in the first stage of the tree. This equation is just a longer version of the weighted average. Each case has its own marginal probability and its own conditional probability for A. You multiply those together for that case's contribution to the overall probability, and then sum the contributions for all cases.

This formula can also be written more compactly using summation notation:

$$P(A) = \sum P(B_i) \times P(A \mid B_i)$$

Where the B_i values represent all possible cases that partition the sample space.

The formula might look intimidating at first, but it simply restates what you already did on the tree. And, if you do find the formula hard to work with, just create a tree to help you solve the problem!

Example: Defect Rates From Three Factories

To see the Law of Total Probability in action, consider a company that buys a part from three factories. The factories do not contribute equally to the supply, and their defect rates differ. Together, they determine the overall probability that a randomly selected part is defective.

Suppose the problem gives you the following information about the factories and their defect probabilities. You need to calculate the total probability for defects (P(Defect)).

Factory 1:

- P(F1) = 0.30
- P(Defect | F1) = 0.02

Factory 2:

- P(F2) = 0.20
- P(Defect | F2) = 0.015

Factory 3:

- P(F3) = 0.50
- P(Defect | F3) = 0.06

Notice that the problem asks us to find the marginal probability of a defect (P(Defect)), but it gives us different marginal probabilities, followed by a series of conditional probabilities for defects. That pattern is a hallmark of LOTP problems!

Here, the three factories are our cases and their marginal probabilities indicate the proportion they contribute to the part supply. There are basically two ways to approach this problem.

If you're comfortable with the formula, you can just go straight to using it. Multiply the appropriate pairs of case marginal probabilities and defect conditional probabilities. Then sum those products.

Or, you can use the tree to organize the information, calculate the necessary values, and then sum the terminal nodes that have defect outcomes.

We'll do both!

The tree shows all the previous marginal and conditional probabilities in their proper locations. The terminal nodes display the joint probabilities I calculated using the multiplication rule. **Note**: The no defect branches are the complement of the defect probability.

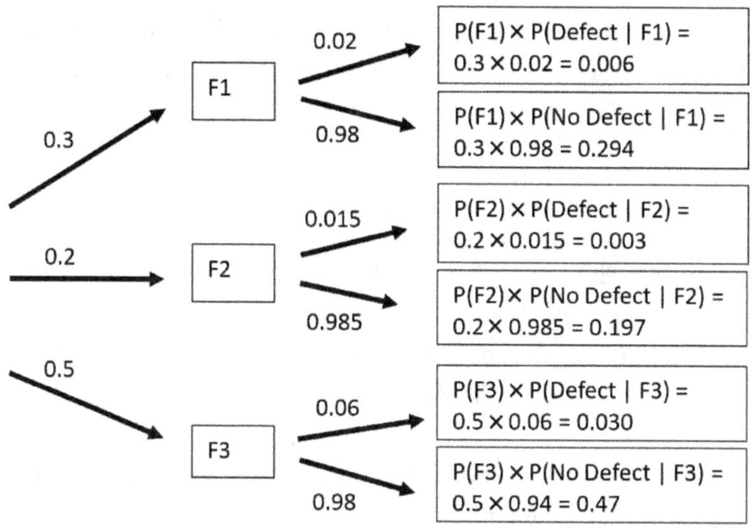

Using the tree, just sum the probabilities for the three branches that end with the defect outcome. Here, a conditional probability that begins with Defect before the bar indicates the outcome is a defective part. Just sum the probabilities in those three terminal nodes that end with defects.

Or, plug those three values into the equation.

$$P(\text{Defect}) = P(F1) \times P(\text{Defect} \mid F1) + P(F2) \times P(\text{Defect} \mid F2) + P(F3) \times P(\text{Defect} \mid F3)$$

$$= (0.3 \times 0.02) + (0.2 \times 0.015) + (0.5 \times 0.06)$$
$$= 0.006 + 0.003 + 0.030$$
$$= 0.039$$

So, the overall defect probability is P(Defect) = 0.039, or 3.9%.

This example also shows that the third factory has a strong influence on the final answer. Factory 3 not only has the highest defect rate (6% vs. ~2%), but it also supplies half of all parts. Because it appears so frequently in the supply chain, its defect probability weighs heavily in the overall result. That is the weighted-average idea behind the Law of Total Probability. Cases that occur more often contribute more to the final probability.

When to Use This Method

When you face a probability problem, it is not always obvious which method to use. The Law of Total Probability helps when an event can happen in several different ways, and you need to account for all of them to find the overall probability. Here are some reliable clues that tell you the LOTP is the right tool.

The problem describes different groups, cases, or categories, but asks for an overall probability.

Look for mutually exclusive groups, such as phrases like "Factory A, Factory B, and Factory C," "three customer types," or "two possible sources." A part can only come from one factory, a customer belongs to one category, and so on. Then you might see wording that suggests finding an overall probability, like "What is the probability a part is defective?" or "What is the chance the customer is satisfied?" If the information is organized by groups or cases and you need to find an overall probability, LOTP is usually the correct approach.

You are asked for a marginal (overall) probability, but the problem gives you a *different* marginal probability plus several conditional probabilities.

This pattern shows up all the time. For example, the problem might ask you for the overall defect rate, but the only marginal probabilities it provides are the production shares for each factory. In other words, you know P(B1), P(B2), P(B3), and the conditional probabilities P(A|Bi), but not P(A). Finding that missing marginal probability is exactly what the Law of Total Probability solves for by combining probabilities across the different cases.

The situation naturally fits a tree diagram.

If you can imagine the process as a first split into cases and a second split into outcomes, LOTP connects directly. Even if you skip drawing the tree, recognizing the branching structure is enough to point you toward the method. It uses information you already have to produce the overall probability you need.

The Law of Total Probability moves forward, from cases to outcomes. Bayes' Theorem works in the opposite direction, starting with an observed outcome and working back to the most likely case.

Bayes' Theorem: Intuition First

In many probability problems, we start with information that describes how likely an outcome is *within* each group or situation. These probabilities usually appear in the form P(A | B): the probability of event A given that case B has already occurred.

These "forward" conditional probabilities are common because they match how we typically collect data. Researchers, engineers, and analysts usually begin by identifying a group and then observing outcomes within that group. For example, a medical study begins by identifying patients with a disease and then records how often their test results are positive. Or, quality analysts assess parts already produced by Factory 1 to determine its defect rate.

Forward probabilities also match the natural order of events. First, the case or condition is established, and then the outcome follows. A patient has a disease, and then the test result appears. A machine produces an item, and then the inspection determines whether it's defective. Because the sequence is intuitive and the data are easy to collect, most real-world information takes the following form:

P(Outcome | Case)

Given the case, what is the probability of the outcome? For example, given a patient has the disease, what is the probability of a positive test result?

Note: In Bayes' Theorem, these cases represent competing explanations for the observed outcome. In later sections, we will refer to them as hypotheses, which is the standard term in Bayesian reasoning.

Unfortunately, these forward probabilities often don't answer the questions people actually care about. Outside of controlled studies and experiments, the questions we face in everyday situations usually run in the opposite direction. In real-world contexts, we often observe an outcome first and then want to know which case is most likely responsible. What we care about often takes the following form:

P(Case | Outcome)

Given the outcome, what is the probability of a specific case? For example, given a patient has a positive test result, what is the probability they have the disease?

Switching the order of a conditional probability asks a completely different question.

You first saw this idea with contingency tables, where switching the order in a conditional probability changed the restricted denominator. Moving from P(A | B) to P(B | A) meant shifting from a column total to a row total (or vice versa). Bayes' Theorem formalizes this same idea when the groups come from real-world contexts rather than a table of observed counts. It gives us a systematic way to reverse the conditional probability when the situation calls for it.

Bayes' Theorem allows us to find probabilities like P(Case | Outcome) when they are more useful. Statisticians refer to this ordering as "reverse" probabilities because they reverse the direction of conditioning. Even though the case or condition comes before the outcome in real life, we often observe the outcome first and then investigate which case applies (e.g., a patient observing a positive test result). That is when you need Bayes' Theorem.

Calculating these reverse probabilities depends on how strongly each case predicts the outcome and also how common each case is in the first place. As a result, the reverse probability cannot usually be measured directly. Instead, it must be inferred by combining the forward conditional probabilities with information about the frequency of each case.

Bayes' Theorem provides the method for finding these reverse probabilities. It takes the forward information we can measure and uses it to answer the reverse question. The following examples show why this reversal matters so much.

Example 1: The Noisy Alarm

Imagine you live in a building with a fire alarm that sometimes sounds even when nothing is wrong. Through testing, the manufacturers know the probability that the alarm sounds when there really is a fire:

P(Alarm | Fire)

If there is a fire, what is the probability that the alarm sounds?

This probability is easy to measure because researchers can set up tests beginning with a known fire condition and observing whether the alarm detects it.

But when you hear the alarm go off, that's not the question you need to answer. You want to know the following:

P(Fire | Alarm)

If you hear an alarm, what is the probability that there is an actual fire?

These two probabilities are not the same. If the alarm sometimes triggers for harmless reasons (burnt toast, steam, or mechanical glitches), relying on P(Alarm | Fire) to judge your danger could be misleading. You might panic unnecessarily or, worse, ignore a real emergency.

This mismatch happens because you are trying to determine which case produced the outcome. The alarm is the observed event, and you want to know whether a fire is likely to have caused it. That is a reverse probability.

Example 2: Medical Testing: Sensitivity vs. "What Does My Result Mean?"

Medical tests perfectly illustrate the need to reverse probabilities. A test's sensitivity is written as P(Positive | Disease).

Given a patient has the disease, what is the probability they will test positive?

Researchers measure this by starting with people who have the disease and recording how often the test detects it. That is the forward direction, and it describes how the test behaves under controlled conditions.

But when your test comes back positive, your real question is very different:

P(Disease | Positive)

Given a positive test result, what is the probability I have the disease?

This probability depends not only on the test's sensitivity and false-positive rate but also on the disease's prevalence in the population. Even a highly sensitive test can yield many false positives among healthy people when the disease is rare. If you confuse P(Positive | Disease) with P(Disease | Positive), you might assume that a positive test almost guarantees disease, even when the true chance is small.

This confusion is widespread, and it stems from using the forward probability when the situation requires the reverse one. We will return to this example later when we discuss the false positive paradox.

Example 3: Credit Card Fraud Alerts: "Is This Really Fraud?"

Suppose your bank texts you saying, "Unusual activity detected on your card." Banks design fraud-detection systems to catch suspicious transactions, and these systems are usually evaluated using forward probabilities, such as:

P(Alert | Fraud)

Given a fraudulent charge, what is the probability that the system issues an alert?

Banks measure this in controlled settings by starting with known cases of fraud and checking whether the system recognizes them. That makes P(Alert | Fraud) straightforward to collect.

But when *you* receive an alert, the question you care about is completely different:

P(Fraud | Alert)

Given that I received this alert, what are the chances of a fraudulent transaction?

Those two probabilities are not the same. The fraud alert system might generate many alerts for legitimate transactions simply because it is tuned to be cautious. If fraud is relatively rare (and it usually is), then a large proportion of alerts will not represent real fraud. Relying on P(Alert | Fraud) would give you the mistaken impression that an alert almost guarantees real fraud.

The correct probability depends on three things:

1. How often fraud actually occurs (the base rate).
2. How often the system catches fraud (P(Alert | Fraud)).
3. How often the system gives false alarms (P(Alert | No Fraud)).

This is a classic situation in which you observe the outcome first (the alert) and want to identify the most likely case: real fraud or a harmless transaction. Without reversing the probability, you cannot answer the question that matters. I will develop this example in more detail in the next section and show exactly how to calculate P(Fraud | Alert).

These examples share the same core idea. In each situation, the original information we have comes from known cases and observing outcomes. Those forward probabilities are useful for designing systems and understanding how they behave under controlled conditions.

But in everyday life, we face the opposite problem. We often observe the outcome first and want to know which case is most likely responsible. To interpret these situations correctly, we must reverse the conditional probability from P(Outcome | Case) to P(Case | Outcome).

In the Bayesian context, you'll often hear about "new evidence," which refers to the observed outcome itself. These outcomes appear at the end of the probability tree, such as the positive test result. This evidence is what triggers the need to reverse the conditional probability and determine how observing it affects the probability of each case.

Now that I've observed a positive test result, what is the probability I have the disease? The test result is the "new evidence" in this context.

Reversing Probabilities in a Tree

Now that you've seen why we often need to reverse conditional probabilities, we can introduce the formal rule that makes this reversal possible. Bayes' Theorem allows us to take a known forward probability, P(Outcome | Case), and calculate the reverse probability, P(Case | Outcome).

As we did with the Law of Total Probability, I'll start with a probability tree and connect it to the formula for Bayes' Theorem. It helps to see where it comes from. The structure of a two-stage probability tree contains all the pieces we need. In fact, Bayes' Theorem follows directly from the same multiplication and addition rules you used earlier with probability trees.

Like LOTP, the first split represents distinct groups, but in Bayes' Theorem we interpret them as potential explanations of the observed outcome we are trying to understand.

Consider a situation with two possible cases: C1 and C2. Each case has its own marginal probability, and each one can lead to different Outcomes. We can represent this structure with a probability tree.

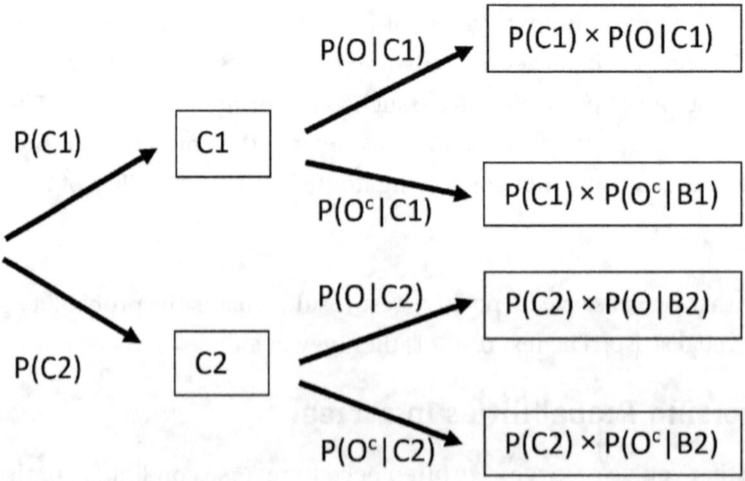

For brevity, I've denoted our case of interest as C1 and the outcome of interest as O. Note O^c or O complement, indicating "not the outcome of interest." This tree represents the "forward" probabilities that come from how the researchers collect the data.

Stage 1 contains the marginal probabilities, which are the base rates for each case. They indicate how common each case is before we observe any new evidence. These base rates play a central role in Bayesian reasoning and will become especially important in the next chapter.

For example, what proportion of all bank charges are fraudulent? What proportion of the population has the disease?

Stage 2 contains the conditional probabilities, which show how well each case predicts the outcome. These conditional values tell us, "If this case occurred, how likely is the outcome?"

For example, when a charge is fraudulent, the algorithm flags it 99% of the time. When the disease is present, the test detects it 95% of the time.

Together, these forward probabilities summarize both how frequently each case appears overall and how strongly each one tends to produce the observed outcome.

Now, we need to reverse the conditional probability from P(O | C1) to P(C1 | O).

For starters, recall from Chapter 3 that a conditional probability is a joint probability divided by a marginal probability. For our problem, it takes the form:

$$P(C1 \mid O) = P(C1 \cap O) / P(O)$$

This formula shows that our reverse probability solution is a ratio with the following components:

Numerator: The joint probability P(C1 ∩ O) is the probability that our specific case of interest (C1) and the outcome of interest (O) occur together. In a probability tree, this value corresponds to the joint probability in a single terminal node that you calculate using the multiplication rule.

Denominator: The marginal probability P(O) relates to the overall probability that our outcome of interest (O) occurs. It is the sum of all paths that lead to it. This sum includes the path through the case of interest (C1) in the numerator, along with all paths through other cases that lead to this outcome. In a probability tree, this process is the same as recovering a marginal probability by summing all branches with the same outcome using the Law of Total Probability.

In a nutshell, the numerator is the joint probability for a single node ending with our *specific* case of interest (C1) and the outcome of interest (O).

The denominator is the total probability across all branches ending in the outcome (O) of interest. It includes all possible cases, which are C1 and C2 in the example tree.

Note that the numerator's terminal node is always included in the denominator's sum. That situation ensures the final value is a proper probability between 0 and 1.

Fortunately, the tree already contains the information needed to calculate it. Let's pull the probabilities from the tree.

The top terminal node represents the joint probability of our case and outcome of interest occurring together. That's our numerator:

$P(C1) \times P(O \mid C1)$

The two terminal nodes that represent the total probability of the outcome of interest occurring are the top branch and the third from the top. You can tell because the conditional probability begins with O before the bar. These two nodes represent all paths reaching our outcome of interest. Here's our denominator:

$P(C1) \times P(O \mid C1) + P(C2) \times P(O \mid C2)$

So, let's put the entire ratio together:

$$P(C1 \mid O) = \frac{P(C1) \times P(O \mid C1)}{P(C1) \times P(O \mid C1) + P(C2) \times P(O \mid C2)}$$

This expression might look complicated at first, but the tree makes each part easy to understand. The numerator focuses on the *specific* case and outcome you care about. The denominator sums all the probabilities for *all* cases that produce the outcome.

Another way to explain it is that Bayes' Theorem compares how well one specific case explains the outcome relative to all cases that produce it.

After you see how Bayes' Theorem arises naturally from a probability tree, the formula becomes far less mysterious.

Bayes' Theorem Formula

Now that we have developed Bayes' Theorem directly from a probability tree, we can write the result in a compact algebraic form. The tree helps you see the idea for a simple case, while the formula extends that same logic to every possible arrangement of cases and outcomes.

The example tree we used had only two cases and an outcome of interest, and the calculations fit that specific structure. But probability trees can have any number of cases, and each case can lead to multiple outcomes. Bayes' Theorem must work for all such cases, not just the one shown in the diagram. The full formula captures this more general situation.

Additionally, while we have written Bayes' Theorem using descriptive terms like "case" and "outcome" for clarity, most textbooks express the formula using letters. In that notation, the observed outcome is labeled A (e.g., fire alarm sounds) and the possible underlying cases are labeled B_1, B_2, ..., B_k (e.g., fire, steam, malfunction, etc.).

Written this way, Bayes' Theorem becomes:

$$P(B_j \mid A) = \frac{P(B_j) \times P(A \mid B_j)}{\sum P(B_i) \times P(A \mid B_i)}$$

Where:

- A is the outcome of interest.
- B_i represents all possible cases.
- B_j is the case of interest.

This is the same formula as before, just using more compact symbols. The ratio is the joint probability of the case of interest and outcome of interest occurring together, divided by the marginal probability of the outcome of interest.

The numerator focuses on the single path that links our case of interest (B_j) to the outcome (A). It assesses the overall frequency of B_j and how strongly it predicts A.

In the denominator, it's the LOTP again. Each possible case B_i receives a weight based on how common it is (its marginal probability) and how strongly it predicts the observed event A (conditional probability). That gives us the familiar pattern of a marginal × conditional probability for each path to the outcome of interest.

The denominator sums the probabilities for all possible cases that produce the outcome.

In the next section, we will apply Bayes' Theorem to an example and walk through each step.

Example: Fraud Alerts

Let's see how Bayes' Theorem works in a real-world context involving credit card fraud alerts.

Suppose your bank uses an automated system to flag suspicious transactions and send alerts. Given that you receive an alert, what is the probability that the transaction is truly fraudulent?

You are given the following information and want to find P(Fraud | Alert):

- P(F) = 0.001 (The probability of a fraudulent transaction.)
- P(A | F) = 0.99 (If a transaction is fraudulent, the system issues an alert 99% of the time. Probability of a true positive.)
- P(A | NF) = 0.01 (If a transaction is not fraudulent, the system still issues an alert 1% of the time. Probability of a false positive.)

Notice how the problem gives us a base rate for fraud and two forward, conditional probabilities. However, it asks for the reverse conditional probability, P(F | A), which is conspicuously absent! That is a classic Bayes' Theorem setup.

There are basically two ways to approach this problem.

- If you are comfortable with the formula, you can go straight to Bayes' Theorem and plug in the values.
- Or you can use a probability tree to visually organize the information, calculate the joint probabilities, and then use the values in the appropriate terminal nodes.

We'll do both.

The following tree shows all the marginal and conditional probabilities in their proper locations, making it easy to see which paths end with an alert.

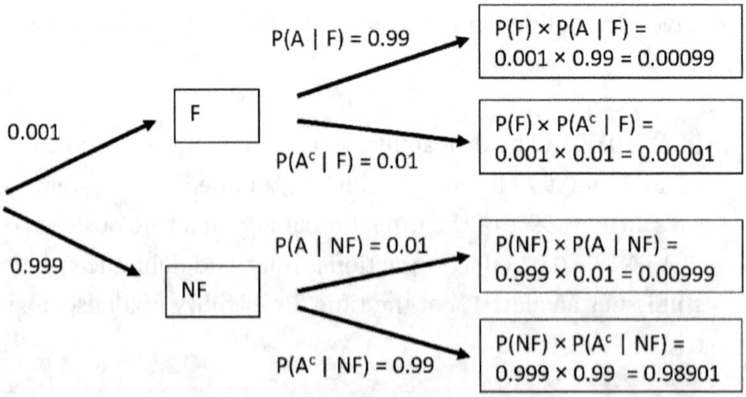

Using trees to solve these problems is helpful in several ways. First, you'll often need to find probability values that aren't given in the problem. For instance, branches at each split must sum to 1, which allows us to find P(NF) even though it is not given. Second, they help you place the correct values in the numerator and denominator of Bayes' Theorem.

To find $P(F \mid A)$, take the joint probability for the path where Fraud (F) occurs and an Alert (A) follows, which corresponds to the branch labeled $P(A \mid F)$. That's the top terminal node in the previous tree. Use that value in the numerator for $P(F \cap A)$.

Then, use LOTP to find the overall probability of an alert. Add the joint probabilities in the two terminal nodes that end with an alert (the first and third nodes from the top). Use that sum in the denominator for P(A).

You can read these values from the tree or plug them into the Bayes' Theorem equation:

$$P(F \mid A) = P(F \cap A) / P(A)$$
$$= P(F) \times P(A \mid F) / [(P(F) \times P(A \mid F) + P(NF) \times P(A \mid NF)]$$
$$= (0.001 \times 0.99) / [(0.001 \times 0.99) + (0.999 \times 0.01)]$$
$$= 0.00099 / (0.00099 + 0.00999)$$
$$= 0.00099 / 0.01098$$
$$= 0.090$$

So, even though the system is very good at catching fraud when it occurs, only about 9% of alerts correspond to truly fraudulent transactions. This happens because fraud is extremely rare. As a result, most alerts (91%) are false positives produced by the far more frequent legitimate transactions.

Example: Factory Defects Revisited

In an earlier example, we used the Law of Total Probability to find the overall probability that a randomly selected part was defective. We treated the factories as cases and the defect status as the outcome, and we found:

$$P(\text{Defect}) = 0.039$$

Now, suppose we change the question. Imagine we randomly draw a part and discover that it is defective. There are no markings to show which factory produced it. Instead of asking, "What is the probability a part is defective?" we now ask, "Given that this part is defective, which factory is it most likely from?"

Using Bayes' Theorem, we need to calculate P(F1 | Defect), P(F2 | Defect), and P(F3 | Defect). Then determine which conditional probability is largest.

The following tree summarizes all the necessary information from the original LOTP problem:

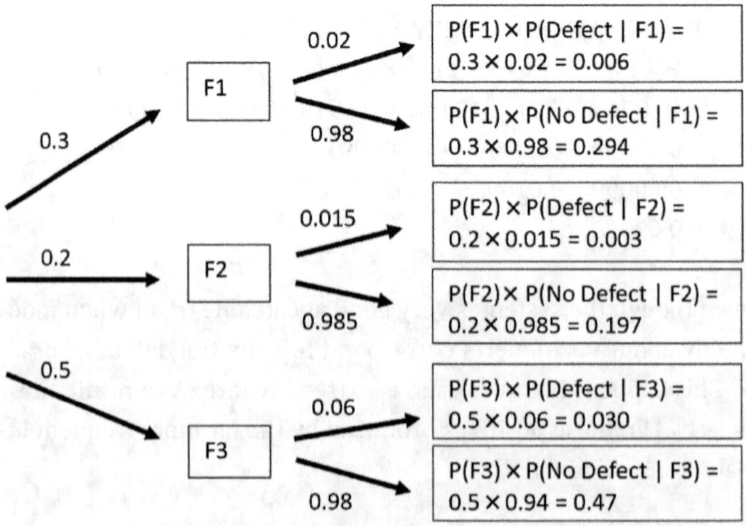

From the tree, we know the joint probabilities for each factory and defect. We'll use them in the numerators for each factory:

- P(F1 ∩ Defect) = 0.006
- P(F2 ∩ Defect) = 0.003
- P(F3 ∩ Defect) = 0.030

In the LOTP section, we solved for the overall defect rate: P(Defect) = 0.039. We need that for all the denominators.

Bayes' Theorem now gives the reverse conditional probabilities:

P(F1 | Defect) = P(F1 ∩ Defect) / P(Defect)
= 0.006 / 0.039 = 0.154

P(F2 | Defect) = P(F2 ∩ Defect) / P(Defect)
= 0.003 / 0.039 = 0.077

P(F3 | Defect) = P(F3 ∩ Defect) / P(Defect)
= 0.030 / 0.039 = 0.769

So, given that the part is defective, there is a 15.4% chance it came from Factory 1, a 7.7% chance it came from Factory 2, and a whopping 76.9% chance it came from Factory 3.

Hence, when you find a defective part, it is by far most likely to come from Factory 3.

Note that these probabilities sum to 1 because they represent all possible sources of a defective part. After we know a part is defective, it must have come from exactly one of the three factories. Bayes' Theorem distributes the total probability across those mutually exclusive possibilities.

This example shows a different use of Bayes' Theorem. Before, we used it to find the probability of a particular case given an outcome. Here we've extended it to find the most likely case out of several possible cases. We started with the outcome (a defective part) and identified the most likely case (Factory 3).

There is an important symmetry between this example and the LOTP example that came before it. In the LOTP problem, we started with the cases (the factories) and used the tree to find the overall probability of the outcome (a defect).

In the Bayes' Theorem problem, we use the same information, but reversed the direction of reasoning. We started with the outcome (a defect) and used the tree and the formula to determine which case is now most likely.

Both methods rely on the same joint probabilities and the same tree structure. The difference lies in *how* we use that structure. LOTP adds together all branches that end in a specific outcome to find its total probability. Bayes uses those branch probabilities to find the probability of the underlying cases after the outcome is known.

Put simply, LOTP moves forward from cases to outcomes, while Bayes moves backward from outcomes to cases.

When to Use This Method

When you face a probability problem, it is not always obvious which method to choose. Bayes' Theorem is the right tool when the problem asks you to reverse a conditional probability. That is, when the probability you want runs in the opposite direction from the information you are given. Here are the clues that tell you Bayes' Theorem is the correct approach.

You are asked to diagnose a case based on an observed result.

Bayes' Theorem is the standard method whenever a problem involves:

- Diagnosing (disease given a positive test)
- Detecting fraud (fraud given an alert)
- Identifying a source (factory given a defect)

If the problem sounds like:

"We've seen this result. What does that say about what caused it?"

...that's a Bayes problem.

You know the base rates for each case and the conditional probabilities within each case.

If the problem supplies:

- How common each case is (base rates): P(B1), P(B2), ...
- How strongly each case predicts the outcome: P(A | B1), P(A | B2), ...

...and then asks:

- Which case is most likely, given the outcome?
- What is $P(B_j \mid A)$?

That is exactly the structure that Bayes' Theorem is built for.

In other words, if you know how likely the outcome is within each case, but you need the probability of the case after observing the outcome, Bayes is the method to use.

Common Pitfalls with Bayes' Theorem

Confusing forward and reverse probabilities
P(Outcome | Case) is not the same as P(Case | Outcome).

Example: A test detects 99% of diseased patients (P(Positive | Disease)), but that does not mean a positive result implies a 99% chance of disease (P(Disease | Positive)).

Ignoring base rates
Even strong conditional probabilities can mislead when the underlying event is rare. In other words, don't focus on only $P(A \mid B)$ and forget how rare B is.

Example: Fraud alerts can be highly accurate for fraudulent transactions, but because fraud is rare, most alerts are triggered by legitimate transactions.

Treating Bayes' Theorem as just a formula
Bayes' Theorem is not just plug-and-chug. It reflects the problem's structure. Use tree diagrams both to understand the problem's logic and to find the required probabilities that are not given.

Forgetting the denominator includes all cases
Students often include only the case of interest and ignore others.

Example: When calculating P(Fraud | Alert), the denominator must include alerts from both fraudulent and non-fraudulent transactions.

Review and Next Steps

This chapter focused on using two powerful tools to do the following:

- Combine probabilities from different cases using the Law of Total Probability.
- Reverse conditional probabilities using Bayes' Theorem to find probabilities for cases.

You learned how to use these tools with probability trees and how they apply to real situations—screening tests, fraud alerts, alarms, and quality control in factories.

Most importantly, you learned that forward information (P(Outcome | Case)) rarely answers the real-world questions we face. To interpret cases correctly, you often need to reverse the conditional probability: P(Case | Outcome). Bayes' Theorem gives you the method for making this reversal.

In the next chapter, we will step beyond the mechanics and focus on how these tools support Bayesian reasoning. You will see how ideas like priors, evidence, and posteriors form the foundation of this methodology, and how misunderstandings of forward and reverse probabilities lead to classic cognitive errors such as the base-rate fallacy and the prosecutor's fallacy.

Chapter Summary

- **Forward vs. reverse conditional probabilities:** P(Outcome | Case) is not the same as P(Case | Outcome). They answer different questions.
- **Probability trees:** Trees help you visualize cases and their conditional probabilities, calculate joint probabilities, and see

how LOTP and Bayes' Theorem arise naturally from the structure.

- **Law of Total Probability:** Combines forward conditional probabilities across all relevant cases for a specific outcome to recover a marginal (overall) probability.
    - $P(A) = \sum P(B_i) \times P(A \mid B_i)$
- **Bayes' Theorem:** Reverses the conditional probability by combining marginal and conditional probabilities to produce a reverse conditional probability that answers a real-world question.
    - $P(B_j \mid A) = \frac{P(B_j) \times P(A \mid B_j)}{\sum P(B_i) \times P(A \mid B_i)}$

Chapter 7 Practice Problems

Question 1: A streaming service offers three subscription plans: Basic, Standard, and Premium. Each plan has a different probability that a user watches at least one show per day. The company provides:

- The proportion of users on each plan.
- The probability that a user watches a show given their plan.

You are asked to find the overall probability that a randomly selected user watches at least one show per day.

Should you use the Law of Total Probability or Bayes' Theorem? Briefly explain why.

Question 2: A wildlife researcher studies nesting success for a bird species across three habitats: forest, grassland, and wetlands. The researcher records how nests are distributed across these habitats (H_1, H_2, H_3) and how likely a nest is to successfully produce chicks within each habitat.

Answer the following questions:

A. Identify the cases in this problem.
B. Identify the outcome of interest.
C. Write the conditional probabilities that represent the probability of a nest successfully producing chicks for each habitat.
D. Let S represent the event that a nest successfully produces chicks. Write the Law of Total Probability expression for P(S) (do not calculate).

Question 3: A company screens job applicants using an automated assessment designed to identify highly qualified candidates. Only a small proportion of applicants are truly highly qualified. The test is very accurate: it correctly identifies most qualified applicants and correctly rejects most unqualified applicants.

After running the assessment, a manager sees that an applicant was flagged as highly qualified.

Explain why the probability that the applicant is truly highly qualified might still be lower than expected, even though the test is highly accurate.

Question 4: A university offers students three ways to contact academic advising. Let the events be defined as follows:

- M_1 = the student used drop-in advising.
- M_2 = the student scheduled an appointment.
- M_3 = the student used email advising.
- R = the student's issue was resolved.

You are given the following information:

- $P(M_1) = 0.30$
- $P(M_2) = 0.45$
- $P(R \mid M_1) = 0.70$
- $P(R \mid M_2) = 0.85$
- $P(R \mid M_3) = 0.60$

Using this information, answer the following:

- **A.** Complete a probability tree for this situation, including all missing probabilities.
- **B.** Use the Law of Total Probability to find P(R), the overall probability that a student's issue is resolved.
- **C.** Given that a student's issue is resolved, which advising method is most likely to have been used? Use Bayes' Theorem to find $P(M_1 \mid R)$, $P(M_2 \mid R)$, and $P(M_3 \mid R)$.

Question 5: A security system scans files for malware. The system is highly accurate: if a file contains malware, it flags it 98% of the time. If a file does not contain malware, it incorrectly flags it 3% of the time. Only 1% of all files actually contain malware.

A file is flagged by the system. What is the probability that the file truly contains malware?

CHAPTER 8

Foundations of Bayesian Reasoning

The ability to think clearly is one of the most valuable skills. Clear thinking leads to better understanding, better decisions, and better outcomes. But achieving clarity is challenging because we live in a world filled with uncertainty. We often rely on assumptions, impressions, or incomplete information. Our beliefs rarely align with reality perfectly, and this gap can lead to misjudgments, blind spots, and poor decisions.

Bayesian reasoning helps close that gap. It provides a systematic method for updating your beliefs when new information arrives, not reactively or emotionally, but deliberately. Each credible piece of evidence can refine your understanding. Instead of treating uncertainty as a barrier, Bayesian reasoning treats it as something we can manage, quantify, and refine.

This approach isn't limited to statisticians or scientists. In everyday life, whenever you learn something new and adjust your expectations, you're already doing a form of Bayesian updating. By understanding priors, likelihoods, and posteriors, you can calibrate your thinking.

You learn how strongly evidence should shift your beliefs and how much weight to give new information. In this chapter, we make that intuitive process explicit and show how it becomes a coherent way of reasoning.

The central idea behind Bayesian reasoning is not new. It began with a simple question: How should someone update a belief about a hypothesis when new information appears? In the 1700s, Thomas Bayes proposed a way to formalize that process, and Pierre-Simon Laplace later expanded it into a general framework for scientific inference. Their insight was that probabilities can also represent uncertainty about the world, not just long-run frequencies.

Earlier approaches to probability relied on counting equally likely outcomes in games of chance. The frequentist view defines probability as the long-run behavior of repeated experiments. Bayesian reasoning instead asks: *Given what we know, how likely is this explanation?*

This shift made Bayesian ideas especially useful in fields with few repeated trials, such as medicine, law, finance, and engineering. Each time you make a prediction, diagnose a problem, interpret evidence, or combine old and new knowledge, you are essentially doing Bayesian updating.

Modern applications make this framework even more important. Spam filters, fraud detection systems, weather forecasts, recommendation algorithms, and medical decision systems all use Bayesian ideas. They continuously update probabilities as new data arrive. Humans do the same thing intuitively; when we get new information, we adjust our expectations. Bayesian reasoning simply provides the mathematical structure that explains how this updating should work and highlights where intuition often goes wrong.

The ideas in this chapter might feel broader than the calculations you have seen so far. That's because Bayesian reasoning goes well beyond the basic uses of Bayes' Theorem. However, the core structure is the same. The tools in Chapter 7 contain the essential logic of Bayesian thinking.

Advanced Bayesian methods build on this base. They use more complex models and repeated updates, but follow the same underlying principles. The goal of this chapter is to help you spot the structure and see how these ideas apply more broadly.

This chapter builds on the mechanics you learned in Chapter 7 and shows how those computations become a broader way of interpreting uncertainty. These ideas teach you how to think about evidence:

- How strongly does each case predict what you saw?
- How common is each case to begin with?

These pieces come together to produce a revised probability that answers a real-world question: Given what you observed, what is the most likely explanation?

This chapter also introduces several classic cognitive errors, such as the base rate fallacy and the prosecutor's fallacy. These mistakes arise when people confuse forward and reverse probabilities or ignore key parts of the evidence. Understanding the Bayesian logic behind these errors helps you avoid them and interpret uncertain information more accurately.

Hypothesis & Evidence Notation

In many textbooks, Bayes' Theorem is written in a more abstract form:

$P(H \mid E)$

where H represents a hypothesis and E represents the evidence you observed.

This notation is equivalent to the case/outcome framework used in this book:

- The hypothesis is any possible explanation (the case).
- The evidence is what you observed (the outcome).

In this framework, we will refer to three key quantities:

- The prior, P(H).
- The likelihood, P(E | H).
- The posterior, P(H | E).

These quantities correspond to the components I will define in the following section:

- The **prior** tells you how plausible each hypothesis was before observing any evidence.
- The **likelihood** tells you how well each hypothesis explains the outcome you observed.
- The **posterior** balances both plausibility *and* how well the hypothesis explains the evidence.

This notation is especially useful in scientific or analytical settings where the "cases" are not literal cases but competing explanations, models, or theories. In short, it is a field-neutral way to express Bayesian reasoning: starting with priors, evaluating how well each hypothesis predicts the evidence (through the likelihood), and updating to a posterior.

Priors, Likelihoods, and Posteriors

These terms don't introduce new mathematics. Instead, they simply give names to the pieces of the updating logic that Bayes, Laplace, and modern statisticians use to connect evidence with belief.

Bayes' Theorem combines the base rate of each case and the strength of the evidence for that case to produce the reweighted, or posterior, probabilities. It works by combining the following three components.

Prior probability

The prior probability, or prior, is your starting point. It is the probability of a case, explanation, or hypothesis *before* observing the outcome. In earlier examples, the prior was the base rate: the proportion of people with a disease, the share of transactions that are fraudulent, or the percentage of parts from a particular factory.

In the factory-defect scenario, the prior for Factory 3 was its base rate of supplying 50% of all parts. That value came before considering whether a specific part was defective. It simply reflected how common Factory 3's output was in the overall supply chain.

Likelihood

One part of Bayesian reasoning asks a simple question: If this explanation were true, how well would it explain what I observed? In Bayesian reasoning, P(Evidence | Hypothesis) is the likelihood. It tells you how consistent the observed evidence is with a particular hypothesis. If a hypothesis makes the observed evidence more expected, it has a higher likelihood.

In the fraud-alert example, ask: If a transaction is actually fraudulent, how likely is it to trigger an alert? That is the likelihood, P(Alert | Fraud). This value tells us how well the "fraud" hypothesis accounts for the alert we observed. A higher likelihood means the alert is more expected when the transaction is fraudulent, indicating the fraud hypothesis better explains the alert.

Posterior probability

The posterior probability is the updated probability of a hypothesis after observing the evidence. In notation, it is P(Case | Outcome), or more generally P(Hypothesis | Evidence). The posterior shows how the evidence reweights the original base rates for the cases. In Bayesian language, this is where "the new information updates your prior belief."

The likelihood asks, "If this hypothesis were true, how likely would this evidence be?" The posterior instead asks, "After seeing this evidence, how plausible is this hypothesis?" The posterior helps you compare competing explanations and determine which is most plausible given the evidence. In practice, this often means looking for the hypothesis with the highest posterior probability among the alternatives.

For example, in the fraud-alert scenario, the prior probability of fraud was very small: P(Fraud) = 0.001. After observing an alert, Bayes' Theorem produced a posterior probability of P(Fraud | Alert) = 0.090. In Bayesian terms, the evidence (the alert) updated the prior from 0.001 to 0.090.

Intuitively, you know that the fraud probability for a flagged transaction *should* be higher than the average fraud rate for all transactions, but how much higher? Bayesian reasoning quantifies that updated probability.

A very different pattern appears in the factory-defect scenario. Factory 3 supplied half of all parts, giving it a prior probability of P(Factory 3) = 0.50, and it also had the highest defect base rate. After observing that a part was defective, Bayes' Theorem produced a posterior probability of P(Factory 3 | Defect) = 0.769. Here, the evidence updated the prior from 0.50 to 0.769. That's a more dramatic increase than in the fraud example.

Without doing the analysis, you can guess that a defective part probably came from Factory 3 because it produces half of all parts. Bayesian reasoning quantifies that probability.

These two examples show how priors and likelihoods interact. When the prior probability is extremely small (e.g., fraud), even strong evidence might produce only a moderate shift. But when the prior is large, and the evidence strongly supports it, as with Factory 3, the posterior can increase dramatically. The size of the update depends on both where you start and how well each possible case predicts the observed outcome.

You've already been using this structure informally. Now you can attach the standard Bayesian terms to it.

Caution: A large likelihood does not automatically imply a large posterior, because the posterior also depends on the prior and on how well competing hypotheses explain the evidence.

One of the strengths of Bayesian reasoning is that it handles new information as it arrives. You don't need all the evidence at once. Instead, you update your belief incrementally. The posterior from your first update becomes the prior for your next one. This creates a running, calibrated view of uncertainty that evolves naturally over time. In everyday life, we already think this way. Each new observation adjusts our expectations. Bayesian reasoning simply gives this intuitive process a clear structure, helping us avoid overreacting to weak evidence or ignoring strong evidence when it appears.

Bayesian reasoning turns probability into a way of thinking about uncertain situations, not just a set of computational steps. As you step back from the mechanics, notice that the tools from Chapter 7 serve a much broader purpose. Those calculations are not isolated procedures. Instead, they form the backbone of how we interpret information in the real world. This chapter moves you from 'plug this into a formula' toward 'this is how evidence reshapes our understanding.'

This framework of priors, likelihoods, and updating posteriors underlies every Bayesian calculation you will encounter.

How to Interpret Bayesian Answers in Practice

Bayesian updating appears in many everyday situations, even when we don't label it formally. We start with an initial expectation, observe new information, and adjust our conclusions. The following examples show how the same logic behind priors, likelihoods, and posteriors appears in familiar contexts.

Diagnosing car trouble begins with a prior belief based on experience. Common issues, such as a weak battery or loose connections, are more likely than major mechanical failures. The initial assumption is updated as new evidence becomes available, such as dashboard warning lights, unusual engine sounds, or diagnostic codes. The conclusion shifts toward the highest probability case that best explains the observed symptoms.

Evaluating a job applicant also follows this process. An initial belief forms from the résumé, credentials, and first impressions, which suggest an expected level of skill or fit. New evidence then comes from interview responses, sample work, communication style, and references. These pieces of information refine the original expectation, leading to an updated assessment of the applicant's qualifications and how well they match the role.

Interpreting weather-related risks works in the same way. A prior belief comes from seasonal climate patterns, which suggest typical levels of risk for a given time and place. As new data arrives, such as radar images, temperature changes, and shifts in atmospheric pressure, the initial expectation is revised. The updated forecast reflects this new information, placing greater emphasis on the data that best explains current conditions.

When you use Bayes' Theorem in real-life situations, it helps to ask yourself a few guiding questions:

1. What was the prior?

Was the case common, rare, or somewhere in between? A tiny prior heavily constrains any posterior, no matter how compelling the evidence looks.

For example, in the fraud alert scenario, fraud was extremely rare, so even a highly accurate test yielded a low probability that the transaction was fraudulent.

2. How strong is the evidence?

A good test or reliable indicator can make the observed evidence much more probable under one hypothesis than under competing alternatives. But even strong evidence cannot fully overcome a very small prior.

3. How big was the change from prior to posterior?

Bayesian updating isn't just about the number. It's about understanding the *shift*. Did the new information meaningfully change your belief, or did it only nudge the probability slightly?

4. Does the result answer the question you care about?

Different conditional probabilities answer different questions. Some describe how a system or test behaves, while others describe what a result means for a specific case.

For example, a test might do a good job detecting a condition when it is present, which helps evaluate how well the test performs. However, what matters to a patient is something different: if the test result is positive, how likely is it that they actually have the condition? This is the reverse conditional probability.

Be careful to match the probability you calculate to the question you are trying to answer.

Deeper Considerations in Bayesian Reasoning

Bayesian reasoning provides a powerful framework for updating beliefs, but it also raises several practical and philosophical questions. These issues do not change how the math works, but they do influence how Bayesian ideas are used in scientific practice and everyday thinking. Understanding these considerations will help you interpret Bayesian results more effectively and recognize both the strengths and limits of this approach.

Subjective and Objective Priors

In the examples, priors came from clear sources: known base rates, historical frequencies, or the structure of the scenario itself. In practice, priors can also come from expert knowledge, reasonable assumptions, or past experience. These are often called subjective priors, and they play a much larger role when data are limited.

Some statisticians welcome subjective priors because they allow the analysis to incorporate meaningful background information. Others take a more cautious view, especially in scientific research, because different analysts may choose different priors and reach different conclusions.

This tension reflects a long-standing debate in statistics. Even if you favor the frequentist perspective, it is useful to understand how Bayesian reasoning handles situations where information is scarce. The influence of the prior typically weakens as more informative data accumulate.

Whether the prior is subjective or objective, transparency is crucial. Readers should always know where a prior came from and how strongly it influenced the results.

Limits of Bayesian Updating

Bayesian updating is only as reliable as the information that goes into it. If the evidence is flawed (unreliable measurements, biased data, or poorly calibrated tests), the update will simply carry those flaws forward. Bayesian reasoning cannot "fix" bad evidence.

Another important limitation arises when data are extremely scarce. In these situations, the posterior can remain heavily influenced by the prior because there is not enough information to pull the belief in a new direction. This is not a defect of the method but rather a reflection of genuine uncertainty.

For example, consider the early days of a new infectious disease. With only a handful of observed cases, risk estimates necessarily rely on expert assumptions about transmission or severity. As more data become available, each new observation shifts the posterior, and the influence of the initial prior fades.

Bayesian updating mirrors the way knowledge stabilizes as evidence accumulates.

Why Humans Often Update Incorrectly

Bayesian reasoning describes how beliefs *should* change when new information arrives. In practice, people update their beliefs in inconsistent ways. We often give too much weight to vivid or emotionally charged evidence and too little weight to base rates. Sometimes we ignore strong evidence because it conflicts with an existing belief; other times we overreact to a weak clue and change our view too quickly. These tendencies reflect natural cognitive shortcuts, but they also reveal why a structured approach to uncertainty is so valuable.

Bayesian reasoning provides a framework for recognizing when our intuition is leading us astray. It highlights the need to balance prior information with the strength of the evidence and clarifies how much updating is reasonable given what we already know.

In the next section, we look at several classic cognitive errors that arise when people confuse the direction of the conditional probability or ignore priors altogether.

Classic Cognitive Errors in Conditional Probability

Many classic reasoning errors occur when people update their beliefs informally, skipping parts of Bayesian logic. These mistakes are surprisingly common in everyday reasoning, news reports, business decisions, and even courtrooms.

One of the running themes of this book is that learning probability is not just about calculating numbers. It also helps you recognize situations where intuition can mislead you. Conditional probabilities create traps because the forward conditional P(Outcome | Case) is often easier to observe and feels more natural, while the reverse conditional P(Case | Outcome) is often the probability we need. When people fail to distinguish between these directions or forget to incorporate base rates, they can reach conclusions that sound plausible but are completely wrong.

This section looks at several classic errors that arise from this confusion. We begin with the base rate fallacy and the closely related false positive paradox, both of which show how ignoring prevalence can make highly accurate tests seem misleading. After that, we examine the Prosecutor's Fallacy, a legal error that confuses P(Evidence | Innocent) with P(Innocent | Evidence). These examples reinforce why conditional probability matters and why the careful reversal provided by Bayes' Theorem is essential.

Base Rate Fallacy and the False Positive Paradox

The base rate fallacy is a cognitive bias that occurs when someone gives too much weight to case-specific details and overlooks the underlying probability of the outcome in the broader population. People are naturally drawn to stories and vivid details, so they often rely on similarity judgments rather than statistical likelihoods. This process makes base rate neglect intuitively appealing but systematically misleading.

Why does this happen? It reflects our brain's preference for narrative over numbers, for compelling details over cold, hard statistics.

Bayes' Theorem forces you to incorporate base rates before updating anything based on new evidence. When people commit the base rate fallacy, they skip that step. They focus on how strongly the evidence points toward a particular outcome and ignore how common the underlying condition is. As a result, the reverse probability P(Disease | Positive) becomes distorted, especially when the condition is rare.

Let's see an example of that in action. The false positive paradox is a classic manifestation of base-rate neglect in medical testing. In this example, we'll look at a medical test scenario. What does a positive test result really mean?

Suppose a patient gets a positive test result for a serious condition. Can you successfully merge case-specific and base-rate information to find the correct answer?

The tools in the previous chapter tell you how to solve this type of problem and not be tripped up by the false positive paradox. However, we'll revisit this type of problem here for several reasons.

First, you and your loved ones are likely to receive positive medical testing results at some point. What do you need to know to evaluate their meaning?

Secondly, I want to look at the false positive paradox from a different angle to help you understand *why* it happens. It's one thing for Bayes' Theorem to tell you the correct but unintuitive answer behind the paradox. It's another thing to truly understand why it occurs.

Here are the details:

Case-specific information: This test is highly sensitive and reasonably specific. It correctly identifies 95% of people with the disease and correctly rules out 95% of those without.

- P(Positive | Disease) = 0.95 (sensitivity)
- P(Negative | No Disease) = 0.95 (specificity)

Because specificity is 0.95, the false positive rate is 1 – 0.95 = 0.05:

- P(Positive | No Disease) = 0.05 (false positives)

Base rate information: 1 in 1,000 people have this condition.

- P(Disease) = 0.001 (prevalence rate)
- P(No Disease) = 0.999

Given this information, what is the probability that the patient with the positive test result has the condition?

The most common answer is 95%, which sounds logical given the test's high accuracy.

In reality, the correct answer is less than 2%. If that's a big surprise, it's because you fell victim to the base rate fallacy!

Specifically, you failed to incorporate the base rate of the disease's occurrence in the population, which has a probability of 0.001.

Solving the Base Rate Fallacy

We need to find the following:

$$P(Disease \mid Positive) = \frac{P(\text{Disease} \cap \text{Positive})}{P(Positive)}$$

You can start with a probability tree diagram for this scenario but, for brevity, I'll go to straight to finding the values to plug into the Baye's Theorem ratio.

First, calculate the numerator of Bayes' Theorem:

P(Disease ∩ Positive) = P(Positive | Disease) × P(Disease)
= 0.95 × 0.001
= 0.00095

Next, calculate P(Positive) in the denominator using the Law of Total Probability:

P(Positive) = P(Positive | Disease) × P(Disease) + P(Positive | No Disease) × P(No Disease)

= (0.95 × 0.001) + (0.05 × 0.999)
= 0.00095 + 0.04995
= 0.05090

Now apply the Bayes' Theorem ratio:

P(Disease | Positive) = 0.00095 / 0.05090
= 0.0187
= 1.87%

The prior probability of 0.1% increases to a posterior probability of about 1.87%, far from the intuitive 95%.

A posterior probability, such as P(Disease | Positive), does not describe a property of the test or a percentage of this individual. Instead, it describes a group. Among people with a positive result and under the assumed conditions, this proportion would have the disease.

Why is the Posterior So Small?

Bayes' formula gives us the correct answer, but it doesn't show *why* the result is so small. To see the intuition behind the false-positive paradox, we'll apply the probabilities to a fictional population and examine the raw numbers. This approach better emphasizes how the base rate affects the ratio of true to false positives.

Let's apply the previous probabilities to a population of 1,000,000 people.

Cases and True Positives

First, find the number of people who actually have the disease:

1,000,000 × 0.001 = 1,000 cases.

Then apply the sensitivity to the cases:

1,000 × 0.95 = 950 true positives.

Non-Cases and False Positives

Next, calculate how many people do **not** have the disease: 1,000,000 × 0.999 = 999,000 non-cases.

Now, applying the 0.05 false positive rate to the non-cases gives: 999,000 × 0.05 = 49,950 false positives.

Putting It All Together

Wow! There are far more false positives than true positives, which explains why the base rate fallacy gives us an extremely biased view.

Types of Positive Results	Number
True	950
False	49,950
Total	50,900

Most of the 1,000 people who actually have the disease receive a true positive result (950). However, among the 999,000 people without the disease, a huge number (49,950) receive false positives. The false positive rate is low, but because non-cases are so common, the test produces far more false positives than true positives.

Only 950 out of 50,900 total positives are true. That's 950 / 50,900 = 1.87%.

Unsurprisingly, this result matches the Bayes' Theorem calculation, but it shows how the false positives swamp the numbers because there are so many non-cases. Very accurate tests can produce many false positives when the underlying condition is rare. Hence, it's the false positive paradox.

Linking Medical Testing to Bayes' Theorem

This example highlights how several common testing measures naturally fit within the framework of conditional probability and Bayes' Theorem. Here, I'll define testing accuracy terms and show how each one maps onto the conditional probabilities we've been using.

Each measure answers a different question depending on which event you condition on.

- **Sensitivity: P(Positive | Disease)**: The probability that the test correctly identifies someone who truly has the disease. It measures the test's ability to detect real cases.
- **Specificity: P(Negative | No Disease)**: The probability that the test correctly identifies someone who does *not* have the disease. It evaluates the test's ability to rule out non-cases.

Both are forward conditional probabilities describing how the test behaves under known conditions. These measures are essential when developing, validating, or comparing diagnostic tests.

But from a patient's point of view, different probabilities matter:

- **Positive Predictive Value (PPV): P(Disease | Positive)** The probability that a person truly has the disease, given a positive test result.
- **Negative Predictive Value (NPV): P(No Disease | Negative)** The probability that a person truly does not have the disease, given a negative test result.

PPV and NPV depend on both accuracy *and* prevalence. When the condition is rare, PPV can be low even with an excellent test. The relationship among these measures becomes clear through Bayes' Theorem, which converts the test's operating characteristics into the reverse probability that the patient cares about.

In the false positive paradox example we just covered, the sensitivity of the test was 0.95. The test is good at detecting the condition when a patient truly has it. However, the PPV was only 0.0187. From a patient's perspective, a positive test result provides little evidence that they actually have the condition in this scenario.

Clinical Decision Implications: Sensitivity vs. PPV

These calculations don't just make good test questions. They have real-world implications. Mammography is a well-studied test for breast cancer and provides a clear example of the difference between sensitivity and positive predictive value.

Mammograms typically have a sensitivity of 80–90%, depending on age and breast tissue density. In practical terms, this means the test detects 8 or 9 out of every 10 cancers when they are present. Sensitivity reflects the test's inherent ability to identify disease. That information helps evaluate the test itself.

Positive predictive value, however, answers a different question: "Given a positive result, what is the chance that the patient actually has cancer?" PPV behaves very differently because it depends heavily on prevalence, resulting in low PPV in broad screening programs but much higher in diagnostic contexts.

In population-wide screening for breast cancer, where very few women actually have breast cancer, PPV can be quite low. Studies show that only about 5–10% of women recalled for further testing after a screening mammogram are ultimately diagnosed with cancer. The low base rate makes false positives common, which pulls PPV down.

In contrast, diagnostic mammography, performed when there is already a reason for suspicion, such as a palpable lump, has a much higher PPV. In these settings, published studies often report PPV values in the 60–70% range. In this setting, the disease prevalence is higher, so a positive result is far more likely to indicate cancer.

In short, what a positive result means depends on the patient's context. Further, clinicians, researchers, and policymakers should rely on sensitivity and specificity when evaluating test performance and on PPV and NPV when interpreting test results for individual patients.

The Prosecutor's Fallacy

The Prosecutor's Fallacy occurs when someone treats P(Evidence | Innocent) as if it were P(Innocent | Evidence). As you know by now, these probabilities answer different questions and can differ dramatically. This fallacy is a real-world example of what happens when those two probabilities are mistakenly treated as the same.

This fallacy commonly appears in courtrooms when DNA evidence is presented. Jurors are often told something like, "Only 1 in a million people would match this DNA profile." That statement refers to P(Match | Innocent) = 0.000001.

This probability tells us how often a randomly chosen innocent person would, by chance, match the evidence. It is a *forward* probability. It says nothing about how likely a person is to be innocent or guilty. The question the court cares about is the reverse conditional:

P(Innocent | Match)

Given that the defendant matches the DNA, what is the probability that they are innocent?

If the defendant matches the DNA, does that mean there is only a 1-in-a-million chance they are innocent?

Not at all.

Imagine a population of 10 million people. With a match rate of 1 in 1,000,000, we expect about 10 innocent matches by sheer coincidence. A DNA match identifies a *pool* of potential suspects, not a single culprit.

Without the proper reversal, the argument sounds compelling but is logically flawed. If investigators had no prior information about who committed the crime, the DNA match simply tells us the defendant is one of a small group of potential matches.

A DNA match must be evaluated alongside other evidence that narrows the suspect pool, such as location data or independent forensic findings. A match becomes meaningful only when combined with other information and it cannot be the entire case. Bayes' Theorem provides a coherent structure for combining these pieces. It starts with the proportion of people who could match (the base rate), and then updates as new information accumulates.

Review and Next Steps

In this chapter, you moved beyond the calculation mechanics of Bayes' Theorem and learned how to use Bayesian reasoning as a way of interpreting uncertainty in real-world situations. You saw how priors, likelihoods, and posteriors fit together to form a coherent framework for updating beliefs, one piece of evidence at a time. You also learned that evidence must be weighed against prior information, that updates can be sequential, and that priors can come from either objective data or reasonable assumptions. These ideas help you understand how Bayesian reasoning calibrates your thinking rather than simply producing a number.

You also explored several deeper considerations, including the role of subjective priors when data are scarce and the limits of Bayesian updating when evidence is weak or unreliable. These concepts prepare you to recognize why people sometimes update their beliefs incorrectly and why structured reasoning is essential. Finally, you examined the base rate fallacy and the prosecutor's fallacy, which are common cognitive errors that arise when people misuse probability.

In the next chapter, we shift from reversing probabilities to counting outcomes. Combinatorics provide the tools to count the number of possible outcomes in complex situations.

Chapter 8 Practice Problems

Question 1: A cybersecurity system monitors login attempts. Historically, only 0.2% of login attempts are malicious. When a login attempt is malicious, the system flags it 90% of the time. When a login attempt is legitimate, the system incorrectly flags it 8% of the time.

A particular login attempt is flagged.

Answer the following. Use the proper probability notation where needed.

- **A.** Identify the prior probability.
- **B.** Identify the likelihood(s).
- **C.** State the meaning of the posterior probability in words (do not calculate it).
- **D.** Explain, in words, what the likelihood tells you in this context.

Question 2: A company manufactures products using three suppliers. Supplier A provides most parts but has a low defect rate. Supplier C provides fewer parts but has a higher defect rate.

A defective product is discovered.

Explain how observing the defect should change your belief about which supplier produced the product. In your answer, describe the roles of both the prior and the likelihood.

Question 3: A rare software bug occurs in 1 out of every 5,000 programs. A diagnostic tool correctly detects the bug 95% of the time when it is present and incorrectly flags 4% of programs that do not have the bug.

A program is flagged by the tool. A developer says: “The tool is highly accurate, so there’s about a 95% chance this program has the bug.”

A. Explain why this conclusion is incorrect.
B. Identify the mistake in terms of Bayesian reasoning.
C. Without calculating, explain whether the true probability is likely to be high, moderate, or low, and why.

Question 4: A test result leads to the conclusion:

P(Fraud | Flagged) = 0.12

Explain what this probability means in plain language and the group it refers to.

Question 5: A company screens job applications for potential fraud. Historically, about 1% of applications contain false information.

A new screening system flags applications that appear suspicious. You are told that an application has been flagged. After reviewing the data, an analyst reports:

P(Fraud | Flagged) = 0.18

Answer the following. Use the proper probability notation where needed.

A. What was the prior probability in this scenario?
B. What is the posterior probability?
C. Describe how the belief changed from prior to posterior.
D. In words, explain what caused this change. Your answer should refer to both the prior and the likelihood.

CHAPTER 9

Counting Methods for Probability

Counting matters in probability. Many probability problems ultimately come down to a simple idea:

Probability = number of outcomes of interest ÷ total number of possible outcomes.

This formula is taught early in probability courses. Yet, real problems often make its use challenging. The challenge is often counting the correct number of outcomes for the numerator and denominator rather than the formula itself.

That is where counting methods become essential.

In simple cases, you can list all outcomes and count them. For example, flipping a coin twice or rolling a die once makes listing easy. However, as problems grow even slightly more complex, this approach quickly becomes impractical or impossible. At this stage, you need systematic ways to count outcomes without listing them all.

Earlier in this book, you learned to think about probabilities in terms of sample spaces and events. Counting methods build directly on that idea. Instead of listing the sample space explicitly, counting methods let you measure its size efficiently.

In probability, we often work with very large sample spaces and then focus on specific subsets within them. Here, the aim isn't to describe each outcome individually, but rather to count how many satisfy the stated conditions.

In this chapter, you're likely to be amazed by how quickly outcomes can grow to surprisingly large numbers for seemingly simple scenarios. For large sample spaces, counting methods let you quickly calculate the total number of outcomes and subsets without creating long, cumbersome lists.

Think of this process as drawing subsets from the sample space and counting outcomes in each category. The probability calculation then follows naturally after you know those counts.

For example, drawing a five-card poker hand selects a subset of five cards from a much larger set of possible hands. There are millions of possible five-card hands, but you don't need to list them to answer probability questions. Instead, use counting methods to determine how many hands meet specific criteria, such as being a royal flush or a full house, and how many total hands are possible. You then enter those values into the numerator and denominator, respectively, to find the probability of a particular hand.

The same idea appears in many settings:

- The number of possible lottery combinations compared to the one winning combination.
- The number of four-digit PINs compared to those with no repeated digits.

- The number of possible seating arrangements compared to those with specific people sitting together.
- The number of exam answer patterns compared to those that earn a passing score.

In each case, the probability depends on counting outcomes efficiently and accurately.

Counting problems can take many different forms because outcomes often differ in important ways. Some situations require keeping track of order, while others do not. In some cases, outcomes can repeat, but in others they can appear only once. Some problems unfold in multiple stages, with different rules applying at each step. Others include restrictions, such as prohibiting certain items from appearing together.

Counting methods can handle all these situations. By matching the structure of a problem to the appropriate counting tool, you can account for order, repetition, multiple stages, and restrictions in a systematic way. This flexibility allows you to count outcomes accurately and find probabilities without guessing or double-counting.

In the following sections, I introduce two primary counting tools: permutations and combinations. Before exploring their applications, I clarify each tool's role and when to use it.

Permutations vs. Combinations

Permutations and combinations are two distinct ways to take a set of items or options and create subsets. For example, if you have 10 people, how many 3-person subsets can you make? While permutation and combination can be synonyms in everyday language, they have different mathematical definitions.

- **Permutations**: The order of outcomes matters.
- **Combinations**: The order does not matter.

We'll start by understanding the difference between permutations and combinations in greater detail.

In some cases, order matters. For example, on a four-digit lock, if the correct numbers are 8 3 6 2, you can't enter them in any other order (e.g., 6 8 2 3) and expect the lock to open! That's a permutation.

Conversely, the order of options doesn't matter for combinations. Imagine making a pizza with pepperoni (P), ham (H), and mushrooms (M). It doesn't matter if you make a PHM, HMP, or MPH pizza. It's still the same pizza! Thus, it is a combination.

Ironically, "combination locks" are really "permutation locks" because sequence matters!

Determining whether something is a permutation or a combination depends on the conditions specified by the problem. Consequently, solving these types of problems requires you to pay close attention to the wording. A common mistake in counting problems is overcounting the same outcome multiple times by treating different orders as distinct when they are not.

For example, selecting five people for a group where everyone has the same role is a combination, because the order you pick them doesn't matter. However, if you're picking five people and their roles depend on when you select them, it's a permutation because the order makes a difference. For example, the first person might be the leader, the second a note-taker, and so on.

Permutation	Combination
Sequence of numbers for a lock.	Numbers for winning the lottery.
Selecting individuals for a team by position.	Selecting children to be members of a class.
Picking first, second, and third place.	Picking three finalists.

For the same items, permutations will always produce at least as many outcomes as combinations, and typically many more.

Why is that the case? The pizza example shows that PHM, HMP, and MHP are all the same combination. However, if we're using those three letters as a password, they're three distinct permutations.

Now that the concepts are clear, let's see how to count permutations and combinations. Their formulas save you from lengthy, error-prone listings.

Permutations

Permutations are sequences where order matters. For example, 9-6-8-4 is a permutation of a four-digit PIN because the order of numbers is crucial.

This section covers how to calculate the number of permutations with and without repetition. Then, we'll work through a step-by-step example using permutations to calculate a probability.

With Repetition

When the outcomes in a permutation can repeat, we refer to it as a permutation with repetition or with replacement. For example, in a four-digit PIN, you can repeat values, such as 1-1-1-1.

To calculate the number of permutations, take the number of possibilities for each event and then multiply that number by itself X times, where X equals the number of events in the sequence. For example, with four-digit PINs, each digit can range from 0 to 9, giving us 10 possibilities for each digit. We have four digits. Consequently, the number of permutations with repetition for these PINs = 10 × 10 × 10 × 10 = 10,000.

We write this mathematically as n^r.

Where:

- n = number of possible outcomes for each event. For instance, n = 10 for the PIN example.
- r = the number of items in each arrangement (i.e., length of sequence). For example, r = 4 for a four-digit pin.

Imagine that a class of 15 children can choose one cookie from five types: Gingerbread, Sugar, Chocolate Chip, Mint, and Peanut Butter. The order matters because each child represents a distinct position in the sequence. There are enough cookies that they can choose any type. Hence, this problem involves permutations with replacement.

How many cookie permutations are there?

In this example,

- n = 5 because there are five possible cookie choices.
- r = 15 because there are 15 students in the class, making it the size of the permutation.

Thus, there are 5^{15} = 30,517,578,125 permutations with repetition. That's a staggering 30 billion permutations for a simple cookie example!

If you guessed the cookie choice of all 15 children, the probability guessing correctly for all them is 1/30,517,578,125. That assumes you don't have insider knowledge about each child's cookie preference.

You have a better chance of winning the jackpot in the Powerball lottery, as you'll see when we get to combinations!

Without Repetition

When outcomes cannot repeat, we call it permutations without repetition. This situation frequently occurs when you're working with unique physical objects that can occur only once in a permutation. Analysts also call this method permutations without replacement because you're drawing from a pool of objects and not replacing them after drawing one.

Imagine you have 10 different books and want to calculate how many possible ways you can arrange them on a bookshelf. After you place the first book, the second book must be a different book.

For the first book, you have 10 books from which to choose. For the second book, you have nine. Then eight for the third book, and so on. Like before, this process involves multiplying the number of possible outcomes together. However, we must reduce the number of outcomes for each subsequent event.

Mathematically, we calculate the permutations for the book example using the following method:

$$10 \times 9 \times 8 \times 7 \times 6 \times 5 \times 4 \times 3 \times 2 \times 1 = 3{,}628{,}800$$

There are 3,628,800 ways to arrange 10 books on a shelf without repeating books.

Whew! I bet you didn't realize there are so many possibilities with 10 books. I'll stick to alphabetical order!

Using Factorials

When you multiply all the numbers from 1 to n, it's a factorial. In the book example, we multiplied all numbers from 10 to 1. Instead of using the long string of multiplication, you can write it as 10! and read it as 10 factorial.

In general, n! equals the product of all numbers up to n. For example, $3! = 3 \times 2 \times 1 = 6$. The exception is $0! = 1$, which simplifies equations.

Factorials are crucial concepts for permutations without replication. The number of permutations for n unique objects is n!. This number snowballs as the number of items increases, as the following table shows.

Factorial	Permutations without repetition
1	1
2	2
3	6
4	24
5	120
6	720
7	5,040
8	40,320
9	362,880
10	3,628,800

Partial Permutations without Repetition

In some cases, you want to consider only a portion of the possible permutations. In the bookshelf example, we wanted to know the total number for 10 books. But what if we could fit only five of the 10 books on the shelf? How many permutations of five books are possible using our 10 books?

Use the following formula to calculate the number of arrangements of r items from n objects. There are several standard methods that statisticians use to notate permutations without repetition, which I show with the formula.

$$P(n,r) = {}^{n}P_{r} = {}_{n}P_{r} = \frac{n!}{(n-r)!}$$

Where:

- n = the number of unique items. For instance, n = 10 for the book example because there are 10 books.
- r = the size of the permutation. For example, r = 5 for the five books we want to place on the shelf.

This equation works both for the complete and partial sets of permutations without repetitions, depending on the values you enter in the equation. Additionally, r cannot be greater than n because you cannot select more unique items than are available.

For complete sets, n = r, causing the denominator to equal 0!, which equals 1. That makes the entire ratio equal n! for complete sets.

In the book example, we have 10 books, but we can only put 5 on the shelf. The first book still has 10 options. However, for placing the second book, we have only nine options because we already placed one. We have eight options for the third book and so on until we place the fifth book. Mathematically, we'd write this as follows for the five books:

$$10 \times 9 \times 8 \times 7 \times 6 = 30{,}240$$

There are 30,240 permutations for placing five books out of our 10 books on a shelf.

Now, we'll use the formula to calculate this example. Again, we'll use n=10 and r=5.

$${}_{10}P_{5} = \frac{10!}{(10-5)!} = \frac{10!}{5!}$$

$$\frac{10!}{5!} = \frac{10 * 9 * 8 * 7 * 6 * 5 * 4 * 3 * 2 * 1}{5 * 4 * 3 * 2 * 1} = 30{,}240$$

Notice how the 5! cancels itself out in the fraction? That leaves us with the 10 × 9 × 8 × 7 × 6 that we had before.

Here's how the equation works. The numerator counts the total number of permutations of all the unique items. The denominator removes the extra permutations that arise from arranging more than r items, leaving only the number of ways to arrange r items. For the book example, the denominator cancels out permutations with more than five books.

Using a standard form of the notation, we'd write this problem as ${}_{10}P_5$ = 30,240.

Using a Permutation Calculator

Next, we'll use my permutation calculator to find the number of permutations. Find the link in the book's Introduction for all the free resources, including the calculators.

Let's use this calculator to recreate the preceding bookshelf example, where we have 10 books but can only place 5 on the shelf. In the calculator, enter the following and click **Calculate**:

- **n (objects)**: 10
- **r (arrangement)**: 5
- **Choose:** *Without Replacement.*

The calculator displays results that match our hand calculations: 30240. It also shows how to calculate and interpret that answer.

Now, try one yourself. A conference has 12 different speakers. The organizer wants to schedule 4 of them to present in a specific order during a morning session.

How many different ways can the organizer arrange the 4 speakers?

See the correct answer at the end of the permutation section.

Using Permutations to Calculate Probabilities

When you're given a probability problem that uses permutations, you need to follow these steps to solve the problem.

1. Set up a ratio to determine the probability.
2. Determine whether the numerator and denominator require combinations, permutations, or a mix. For this section, we'll stick with permutations.
3. Are they with repetitions, without, or a mix?
4. Both types of repetition require you to identify the n and r to enter into the equations.

Problem: What is the probability that a randomly selected four-digit PIN does not have repeated digits?

This question builds on several of the examples in this post.

First, we need to set up our probability ratio. In this example, use the following ratio for the events of interest and the total number of events.

$$Probability = \frac{Permutations\ of\ interest}{Total\ number\ of\ permuationas}$$

Numerator

Let's tackle the numerator. We need to find the number of four-digit PINs that do not have repeating digits. That's a permutation because order matters, and it's without replication because we can't have repeats. Let's identify the n and r. We'll use n=10 because 10 digits are available for the first item and r=4 because we're discussing four-digit PINs.

Let's enter that into the equation for permutations without repetition to calculate the numerator:

$${}_{10}P_4 = \frac{n!}{(n-r)!} = \frac{10!}{6!} = \frac{10*9*8*7*6*5*4*3*2*1}{6*5*4*3*2*1} = 5{,}040$$

Denominator

For the denominator, we need to calculate all possible permutations for four-digit PINs with repeats. We need to enter our n and r into the equation for permutations with repeats.

$n^r = 10^4 = 10{,}000$

Consequently, the probability of a four-digit PIN with no repeating digits equals the following:

$$\frac{{}_{10}P_4}{10^4} = \frac{5{,}040}{10{,}000} = 0.504$$

Just over half of all four-digit PINs have repeated values.

Next, let's see how to calculate combinations.

CALCULATOR SOLUTION: For ${}_{12}P_4$, when you arrange 4 items from 12 objects without replacement, there are 11,880 possible permutations.

Combinations

Combinations are sequences of outcomes in which the order does not matter. For example, suppose we're randomly drawing team members where all members have the same role. Using letters to signify team members, C, J, M is equivalent to M, J, C. Either way, you have the same people on the team. Consequently, the order of the letters does not matter because you have the same team! These three letters form one combination.

In this section, I'll show you how to calculate the number of combinations with and without repetition and teach you the standard notation for combinations. I'll also work through example problems so you can see the process in action and how to use it to solve probability problems. You'll learn how to calculate the probability of winning the jackpot in the Powerball lottery!

With Repetition

When the outcomes in a combination can repeat, you have repetition. When ordering pizza, if you can choose a topping multiple times, such as double or triple ham, we have combinations with repetition. It's also known as combinations with replacement.

To calculate the number of combinations with repetitions, use the following equation:

$$_{(n+r-1)}C_r = \frac{(r+n-1)!}{r!(n-1)!}$$

Where:

- n = the number of options.
- r = the size of each combination.

Allowing repetition effectively increases the number of available selections, which is why we use n + r − 1 instead of n.

In addition to the standard notation, there's a standard way to verbalize combinations. For (n + r - 1)Cr, you say, "n choose r."

Let's work through an example of the number of combinations with repetition. Imagine we own a pizza restaurant and have six toppings for customers to choose from. We have a special offer for pizzas with three toppings. How many potential combinations fall under our special? Because we're allowing repetition, customers can get triple-bacon pizzas if they want to risk a hospital visit!

To solve these types of problems, you need to determine the n and r for the equation. The number of possible outcomes equals the number of toppings, which is 6. Each combination has a size of three because that's what we're allowing for our special. So, n = 6 and r = 3.

Using the standard notation for combinations of (n + r - 1)Cr, where 6 + 3 − 1 = 8, we'd write this as ${}_8C_3$ and say "8 choose 3."

Inputting those numbers into the equation:

$$\frac{(3+6-1)!}{3!(6-1)!} = \frac{8!}{3!5!} = \frac{40{,}320}{720} = 56$$

${}_8C_3 = 56$.

There are 56 combinations of three toppings when you have six topping choices and allow repeats.

Without Repetition

When the outcomes cannot repeat, you are working with combinations without repetition. These calculations use permutations without repetition as a starting point.

In the following equation, the first term counts the number of permutations in which the order matters. The second portion reduces that number because the order does not matter for combinations. The product of the two produces the final equation.

$$\frac{n!}{(n-r)!} \times \frac{1}{r!} = \frac{n!}{r!(n-r)!}$$

In the simplified form:

$${}_{n}C_{r} = \frac{n!}{r!(n-r)!}$$

Where:

- n = the number of possible selections at the start.
- r = the size of each combination.

For example, imagine a process that selects five people from a pool of 30 to be on a team. The order in which the process draws team members does not matter because everyone has the same role. Suppose a team comprises people A, B, C, D, E. That is equivalent to a team where the same people were drawn but in a different order. Additionally, because the process can select a person only once, there is no repetition.

In this example, n = 30 because we're drawing from a pool of 30 candidates. R = 5 because the team can have only five members. Inputting those numbers into the equation:

$$\frac{30!}{5!(30-5)!} = 142{,}506$$

Using the standard notation: ${}_{30}C_{5} = 142{,}506$.

There are 142,506 combinations of five-person teams when you draw from a pool of 30 individuals.

This formula for combinations without repetition is a crucial part of the binomial distribution, which we'll cover in Chapter 11.

Using a Combination Calculator

Let's use my combination calculator to find the number of permutations. Find the link in the book's Introduction for all the free resources, including the calculators.

Let's use this calculator to recreate the preceding bookshelf example, where we have 30 candidates but can select only 5 for the team. In the calculator, enter the following and click **Calculate**:

- **n (objects)**: 30
- **r (sample)**: 5
- **Choose:** *Without Replacement*.

The calculator displays results that match our hand calculations: 142,506. It also shows how to calculate and interpret that answer.

Now, try one yourself. A science lab has 10 different chemical samples. A researcher selects 3 of them to test together in an experiment. The order of mixing does not matter.

How many different groups of 3 samples can the researcher choose?

See the correct answer at the end of the combination section.

Using Combinations to Calculate Probabilities

In the previous examples, we calculated the number of combinations for different scenarios. However, we didn't use them to calculate probabilities. In this context, a probability is the number of combinations that constitute an event, divided by the total number of combinations.

When you're given a probability problem using combinations, you need to determine several things to solve the problem.

1. Set up a ratio to determine the probability.
2. Determine whether the numerator and denominator require combinations, permutations, or a mix. For this section, we'll stick with combinations.
3. Are these combinations with or without repetitions?
4. Both types of combinations require you to identify the n and r to enter into the equations.

Let's work through an example probability calculation using combinations!

Problem: What is the probability of winning the basic game and the jackpot in the Powerball lottery?

The classic example of probability using combinations without repetition is a lottery in which machines randomly draw balls with numbers from a pool. The order of the numbers does not matter. You just need to match the numbers. Because each ball can be drawn only once, it is without repetition.

For this example, we'll look at the Powerball lottery, which is a lottery offered in 45 states in the United States. In the basic game, there are 69 white balls in a machine that randomly selects five of the balls. The following ratio defines the probability of winning:

$$Probability = \frac{Winning\ combination}{Total\ number\ of\ combinations}$$

The numerator is simple. There is one winning combination without repetition. You could use the equation to calculate ${}_5C_5$, which equals the number of combinations where you have five numbers in the winning combination, and there are five winning numbers from which to choose. But it makes sense intuitively. There's just one winning combination.

For the denominator, you need to calculate ${}_{69}C_5$, which equals the number of combinations when you draw five numbers from a total of 69 numbers. Let's enter these numbers into the equation:

$$\frac{69!}{5!(69-5)!} = 11{,}238{,}513$$

$${}_{69}C_5 = 11{,}238{,}513$$

When you draw five numbers out of 69 without repetition, there are 11,238,513 combinations.

Using the standard notation, the probability of winning the basic game in Powerball is the following:

$$\frac{{}_{5}C_5}{{}_{69}C_5} = \frac{1}{11{,}238{,}513} = 0.000000089$$

You have a 1 in 11,238,513 chance of winning the basic game in the Powerball lottery.

To win the jackpot in the Powerball, you need to win the basic game *and* match one red Powerball number. The Powerballs are 26 red balls in a separate machine that randomly draws one. Let's calculate the probability of winning the jackpot!

For the numerator, there is again one winning combination for the Powerball. For the denominator, we need to calculate the number of combinations for ${}_{26}C_1 = 26$. That's not surprising. There are 26 combinations when you choose one number from 26 options! Thus, the chance of matching the Powerball is the following:

$$\frac{{}_{1}C_{1}}{{}_{26}C_{1}} = \frac{1}{26} = 0.038$$

To include this in the probability calculations for the basic game, we need to use the multiplication rule for independent events and multiply the two ratios.

Using the standard notation for combinations, the probability of winning the jackpot in Powerball is the following:

$$\frac{({}_{5}C_{5})({}_{1}C_{1})}{({}_{69}C_{5})({}_{26}C_{1})} = \frac{1}{292{,}201{,}338} = 0.0000000034$$

You have a 1 in 292,201,338 chance of winning the jackpot in the Powerball lottery.

Think back to the roughly 1 in 30 *billion* chance of guessing the cookie list for 15 kids. Why is that so much harder than the roughly 1 in 300 *million* chance of winning the lottery, even though the lottery involves many more numbers (95 balls)? Guessing the cookie list is literally two orders of magnitude (100X) harder!

For the cookie list, the order matters because you must match the correct cookie to the correct child. In contrast, with the lottery, you only need to match the final set of numbers, not the order in which they are selected. The counting calculations accurately reflect this difference.

Now that you know these tools, here's how to recognize when each one applies.

CALCULATOR SOLUTION: For ${}_{10}C_{3}$, when you sample 3 items from 10 objects without replacement, there are 120 possible combinations.

When to Use These Tools

When students struggle with permutations and combinations, the difficulty is usually not performing the calculations. The difficulty is recognizing which tool applies to a given problem. That decision depends on how the outcomes are defined and what the problem is asking you to count.

The key is to focus on the structure of the situation rather than on surface details or keywords alone.

Step 1: Do You Need Counting at All?

Ask whether the probability problem requires you to determine how many outcomes are possible. If the problem involves:

- Listing all outcomes is impractical.
- A large sample space.
- A probability expressed as "how many ways" something can occur.

Then, counting methods are likely needed.

If the problem already provides probabilities directly, counting might not be required.

Step 2: Does Order Matter?

Ask whether changing the order of the same items creates a different outcome.

Problem statements where order matters often ask how many ways items can be arranged, ordered, or assigned to positions. They can involve:

- Sequences or arrangements.
- Assigned roles or rankings (first, second, third; leader, assistant).
- Passwords, PINs, rankings, or seating arrangements.

In these cases, you are arranging items, and permutations apply. Problem statements where order does not matter often ask how many ways items can be chosen, selected, or grouped. They can involve:

- Selecting a group where all members have the same role.
- Choosing items where only the collection matters.
- Lottery numbers, hands of cards, or team membership.

In these cases, you are selecting items, and combinations apply.

Focus on what defines a distinct outcome in the problem. If changing the order creates a different result, use permutations. If the same items in a different order represent the same result, use combinations.

Step 3: Can Outcomes Repeat?

Next, determine whether repetition is possible. Repetition is allowed when:

- Items can be reused.
- Choices are made with replacement.
- Digits, letters, or options can appear more than once.

This leads to counting with repetition.

Repetition is not allowed when:

- Items are unique.
- Each item can be used only once.
- Choices are drawn from a pool without replacement.

This leads to counting without repetition.

Step 4: Does the Problem Have Multiple Stages or Conditions?

Some probability problems involve more than one step. For example:

- Selecting a group and then assigning roles
- Choosing items first and arranging them later
- Applying restrictions, such as certain items not appearing together

In these cases, different stages of the problem might require different counting tools. A single problem can legitimately involve both permutations and combinations.

The correct approach is to break the process into stages and apply the appropriate tool at each stage.

Now that you've seen the core tools, let's apply them to more complex problems that combine these ideas.

More Complex Probability Problems

The earlier examples in this chapter focused on using permutations or combinations in relatively straightforward ways. In those cases, we just used a single counting tool. In practice, probability problems often add more structure.

The examples in this section build on what you have already learned by introducing additional layers. Some problems require multiple counting tools within the same calculation. Others involve multiple stages, where different rules apply at each step. Still others place restrictions on which outcomes are allowed, which changes how the numerator is counted. In all cases, the goal is the same: count outcomes accurately so you can calculate probabilities correctly.

Example: Comparing a Royal Flush to a Full House in Poker

Suppose you draw a five-card poker hand from a standard 52-card deck. We want to compare:

- The probability of getting a royal flush.
- The probability of getting a full house.

A royal flush is A–K–Q–J–10, all in the same suit. (There is exactly one royal flush per suit.)

A full house is five cards with the pattern three of one rank and two of another rank, such as three 7s and two Kings.

A poker hand is just a set of five cards. The order you receive them does not matter. Also, after a card is drawn, it cannot be drawn again. So, we use combinations without replacement.

The following two equations represent the probabilities of a royal flush and a full house. Note that the denominator is the same for both.

P(Royal Flush) = (number of royal flush hands) / (total number of possible 5-card hands)

P(Full House) = (number of full house hands) / (total number of possible 5-card hands)

Total number of possible five-card hands

The denominator is the number of ways to choose 5 cards from 52:

Total hands = ${}_{52}C_5$ = (52 × 51 × 50 × 49 × 48) / (5 × 4 × 3 × 2 × 1) = 2,598,960

Counting royal flushes

There are 4 suits, and each suit has exactly 1 royal flush.

Number of royal flushes = 4.

So:
P(royal flush) = 4 / 2,598,960
= 1 / 649,740
= 0.000001539

Counting full houses

To count full houses, build the hand in two parts: the three-of-a-kind and the pair.

First, choose the rank for the three-of-a-kind. There are 13 possible ranks (2, 3, 4, ..., Q, K, A), giving 13C1 = 13 choices. Each rank has four cards, one in each suit, so you then choose 3 of those 4 cards, which is counted by 4C3.

Next, choose a different rank for the pair, leaving 12 possible ranks. Again, that rank has four cards, and you choose 2 of them, which is counted by 4C2. By multiplying the number of choices at each step, you get the total number of full house hands.

In summary, here are the steps for calculating the answer:

1. Choose the rank for the three-of-a-kind: 13 choices.
2. Choose which 3 suits out of 4 for that rank: ${}_4C_3 = 4$.
3. Choose the rank for the pair (must be different): 12 choices.
4. Choose which 2 suits out of 4 for that rank: ${}_4C_2 = 6$.

Number of full houses = $13 \times {}_4C_3 \times 12 \times {}_4C_2$
= 13 × 4 × 12 × 6
= 3,744.

So:

P(full house) = 3,744 / 2,598,960
= 0.0014406
= 1 in 694.17.

A royal flush is far rarer than a full house.

- Royal flush: 1 in 649,740
- Full house: about 1 in 694

So, a full house is roughly 936 times more likely than a royal flush.

Example: Arranging Books with a Restriction

A librarian randomly arranges 8 different books on a shelf in a row. Two of the books are a *Dessert Cookbook* and a *Strict Diet Guide*. To avoid sending mixed messages, the librarian does not want those two books placed next to each other.

We will find the probability that a random arrangement of the books keeps the Dessert Cookbook and the Diet Guide separated.

This is an arrangement problem, meaning it involves permutations.

- We are arranging distinct items in a row, so order matters.
- Each book appears exactly once, so this is a permutation without repetition.
- The condition "not next to each other" restricts which permutations are allowed.

To calculate the probability, we use:

Probability = (number of allowed permutations) / (total number of permutations)

Denominator: Total number of possible arrangements

If there were no restrictions, the 8 books could be arranged in:

Total arrangements = 8!
= 8 × 7 × 6 × 5 × 4 × 3 × 2 × 1
= 40,320

That value, 40,320, counts all possible outcomes of the random arrangement and serves as the denominator.

Numerator: Number of allowed arrangements (books not adjacent)

To count the arrangements where the two books are not next to each other, it is easier to subtract the arrangements we want to exclude. Instead of counting all the allowed arrangements directly, we count the simpler disallowed case where the two books are adjacent and subtract it from the total. This is yet another example of how useful complements can be!

Allowed arrangements = Total arrangements – Disallowed arrangements.

First, count the arrangements where the Dessert Cookbook and Diet Guide are adjacent.

If the two books must be next to each other, treat them as a single block (as if they were glued together). Then we are arranging:

- 1 block (Dessert + Diet together)
- 6 other individual books

When you count the pair as one block, that gives 7 objects to arrange:

Adjacent arrangements = 7!
= 7 × 6 × 5 × 4 × 3 × 2 × 1
= 5,040

Within that block, the two books can appear in either order:

Internal arrangements of the block = 2! = 2

So the total number of adjacent arrangements is:

Adjacent arrangements = 7! × 2!
= 5,040 × 2
= 10,080

Now subtract these from the total to find the number of allowed arrangements:

Allowed arrangements (not adjacent)
= 8! − (7! × 2!)
= 40,320 − 10,080
= 30,240

That value, 30,240, is the numerator.

Probability

P(Dessert Cookbook and Diet Guide not adjacent)
= Allowed arrangements / Total arrangements
= 30,240 / 40,320
= 0.75

The probability that the librarian's random arrangement keeps the *Dessert Cookbook* and the *Strict Diet Guide* separated is 0.75, or 75%.

Example: Forming a Committee and Assigning Officer Roles

A department has 12 faculty members, and it will form a 5-person hiring committee. After the committee is selected, it assigns three distinct officer roles: Chair, Secretary, and Presenter.

Assume each stage is random, so all committee selections are equally likely and, after a committee is chosen, all role assignments are equally likely.

We will find the probability that Jim and Carmen are both on the committee and Jim is the Chair.

The department randomly selects the committee from the 12 faculty members, and then randomly assigns the three officer roles among the five committee members. When I write "Jim is Chair," I don't mean the department intentionally selects him for that role. Instead, we are counting the outcomes in which the random assignment happens to give Jim the Chair position.

This problem is multi-stage, which is why it uses more than one counting tool:

- Stage 1 selects the committee members. The order of selection does not matter, and no person can be selected twice. That means we use combinations without repetition.
- Stage 2 assigns officer roles within the selected committee. Here, order matters because Chair, Secretary, and Presenter are different positions. That means we use permutations without repetition.

To calculate the probability, we use the basic counting form:

Probability = (number of favorable outcomes) / (total number of possible outcomes)

The denominator counts all possible outcomes of the entire random process (committee selection and role assignment).

The numerator counts only the outcomes that satisfy the conditions we care about (Jim and Carmen on the committee, and Jim assigned Chair).

Denominator: Total number of possible outcomes

First, count how many committees could be formed:

Total committees = ${}_{12}C_5$
= $(12 \times 11 \times 10 \times 9 \times 8) / (5 \times 4 \times 3 \times 2 \times 1)$
= 792

For each specific committee, we must assign the three distinct roles of Chair, Secretary, and Presenter to three of the five committee members. Because the roles are different, assigning A as Chair and B as Secretary is a different outcome from assigning B as Chair and A as Secretary. That is why we use 5P3.

Total role assignments = ${}_5P_3$
= $5 \times 4 \times 3$
= 60

For every possible committee, there are 60 possible ways to assign the roles, so we multiply the number of committees by the number of role assignments. So, the total number of possible outcomes for the full process is:

Total outcomes = $({}_{12}C_5) \times ({}_5P_3)$
= 792×60
= 47,520

47,520 is the denominator in our probability.

Numerator: Number of favorable outcomes

Now, count how many outcomes meet our conditions: Jim and Carmen are on the committee, and Jim is assigned Chair.

Start with the committee selection stage. If Jim and Carmen must be on the committee, then two of the five seats are already taken. That means we only need to choose the remaining three committee members from the people who are left.

After including Jim and Carmen, there are:

Remaining faculty to choose from = 12 − 2 = 10

The number of possible committees that include both Jim and Carmen is:

Committees containing Jim and Carmen = ${}_{10}C_3$
= (10 × 9 × 8) / (3 × 2 × 1)
= 120

Now move to the role assignment stage for one of those committees. Think of it this way:

- We require Jim to be Chair; that role is already filled.
- That leaves two remaining roles (Secretary and Presenter) to be assigned.

Those two roles must go to two different people chosen from the four committee members other than Jim. Because Secretary and Presenter are different positions, order matters, so the number of assignments is ${}_4P_2$.

Role assignments given Jim is Chair = ${}_4P_2$

$= 4 \times 3$
$= 12$

For each committee that includes Jim and Carmen, there are 12 ways to assign the remaining roles with Jim as Chair, so we multiply the number of such committees by the number of role assignments. Therefore, the total number of favorable outcomes is:

Favorable outcomes = $({}_{10}C_3) \times ({}_4P_2)$
$= 120 \times 12$
$= 1{,}440$

1,440 goes in the numerator of our probability.

Probability

P(Jim and Carmen on the committee and Jim is Chair)
= Favorable outcomes / Total outcomes
= 1,440 / 47,520
= 0.0303

So, the probability is about 3.03%.

This example illustrates how counting methods can be combined across stages to model complex probability processes.

Review and Next Steps

In this chapter, you learned how counting methods make probability calculations possible when listing outcomes is impractical. By identifying when order matters, when repetition is allowed, and how restrictions shape which outcomes are permitted, you saw how permutations and combinations provide a systematic way to count outcomes accurately. You also worked through examples that showed how counting methods can be combined across stages and adapted to handle real constraints.

These ideas are more than isolated techniques. Counting methods describe the *structure* of many probability problems.

- Use counting methods when listing all possible outcomes is impractical, but the total number of outcomes can be determined systematically.
- If changing the order creates a different outcome, permutations apply. If it does not, combinations apply.
- Some problems allow the same item to be used more than once, while others do not.
- A multi-stage problem can legitimately use both permutations and combinations.
- In problems with restrictions, it is often easier to count disallowed outcomes and subtract them from the total using a complement.
- In probability calculations, counting methods determine the numerator (favorable or allowed outcomes) and the denominator (all possible outcomes).

In the next chapter, we shift from counting individual outcomes to modeling random variables, their distributions, and the properties of those distributions. However, we'll encounter counting methods in later chapters when we get to discrete probability distributions, such as the binomial distribution.

Formula Quick Reference

Method	Formula
Permutations with repetition	n^r
Permutations without repetition (full)	${}_nP_r = n!$
Permutations without repetition (partial)	${}_nP_r = n! / (n - r)!$
Combinations with repetition	${}_{n+r-1}C_r = (r + n - 1)! / [(r!(n - 1)!]$
Combinations without repetition	${}_nC_r = n! / [r!(n - r)!]$

Chapter 9 Practice Problems

Question 1: For each of the following scenarios, identify the most appropriate counting method (don't calculate the solution):

- Permutations with repetition
- Permutations without repetition
- Combinations with repetition
- Combinations without repetition

Scenarios:

A. A company generates 4-character product codes using digits 0–9. The same digit can appear more than once, and different orders produce different codes.

B. A student selects 3 courses from a list of 12 electives to take next semester. Each course can be taken only once.

C. A bakery offers 5 types of cookies. A customer chooses 4 cookies for a box, and they can select multiple cookies of the same type.

D. A race assigns gold, silver, and bronze medals to 3 of the 10 runners. Each medal is distinct and no runner can receive more than one medal.

Question 2: A streaming service generates a 3-letter recommendation code for each user using the letters A–Z. Letters can repeat.

What is the probability that a randomly generated code contains no repeated letters?

Question 3: A museum has 9 different paintings and randomly selects 4 of them for inclusion in a special exhibit.

What is the probability that the selected set of paintings includes both a specific Monet painting and a specific Van Gogh painting?

Question 4: A smoothie shop offers 7 different fruits. A customer selects 3 fruits to include in a smoothie. The same fruit can be chosen more than once (e.g., triple strawberry).

What is the probability that a randomly created smoothie includes at least one banana?

Question 5: A music app has a curated list of 14 songs: 8 upbeat tracks and 6 slow tracks.

First, the app randomly selects 5 songs to create a featured playlist. Then, from those 5 songs, it randomly assigns two distinct positions: Opening Track and Closing Track.

What is the probability that the playlist contains exactly 3 upbeat songs and 2 slow songs, and that a specific slow song is chosen as the Opening Track?

CHAPTER 10

Probability Distributions: Modeling Patterns

So far, we have focused on individual events, probabilities, and long-run averages. We've learned how to count outcomes and calculate probabilities. Those tools answer important questions, but they still leave something out. They reveal only parts of the randomness, not the overall pattern.

Often, the goal is not to predict a single outcome, but to understand randomness across many outcomes. We want to know which values are common, which are rare, how outcomes cluster, and how much variation we should expect. Probability distributions provide that broader view. They shift the focus from individual results to models that describe long-run behavior.

In this chapter, you will learn what a probability distribution is and how it models randomness. You will see how random variables map outcomes to numbers, how the expected value represents the average of a random process, and how the standard deviation measures the spread around that average. Throughout the chapter, I use simple probability distributions defined by probability tables rather than named probability models. This approach keeps the emphasis on usage rather than mechanics for this introductory chapter.

This chapter establishes the foundation for what comes next. It introduces the language and ideas of probability distributions without covering specific distributions in detail. Later chapters expand on these ideas by introducing model-based distributions, such as the binomial and normal distributions. Here, the goal is to understand what probability distributions are, why they work, and how they clarify randomness.

Modeling Randomness

When we model randomness, we are not just interested in whether an event happens. We want to describe how outcomes vary, which outcomes are common and which are rare, and what happens in the long run. Random experiments, random variables, and probability distributions work together to make that possible.

Random experiments

Random experiments are the sources of randomness that produce uncertain outcomes. Here, the word "experiment" does not refer specifically to a scientific study. Instead, it refers to any repeatable process that generates outcomes influenced by chance. While these processes are random, their well-defined outcomes still follow stable long-run probability patterns.

For example, random experiments can incorporate coin tosses, die rolls, or repeated measurements on random samples drawn from a population.

In this context, an experimental outcome is unknowable, even in principle, before a trial occurs. For instance, if your experiment involves randomly drawing a person from a population and measuring their height, you cannot know the next height until you select and measure the next person.

Random variables

Random variables record the outcomes of an experiment numerically. These variables don't have one specific value. Instead, chance determines their values for each observation.

Over repeated trials, random variables take on different values representing outcomes from events, categories, labels, or measurements. Variables record these outcomes numerically, enabling quantitative analysis. Think of them as a data column in a spreadsheet for variables like height or coin toss results.

For example, in the height experiment, you'll obtain a value for each subject that you store in a spreadsheet as a column of heights.

Random variables can take two distinct forms: discrete or continuous.

A discrete random variable has distinct values that are countable and finite or countably infinite. This type often arises when you are counting events. For example, discrete random variables include the following:

- The number of heads during a series of coin tosses.
- The number of library books checked out per hour.

A continuous random variable takes on uncountably infinite values within a range. Further, there are infinite values between any two values.

This data type often occurs when you measure a quantity on a scale. For example, continuous random variables include the following:

- Height and weight.
- Time and duration.
- Temperatures.

Probability distributions

Probability distributions give structure to the randomness in a random variable. While the next observed value in a random experiment is unknown, it is not entirely unpredictable because a probability distribution defines the likelihood of all possible values. As a result, these distributions allow you to make predictions about real-world phenomena.

A probability distribution is a mathematical representation that assigns probabilities to the possible values of a random variable. It describes how those values are patterned across many repetitions of the same random experiment.

Rather than focusing on a single outcome, a probability distribution shows which values are common, which are rare, and the overall variation in outcomes. Analysts often display distributions using tables or graphs that summarize this pattern.

This shift in perspective matters. When we ask whether a coin flip will land heads, we are asking about a single outcome. When we ask how often heads appears in the long run, or how variable the results are, we are asking about a distribution. Probability distributions move us from individual events to long-run behavior.

Crucially, a probability distribution does not describe a single dataset. Instead, it represents the pattern of outcomes we would expect if a random process repeats under identical conditions. In practice, we often use data from a sample to estimate this pattern, but the distribution itself describes the underlying process behavior, not just one dataset.

Probability distributions share many of the same descriptive features as histograms. They have a shape, center, and spread. However, the focus for probability distributions is, unsurprisingly, on the probabilities of the outcomes.

Every probability distribution must satisfy two basic conditions. First, each probability must be non-negative. Negative probabilities have no meaning. Second, the probabilities across all possible values must sum to 1. Together, these rules ensure that the distribution accounts for all possible outcomes.

It's important to separate the idea of a probability distribution from formulas and named models. A distribution does not require a binomial formula, a normal curve, or any other mathematical structure. We can define a probability distribution simply by listing possible values and assigning probabilities to them in a table, as long as those probabilities follow the basic rules.

In summary, each concept plays a distinct role. The random experiment introduces uncertainty through fluctuating outcomes; the random variable records the experiment's outcome numerically; and the probability distribution models the long-run behavior of all possible values. Together, they form a complete model of randomness.

Example

Let's build a complete example using the height measurement scenario to show how random experiments, variables, and probability distributions work together.

Suppose you draw a random sample of people and measure each person's height. The random experiment consists of selecting a person at random and measuring their height. The study cannot know the next height until it randomly selects and measures the next person.

Height is the random variable that records a value for each outcome. Each time the study selects another person and measures their height, the variable takes a new value.

As these measurements accumulate, you can graph the distribution of heights using a histogram. This distribution shows the most common heights, the amount of variation among individuals, and the likelihood of different height ranges.

The histogram displays the observed pattern of heights in the middle school girl population.

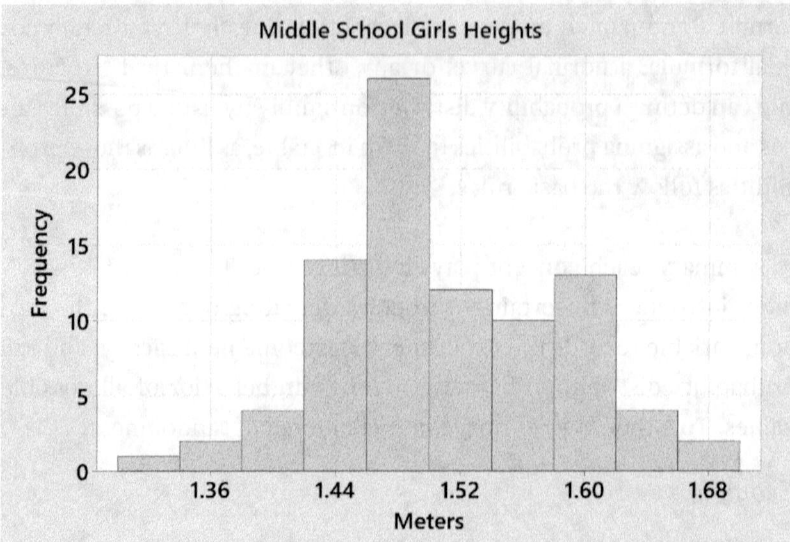

While the graph displays the observed data, the probability distribution describes the underlying pattern that generates those outcomes. It represents the long-run behavior of the random process rather than just the specific sample you see. You use the distribution to find probabilities for specific outcome values or ranges of values. But let's use the histogram to get an idea of what we can learn from a probability distribution.

Imagine you plan to draw a random middle school girl and measure her height, but haven't done so yet. You don't yet know the next height measurement.

Even though the next person's height is unknown, you can use the histogram to identify more and less probable values. Heights near the center of the distribution are more likely than those in the tails. You can also see the entire height range and the extremes where no heights were measured.

All heights are between 1.3m and 1.7m, with heights near 1.5m being the most common. Likelihoods drop off as you move away from the center in both directions. The key takeaway is that while you don't know the next randomly selected girl's height, you can identify the likely range and most probable values.

For the height data, you can use the normal distribution to model it mathematically to obtain precise probabilities, which we cover in a later chapter. However, the histogram gives you an idea of what you can learn.

Let's take a closer look at the general properties of probability distributions.

Probability Distribution Properties

Probability distributions describe how likelihood is allocated across all possible outcomes of a random experiment. The "distribution" in the name refers to the focus being on the global pattern of outcomes rather than a single outcome. However, you can use these distributions to find probabilities for specific outcomes; it's just done in the context of all possible outcomes.

A probability distribution can be specified directly by listing outcomes and their probabilities in table, or through a mathematical function or model that generates those probabilities. Both methods need to follow specific rules to produce valid probability distributions.

This chapter highlights probability distributions defined by probability tables, allowing us to focus on usage and interpretation rather than calculations. Table-based distributions can come from observed data, such as the number of cars per household, or from logical reasoning about a random process, such as die rolls. Later chapters cover model-based distributions that use parameterized functions.

Statisticians use the following notation to describe probability distributions:

P(x) = the probability that the random variable takes the value x.

This notation is often written more explicitly as P(X = x), where X represents the random variable and x is a specific value that it can take. For example, if X is the outcome of a die roll, then P(X = 3) denotes the probability that the random variable takes the value 3.

We'll often use the simpler notation P(x) for convenience. Both notations represent the same idea: the probability that the random variable equals a particular value.

All probability distributions must adhere to the following rules to be valid:

- The sum of all probabilities must equal 1.
- The probability for a value or range of values must be between 0 and 1, inclusive.
- The distribution must include all possible outcomes for the random experiment.

In short, because the total probability for the entire distribution is 1, one of the values must occur for each opportunity. For example, the likelihood of rolling a specific number on a die is 1/6. The total probability for all six values equals one. When you throw a die, you inevitably obtain one of the possible values.

Probability distributions describe the dispersion of outcomes for a random variable. Consequently, the type of variable determines the probability distribution. Just as random variables come in two flavors, discrete and continuous, so do probability distributions. We'll start with discrete distributions because they're simpler to understand.

Discrete probability functions, also known as probability mass functions, can assume a set of distinct values. For example, coin tosses and event counts are discrete functions. These are discrete distributions because there are no in-between values. For example, you can have only heads or tails in a coin toss. Similarly, if you're counting the number of books that a library checks out per hour, you can count 21 or 22 books, but nothing in between.

In this chapter, we work with discrete probability tables. When the discrete distribution has a finite number of outcomes, you can display all of them with their corresponding probabilities in a table. For example, according to a study, the likelihood of the number of cars in a California household is the following:

Number of Cars	Probability
0	0.03
1	0.13
2	0.70
3	0.10
4	0.04

Notice that the table includes all observed values and assigns a probability to each value. Further, the probabilities sum to 1. Collectively, these properties indicate that each household observation will always produce one of the values in the table.

To find probabilities for ranges or multiple outcomes, simply add those values together.

For example, the probability of having 1, 2, or 3 cars is just the sum of the probabilities for those three values: 0.13 + 0.70 + 0.10 = 0.93.

The following table defines the probability mass function for the sum of two dice. The sample space consists of 6 × 6 = 36 possible outcomes for the two dice. To find the probability, you count the number of paired dice outcomes that produce a specific sum and divide by all possible combinations (36). For example, there are four pairs that sum to 5. Hence, the probability of two dice summing to 5 is 4/36.

Sum of 2 Dice	Paired Outcomes	Probability
2	(1,1)	1/36
3	(1,2)(2,1)	2/36
4	(1,3)(3,1)(2,2)	3/36
5	(1,4)(4,1)(2,3)(3,2)	4/36
6	(1,5)(5,1)(2,4)(4,2)(3,3)	5/36
7	(1,6)(6,1)(2,5)(5,2)(3,4)(4,3)	6/36
8	(2,6)(6,2)(3,5)(5,3)(4,4)	5/36
9	(3,6)(6,3)(4,5)(5,4)	4/36
10	(4,6)(6,4)(5,5)	3/36
11	(5,6)(6,5)	2/36
12	(6,6)	1/36

Again, notice how the table includes all possible values, lists only non-zero probabilities, and they sum to 1.

The following graphs show the same information as the tables. They are informative because they display the probabilities for all possible outcomes for both random processes. The bar heights make it easy to determine which outcomes are most likely.

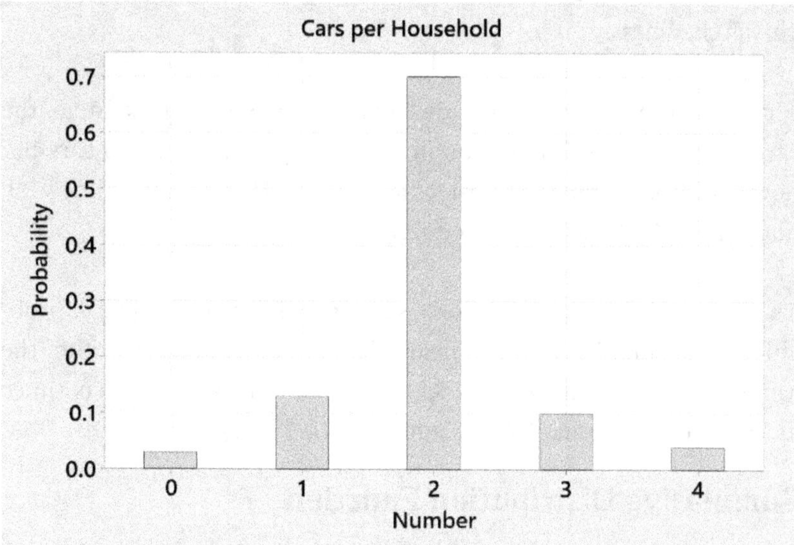

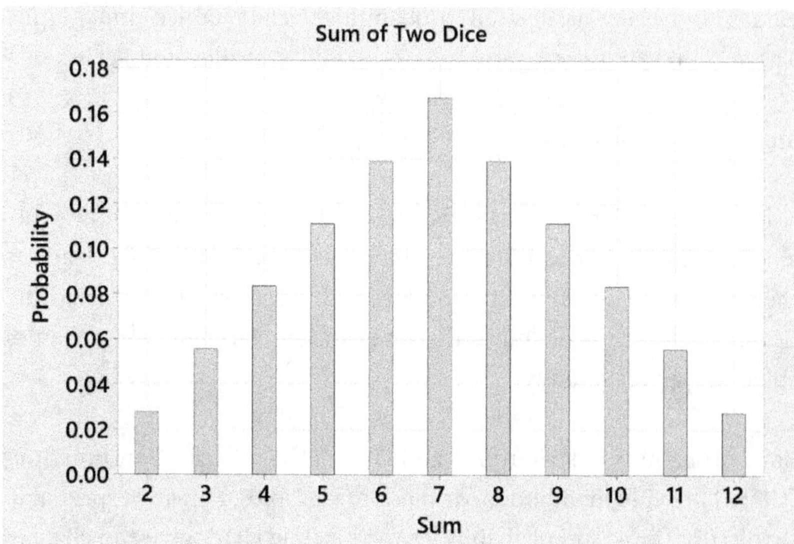

Think back to your basic statistics. Recall that all distributions have several basic properties. These include a central tendency (e.g., mean, median, and mode) and a measure of variability (e.g., standard deviation) that assesses how far observations tend to fall from the center.

You can also assess other distribution properties, such as symmetry versus skewness.

That's all true for probability distributions as well. For example, the car-per-household distribution in California shows that two cars per household is by far the most common (0.70). And that almost all households (93%) have between 1 and 3 cars.

The sum of two die distributions is symmetrical and has a peak probability at a sum of 7. Further, nearly half the outcomes fall within the range of 6 – 8 (5/36 + 6/36 + 5/36 = 16/36 = 0.44). You're six times as likely to roll a sum of 7 as a sum of 12 (0.1875 vs. 0.03125)!

Cumulative Distribution Function

So far, we have described probability distributions by listing possible values and their associated probabilities using tables and graphs. These representations show how probability is allocated across outcomes, value by value. In notation, we represent that using $P(X = x)$, indicating that it's a probability for a specific value (x).

Another way to present the same information is to focus on how probability accumulates as outcomes increase. That perspective leads to the cumulative distribution function, which reorganizes the distribution to show the probability that the random variable is less than or equal to a given value.

Both probability distributions and cumulative distribution functions (CDFs) provide likelihoods for random variables. However, probability distributions find probabilities for x, while CDFs give the chances for $\leq$ x.

CDFs require that the possible values of the random variable have a natural order so they can be accumulated meaningfully. For example, numerical outcomes like heights or dice sums work well, but categorical outcomes without an inherent order, such as blood type or favorite food, do not.

CDFs use the following notation:

$$CDF(x) = P(X \leq x)$$

X is the random variable, and x is a specific value. A cumulative distribution function describes the probabilities of a random variable having values less than or equal to x. It is a cumulative function because it sums the total likelihood up to that point. Its output always ranges between 0 and 1.

These functions are non-decreasing. As x increases, the likelihood can either increase or stay constant, but it can't decrease because negative probabilities do not exist.

Cumulative distribution functions are excellent for providing probabilities that the next observation will be less than or equal to the value you specify.

Let's return to the number of cars per California household probability distribution. To create a cumulative distribution, we form a running total in the last column by adding each probability to the cumulative total above it.

Number of Cars	Probability	Cumulative
0	0.03	0.03
1	0.13	0.16
2	0.70	0.86
3	0.10	0.96
4	0.04	1.00

In the table, notice how the final row must have a cumulative probability of 1. Because all likelihoods in a probability distribution must sum to 1, the highest value must have a cumulative probability of 1.

Cumulative distributions are just a different way of showing the same information as a probability distribution. When you have one type of distribution, you can construct the other type. For instance, given a cumulative distribution, you can create a probability distribution by subtracting out the changes between cumulative values.

Note the relationship between the Probability and Cumulative columns. The Probability column provides the probability for a specific outcome, while the Cumulative column shows the running total of probabilities as they accumulate across outcomes. The increase in the cumulative values from one outcome to the next equals the probability of that outcome.

Cumulative distributions are especially useful for answering probability questions phrased in terms of "at most," "less than," or "no more than." They allow you to find the probability that a random variable falls below a given value just by looking up that value in a table or graph.

CDFs are also a natural tool for working with percentiles, which identify the value below which a specified percentage of outcomes fall. For example, the 90th percentile marks the point below which 90% of the distribution lies. Hence, CDFs are particularly useful for understanding relative standing and thresholds within a distribution.

For example, in the previous table, there is an 86% probability that a household will have two cars or fewer. Alternatively, you can say that a household with two cars is at the 86th percentile.

Just as you can graph probability distributions, you can plot the cumulative distributions. In CDF graphs, as the random variable increases on the horizontal axis, the graph line can only either increase or stay constant, but never go downward.

These graphs are convenient because you can pick out an x-value and quickly find the cumulative probability $P(X \le x)$ for it. Further, steeper sections reflect larger increases in cumulative probability, which correspond to outcomes with higher probabilities.

The following graph displays the CDF for the car probabilities. Notice how the steep rise for two cars corresponds to the large change in the cumulative probability from having 1 car to 2 cars. The flatter portions reflect smaller increases.

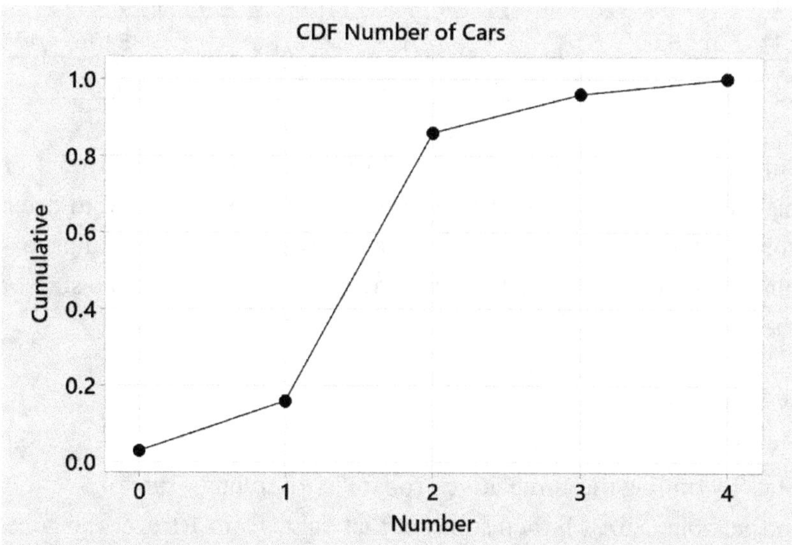

Additionally, you can also use cumulative distributions with complements to find probabilities for "greater than" or "at least" questions by subtracting from 1. And, you can find the probability that a value falls within an interval by subtracting two cumulative probabilities.

Let's work with the sum of two dice probability distribution for these two calculations. I've added the running total, and again, the final row has a probability of 1 (36/36).

Sum	Paired Outcomes	Probability	Cumulative
2	(1,1)	1/36	1/36
3	(1,2)(2,1)	2/36	3/36
4	(1,3)(3,1)(2,2)	3/36	6/36
5	(1,4)(4,1)(2,3)(3,2)	4/36	10/36
6	(1,5)(5,1)(2,4)(4,2)(3,3)	5/36	15/36
7	(1,6)(6,1)(2,5)(5,2)(3,4)(4,3)	6/36	21/36
8	(2,6)(6,2)(3,5)(5,3)(4,4)	5/36	26/36
9	(3,6)(6,3)(4,5)(5,4)	4/36	30/36
10	(4,6)(6,4)(5,5)	3/36	33/36
11	(5,6)(6,5)	2/36	35/36
12	(6,6)	1/36	36/36

Suppose you need to find the probability of rolling a sum of 9 or higher. Cumulative probabilities are $P(X \le x)$, so we'll need to use a complement to solve this greater than or equal to problem. In probability notation, we write this as follows, where S denotes the sum of the two dice:

$P(S \ge 9) = 1 - P(S \le 8)$

Start by finding the cumulative probability of rolling a sum of 8 or less in the table (26/36), then subtract that value from 1 to get the complement. Why use 8? Because dice sums are discrete whole numbers, there are no possible outcomes between 8 and 9. As a result, the complement of "sum ≤ 8" is "sum ≥ 9."

$P(S \ge 9) = 1 - 26/36 = 10/36 = 0.278.$

You have a 0.278 probability of rolling two dice and obtaining a sum of at least 9. Or, you'll get a sum of at least 9 for approximately one-quarter of dice rolls.

You can also use cumulative probabilities to find likelihoods for ranges by subtracting.

Suppose you need to find the probability of rolling a sum between 7 and 9 inclusive, denoted as the following:

$$P(7 \leq S \leq 9) = P(S \leq 9) - P(S \leq 6)$$

Because we want to include a sum of 7, we subtract $P(S \leq 6)$ rather than $P(S \leq 7)$. This removes all outcomes below 7, leaving the sums 7, 8, and 9.

$$P(7 \leq S \leq 9) = P(S \leq 9) - P(S \leq 6) = 30/36 - 15/36 = 15/36 = 0.4167$$

You have a 41.67% probability of rolling a sum of 7, 8, or 9.

CDFs become especially important when we move to continuous distributions, where probabilities are defined over ranges rather than individual values and are naturally expressed as accumulated areas. We'll return to these ideas in a more advanced form when we study continuous probability distributions in Chapter 13.

Tips for Using CDFs for Discrete Distributions

When solving probability problems using cumulative probabilities, the most common mistakes come from misreading the question or writing probability notation that does not match the wording of the problem. Before performing any calculations, take a moment to translate the problem description into a precise probability statement. Pay close attention to phrases such as *at least*, *at most*, *greater than*, *less than*, *between*, and *inclusive*. These words determine whether bounds are inclusive (≤, ≥) or exclusive (<, >).

For discrete distributions, inclusive and exclusive bounds matter because there are gaps between possible values. As a general rule, if a problem asks for values greater than or equal to a, include a in the probability statement. If it asks for values strictly greater than a, exclude a.

When using cumulative probabilities to find ranges, remember that subtracting P(X ≤ a) removes all outcomes up to and including a, while subtracting P(X ≤ a − 1) removes only outcomes below a. Choosing the correct cutoff depends entirely on whether the lower bound is inclusive or exclusive.

Complements can simplify many problems, but they also require careful handling of bounds. When you replace a probability such as "X ≥ a" with its complement, ensure the complementary event truly captures all remaining outcomes. For discrete variables, complements often shift the cutoff value by one unit.

Always check that the probability statement you calculate matches the original question in words. Performing a quick sanity check by listing the specific values you want to include can help confirm that your notation and calculations align with the problem you are trying to solve.

Probability tables, graphs, and cumulative distribution functions describe how probabilities are organized across possible outcomes. They show how likelihood is distributed across values and how it accumulates as those values increase, but they do not summarize the distribution's overall behavior. Expected value and standard deviation provide that summary using just two numbers.

As you might recall from introductory statistics, distributions have central tendencies and variability around their centers. The same is true for probability distributions. To understand how averages emerge and how closely outcomes cluster around the averages over many repetitions, we need additional summary measures. The next sections introduce those measures, beginning with the expected value.

Expected Values (EVs)

The expected value (EV) is the long-run average outcome of a random variable based on its possible outcomes and their respective probabilities. It is the central tendency of the probability distribution for a random process and represents the mean outcome. If a random experiment were repeated many times, the expected value tells us the average result we'd see.

However, the expected value does not necessarily correspond to the most common or most likely outcome. In some distributions, especially those that are skewed or have rare extreme values, the expected value can differ substantially from the most common value.

Caution: The expected value is not a prediction of a single outcome, but a summary of long-run behavior. Additionally, the EV need not be a possible outcome. Families can have an average of 2.2 kids, even though no single family can have that number!

Statisticians denote it as E(X), where E is "expected value," and X is the random variable.

The origins of expected values trace back to the 17th century when mathematicians Blaise Pascal and Pierre de Fermat corresponded about gambling problems. The concept of expected value emerged from these early deliberations on fair bets and equitable decisions.

Expected values help you make informed decisions by understanding the average outcomes of various random experiments and scenarios.

Understanding the general process for calculating the expected value for a discrete random variable is the best place to start. For discrete probability distributions, find the expected value by multiplying each possible value of the random variable by its probability and then summing all those products. Use the following formula:

$$E(X) = \sum_{i=1}^{n} x_i \times P(x_i)$$

Where:

- X is the random variable.
- x_i are the specific values.
- i is the index variable, ranging from 1 to n, over all possible values of the discrete variable.
- n is the number of possible outcomes.
- $P(x_i)$ is the probability of x_i.

This calculation is just a weighted average, where each outcome's weight is its probability. Because the probabilities sum to 1, there is no need to divide.

Probability Tables

For probability tables, this process becomes especially easy.

1. Multiply each outcome by its probability.
2. Sum those values

For the number of cars example, we can use the table to calculate the average number of cars per California household.

Number of Cars	Probability	Outcome × Probability
0	0.03	0.00
1	0.13	0.13
2	0.70	1.40
3	0.10	0.30
4	0.04	0.16
	Sum = EV	1.99

Given the probability distribution, households in California have an average of 1.99 cars. If we randomly selected many households, the average would converge on this value.

Equiprobable Outcomes

When outcomes are equally likely (like rolling a fair die), the expected value formula is a straightforward average of the results because all the weights (i.e., probabilities) are equal.

For example, when you roll a die, each outcome (1 through 6) has an equal 1/6 chance.

$E(X) = (1 + 2 + 3 + 4 + 5 + 6) / 6 = 3.5$

Therefore, the expected value of rolling a die is 3.5. When you roll a die many times, the average will converge on this value.

Games of Chance

Imagine a casino game where you have a 0.5 chance to win $100, a 0.4 chance to win $500, and a 0.1 probability to lose $100.

To find the expected value for the casino game, we'll take each outcome (the winnings and loss), multiply it by its probability, and sum them.

Outcome	Probability	Outcome × Probability
100	0.5	50
500	0.4	200
-100	0.1	-10
	Sum = EV	240

$E(X) = 100 \times 0.5 + 500 \times 0.4 - 100 \times 0.1 = \$240.$

Hence, the expected value for that game is $240. As more people play this game, the average outcome will converge on this value.

Now imagine a casino uses these probabilities. How much would the casino need to charge to ensure it makes money? If the average payout is $240, then charging more than that amount per game produces a positive expected profit in the long run. For instance, if the casino charges $250 to play, the expected long-run profit is $10 per game.

The house always wins in the long run because it knows how much to charge!

Conversely, if the casino charges $250, players are expected to lose an average of $10 per game ($240 – $250 = $-10) over the long run.

Several notes before moving on. The previous calculation method applies to discrete probability tables. The parameter-based probability distributions we cover in the upcoming chapters have their own calculation methods.

Finally, I hope you noticed how I mentioned "converging over the long run" in relation to expected values. The law of large numbers shows why that is important!

Law of Large Numbers

The expected value describes the long-run average outcome of a random experiment, but what does that mean exactly?

The Law of Large Numbers is a cornerstone concept in probability theory. This law states that as a random experiment is repeated many times under the same conditions, the average of the observed outcomes tends to approach the expected value.

Conversely, large deviations from the expected value are more common in the short run than in the long run. When only a small number of trials.have occurred, random variation can dominate, producing averages that differ substantially from the expected value.

For instance, if you flip a coin four times, it's not surprising to get three heads (75%). However, after 100 coin flips, the percentage will be very close to the expected 50%.

This law brings a type of order to random events. For example, if you're talking about flipping coins, rolling dice, or games of chance, you are more likely to observe an unusual sequence of events over the short run. However, as the number of trials grows, the overall outcomes converge on the expected value.

Importantly, this stabilization reflects averaging over many outcomes, not increased predictability of individual results. After a sequence of coin tosses of any length, the probability of the next heads is still 1/2 regardless of the prior outcomes.

Why does this stabilization occur?

When only a small number of outcomes are observed, random fluctuations can easily push the average above or below the expected value because there aren't enough outcomes to offset the deviation. As more outcomes are added, unusually high and unusually low results become increasingly likely to balance out. While individual outcomes remain random, their deviations from the expected value tend to cancel one another when averaged over many trials, resulting in the overall average settling near the expected value.

While there are mathematical proofs for the law of large numbers, I will instead simulate it using my favorite free random sampling program, Statistics101! You can download the scripts for the coin toss and casino examples. See the resource page for this book on my website.

You can perform the simulations yourself and see the results. I include example graphs that I created using these scripts. I exported the data to Excel for prettier graphs, but Statistics101 produces graphs as well. Your simulations won't match mine, but they should follow the same overall pattern.

These simulations work by performing the random experiment many times and keeping a running average. This process produces a sequence of sample means as the sample size increases. If the law of large numbers holds true, we'd expect the sample means to converge on the expected value as the sample size increases. Let's see!

Coin Flipping Example

Let's start with coin flips. For this scenario, we'll define an event as heads in the coin toss. A coin toss is one trial. The law of large numbers predicts that as the number of trials increases, the proportion will converge on the expected value of 0.50. The fluctuating line represents the running proportion of heads. Note that the darker horizontal line is at the expected value of 0.5 heads.

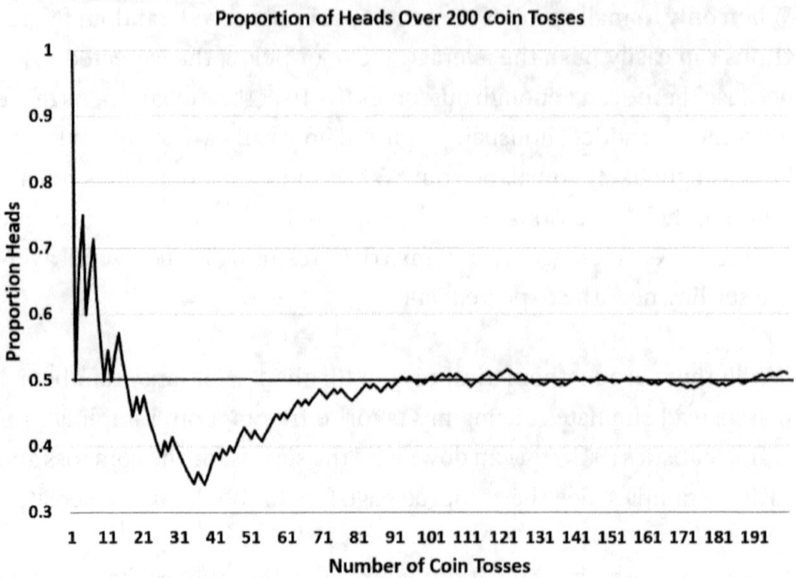

In the short run, early in the sequence, the running average is more erratic, ranging from mostly heads to only 35% heads. Later, the average stabilizes and converges on the expected value (0.5) as the sample size increases.

Casino Example

Now, let's revisit the previous casino example where we calculated an expected value of $240 for the game. The line in this graph represents the running average payout per game. Keep in mind, the casino charges $250 to play this game once. Note the reference line at the expected value of $240.

This graph shows the same pattern as the coin example. Larger deviations from the expected value in the short run, and then the values converge on the expected value as the number of trials increases.

Early in the sequence, the average is around $300 a game, meaning the casino is losing an average of $50 per game. Then the average payout drops to around $170, indicating the casino is making an average of $70 a game!

Finally, the average payout converges on the expected value of $240 a game as the number of games played increases. That is good for the casino, which makes an average profit of $10 per game, but it's not so good for players, who lose an average of $10 per game.

Gambler's ruin in probability theory refers to the long-run tendency for players to eventually lose their money when repeatedly playing games with a negative expected value. Even if players experience occasional short-term wins, the average loss per game gradually works against them over many plays. You can see this process in action in the previous graph.

This example shows how casinos can predict their earnings from games of chance when they have high traffic volume. Their earnings will converge on predictable values over many games. You might beat the house with several lucky hands, but in the long run, the house always wins!

In summary, the law of large numbers is a fundamental principle that brings a degree of predictability to the randomness. From flipping coins and casino games to other long-term probabilities, this law ensures that average outcomes align with theoretical probabilities over a large number of trials.

Finally, notice that the running average for the coin toss stabilizes much sooner than the casino game (approximately 90 coin tosses versus 270 games). This difference occurs because the casino game has much greater inherent variability in its individual outcomes. Larger variability causes the average to fluctuate more from one trial to the next, so it takes longer for the average to settle near the expected value. The next section shows you how to quantify this variability using the standard deviation.

Standard Deviation: Variability Around the EV

The Law of Large Numbers explains how averages stabilize in the long run, but it does not reduce the variability of individual outcomes. As more trials occur, the average becomes more stable, even though individual results continue to vary consistently from one trial to the next. To describe how much individual outcomes vary around the expected value, we use a measure of spread called the standard deviation.

The previous graphs should make the idea of variability between outcomes obvious. In the height histogram, there was a peak at about 1.5m, but a clear range of values around it. Similarly, while 7 has the highest probability for the sum of two dice, you can see the distribution tapering off from this peak. In fact, you have an 81.25% chance of rolling a different sum! While the expected value tells you the long-run average, it doesn't capture the variability around it.

The standard deviation provides that measure. It describes the typical distance between individual outcomes and the expected value. While the expected value identifies the center of a distribution, the standard deviation tells us how tightly or loosely outcomes cluster around that center. Distributions with the same expected value can behave very differently depending on their standard deviation.

Standard deviation is derived from a related quantity, the variance, which measures the average squared deviation from the expected value. Variance plays an important role mathematically, but because it is expressed in squared units, it is harder to interpret directly. For that reason, this section focuses on standard deviation as the primary measure of variability in probability distributions.

Use the following formula to find the standard deviation when a probability table specifies the possible values and likelihoods for a discrete probability distribution.

$$\sigma = \sqrt{\sum_{i=1}^{n}(x_i - \mu)^2 \times P(x_i)}$$

Where:

- σ is the standard deviation of the probability distribution.
- μ is the expected value of the random variable.
- x_i represents the possible values the random variable can take.
- $P(x_i)$ is the probability associated with each value x_i.
- The summation runs over all possible values of the random variable.

The calculation follows four steps:

1. Subtract the expected value from each possible value to find the deviation.
2. Square those deviations to ensure all deviations are positive.
3. Weight the squared deviations by probability to reflect how often each deviation occurs.
4. Take the square root to return the result to the original units.

In the casino game example that follows, I'll walk you through these steps.

Example: Sum of Two Dice

Consider the random experiment of rolling two fair six-sided dice and recording the sum of the two values. Earlier, we listed all possible sums and their probabilities. That table defines a discrete probability distribution for the random variable X, the sum of the dice.

The expected value of this distribution is 7. That tells us the distribution's long-run average. To understand how much individual outcomes typically differ from this average, we calculate the standard deviation.

Carrying out this calculation using the probability table for the sums of two dice yields a standard deviation of approximately 2.4.

This result tells us that individual dice sums typically differ from the expected value of 7 by about 2.4 points. While rolls of 7 are the most common, outcomes like 5, 9, 4, or 10 are also quite typical. Much larger deviations, such as rolling a 2 or a 12, are possible but occur less often.

The standard deviation complements the expected value. The expected value identifies the center of the distribution, while the standard deviation describes how spread out the outcomes are around that center. Together, they provide a clearer picture of what to expect from repeated rolls of the dice, including where outcomes tend to cluster and how much variability exists from one roll to the next.

Example: Casino Game

Now consider the casino game introduced earlier, where the random variable represents the casino's payout to the player. The possible payouts and their probabilities are listed in a probability table, which defines the distribution for this game.

From the earlier calculation, we know the expected value of the payout is $240. That value tells us the long-run average payout per play, but it does not tell us how much individual outcomes typically differ from that average. To understand the payout variability, we calculate the standard deviation using the probability table.

We are given the first two columns in the table, and we previously calculated the expected value ($240). That's all we need.

First, in column 3, calculate the difference between each outcome and the expected value (μ). Next, square those differences to produce column 4. In the final column, multiply each squared difference by the probability of the corresponding outcome. Finally, sum the values in the last column and take the square root of that total. That result ($220) is the standard deviation for this probability distribution.

Outcome (X)	Prob. (P)	X - μ	(X – μ)²	(X – μ)² × P
100	0.5	-140	19,600	9800
500	0.4	260	67,600	27040
-100	0.1	-340	115,600	11560
			Sum	48400
			Sq. Root	220

Note: The sum in the table is the variance (σ^2). The square root is the standard deviation (σ).

The standard deviation of $220 indicates that individual payouts tend to deviate substantially from the expected value of $240. Although the expected value describes the long-run average payout, outcomes from a single play can vary widely around that average. Winning $100 or $500, or losing money altogether, are all realistic possibilities on any given play.

This example highlights how expected value and standard deviation describe different aspects of a probability distribution. In this case, the standard deviation is nearly as large as the expected value itself, indicating substantial variability and risk.

Compared to the sum of two dice, where outcomes cluster relatively tightly around the expected value, the casino game shows a much greater spread. This contrast illustrates how two distributions can differ dramatically in variability even when both have a well-defined long-run average.

Together, these measures provide a clearer picture of the game's behavior. While the average payout converges toward $240 over many plays, individual outcomes remain highly variable, which is exactly what players experience.

Interpreting Standard Deviations

When interpreting standard deviation, it is important to keep its role in mind. Standard deviation describes how much individual outcomes typically differ from the expected value. A small standard deviation indicates that outcomes tend to cluster closely around the average, while a large standard deviation signals that outcomes are more spread out and variable.

Interpret the standard deviation relative to the expected value and the context of the problem. A large standard deviation relative to the expected value suggests substantial volatility and risk, even if the long-run average is well defined. Conversely, a small standard deviation indicates more stable outcomes from one trial to the next.

In this chapter, we calculated standard deviations directly from probability tables to emphasize interpretation. In later chapters, you will see that named probability distributions have their own formulas for calculating standard deviation efficiently. Those results build on the same ideas developed here, but allow you to calculate variability more easily for specific probability models.

Review and Next Steps

In this chapter, you shifted from analyzing individual events to using probability distributions to model outcome patterns generated by a random process. By connecting random experiments, random variables, and probability distributions, you now have a framework for understanding how uncertainty behaves over many repetitions.

You saw how tables, graphs, and cumulative distribution functions can represent discrete probability distributions, each offering a different perspective on the same underlying information. You then explored how distributions behave through key summary measures. The expected value identifies the long-run average outcome, the Law of Large Numbers explains why observed averages stabilize around that value, and the standard deviation describes how much individual outcomes typically vary around the average.

Together, these tools show why randomness can produce unpredictable individual outcomes while still exhibiting stable long-run patterns. They also highlight an important theme: understanding randomness means understanding both its structure and its variability, not eliminating uncertainty.

In the next chapters, we will build on this foundation by studying specific probability models. You will see how named distributions, such as the binomial distribution, provide efficient methods for calculating probabilities, expected values, and standard deviations. The goal of modeling randomness remains the same, but with tools tailored to specific types of random processes.

- Probability distributions model randomness by describing patterns across all possible outcomes rather than focusing on individual events.
- Random experiments generate uncertain outcomes, random variables translate those outcomes into numerical values, and probability distributions summarize how those values behave in the long run.
- Cumulative distribution functions are especially useful for answering "at most" questions and for identifying percentiles, which describe relative standing within a distribution.
- Use complements for "greater than" and differences for "between" when working with cumulative probabilities.
- The expected value represents the long-run average outcome of a random process.
- The Law of Large Numbers says the observed averages tend to stabilize near the expected value as the number of trials increases.
- Standard deviation measures how much individual outcomes typically differ from the expected value.

- Expected value and standard deviation together describe both the center and spread of a distribution, offering a more complete picture of how randomness behaves in practice.

Formula Quick Reference

Measure	Formula
Expected value	$\mu = \Sigma\,[x \times P(x)]$
Standard deviation	$\sigma = \sqrt{(\Sigma\,[(x - \mu)^2 \times P(x)])}$

Chapter 10 Practice Problems

Question 1: A bookstore tracks the number of books a customer buys during a visit. Let X represent the number of books purchased by a randomly selected customer. Based on the probability distribution for X, the expected value is:

E(X) = 1.8 books

Answer the following questions.

A. Explain what an expected value of 1.8 books means in this context.

B. Is 1.8 books a valid expected value even though a customer cannot buy 1.8 books? Explain why or why not.

Question 2: A streaming service tracks the number of hours users watch content per week. Two different user groups have the same average viewing time of 10 hours per week, but different standard deviations:

- Group A: standard deviation = 1.5 hours
- Group B: standard deviation = 6 hours

Answer the following questions.

A. Which group has more consistent viewing behavior? Explain what the standard deviation tells you about that group.

B. Which group is more likely to include users with very low or very high viewing times? Explain why.

Question 3: A delivery company tracks the number of packages that arrive late on a given day. Let X be the number of late packages per day.

x	P(X ≤ x)	P(X = x)
0	0.30	
1	0.55	
2	0.80	
3	0.95	
4	1.00	

Assume these are all the possible values of X.

A. Use the supplied values to fill in the missing values in the table.

B. Verify that your probability distribution is valid.

C. What is the probability that exactly 2 packages are late on a given day?

D. What is the probability that more than 2 packages are late?

E. What is the probability that between 1 and 3 packages (inclusive) are late?

Question 4: A casino offers a slot machine that costs $1.00 per play. The machine has the following payout distribution, where the random variable X represents the amount paid back to the player on a single play.

Payout (X)	Probability
$0.00	0.489
$0.50	0.260
$1.00	0.100
$2.00	0.100
$5.00	0.040
$20.00	0.010
$100.00	0.001

Assume this table includes all possible outcomes for one play of the machine.

- **A.** Verify that this is a valid probability distribution.
- **B.** Calculate the expected value of the payout for one play of the machine.
- **C.** Because the machine costs $1.00 per play, determine the player's expected net gain or loss per play.
- **D.** Based on the expected value, do players tend to make money or lose money in the long run? Explain why and state how much they gain or lose on average per play.
- **E.** Calculate the standard deviation of the payout distribution.
- **F.** Casinos sometimes advertise the payout percentage, which is the percentage of money played that the machine returns to players in the long run. Based on this probability distribution, what is the machine's payout percentage?
- **G.** Based on the expected value and standard deviation, briefly describe the overall behavior of this slot machine in terms of average return and variability.

CHAPTER 11

Probability Distributions for Binary Outcomes

In Chapter 10, you learned how probability distributions describe patterns that emerge from random processes rather than individual outcomes. You also saw how expected value and variability summarize what randomness looks like in the long run. This chapter applies those ideas to one of the most important types of data: binary outcomes.

Binary random variables arise when each observation has only two possible results, such as success or failure, yes or no, or defective or not defective. When these outcomes occur repeatedly under well-defined conditions, probability distributions allow us to describe how counts of successes or waiting times until one or more successes behave across many repetitions of the process.

Up to this point, you've seen probability distributions shown as tables listing every possible outcome and its probability, which works well when the number of outcomes is small and fixed.

In this chapter, we begin working with distributions where a formula defines the probabilities rather than a table. These distributions use a function that generates probabilities after you specify a few key values that describe the process. Statisticians refer to these values as parameters and they summarize how the random process behaves. Plug these parameters into the formula to find the outcome probabilities.

I'll show you how to use the binomial, geometric, negative binomial, and hypergeometric distributions to find probabilities for binary random variables. Each distribution reflects a different way of organizing random outcomes and answers different types of questions. For example:

- How many times is an event likely to occur?
- When is the first instance probable?
- How many opportunities do I need to produce a specific number of events?

However, as we covered previously, the power of probability distributions isn't just the ability to find the likelihood for a particular outcome but their ability to show the center, spread, and overall shape of all possible outcomes. That's what makes them distributions.

While I do cover the formulas and manual calculations, you'll usually use statistical software to do that. Therefore, the real focus is on knowing when and how to use each distribution and interpreting the results. For each binary probability distribution, I start by focusing on the type of information it provides, how to interpret its results, and how to use it to solve probability problems with a distribution calculator.

At the end of each distribution's section, I cover the formula and perform example calculations. That helps you see how these distributions work, rather than leaving them as "black boxes" that spit out answers.

Further, it is important to understand what these distributions represent conceptually. These binary distributions are not models you "fit" to data after the fact. Instead, each distribution applies by design when the data come from a process satisfying certain assumptions. When those conditions are met, the distribution naturally follows from the process's structure.

Binary Data Assumptions

A binary random variable exists when you can place each observation into only two categories. These two values indicate whether an event occurred or whether an item has a particular characteristic. For example, an inspection process produces binary pass/fail results. When a customer enters a store, there are two possible outcomes: sale or no sale. In a flu vaccination study, the outcomes for each participant are either "infected" or "not infected."

To be a binary variable, the two outcomes must be mutually exclusive. Each observation must fall into one, and only one, of the two categories. Outcomes cannot be classified as both categories or neither category.

At a basic level, binary data allow you to calculate proportions and percentages easily. What proportion of items pass inspection? What percentage of customers make a purchase? What proportion of people become infected with the flu?

As discussed in Chapter 10, probability distributions are based on random experiments, which are repeatable processes where each trial produces an outcome that can't be predicted in advance. While future outcomes are unknowable, binary random experiments follow a defined structure that produces a long-run pattern of outcomes. The assumptions in this section specify that structure.

To use the binomial, geometric, negative binomial, and hypergeometric distributions, the data must arise from a well-defined sequence of trials that satisfy the following assumptions.

There are only two possible outcomes per trial. For example, accept or reject, sale or no sale, etc.

Each trial is independent (except for the hypergeometric distribution). The result of one trial does not affect the results of another trial. For instance, when flipping a coin, the outcome of a coin toss doesn't influence the next coin toss.

The probability remains constant over time (except for hypergeometric). In some cases, this assumption is valid based on the physical properties, such as flipping a coin. In other settings, the probability might change over time. When that is a concern, tools such as a P chart (a type of control chart) can help assess whether the probability remains stable. For example, the likelihood that a process produces defective products might change over time.

In addition, the definition of a trial and what counts as a success must be fixed in advance and applied consistently throughout the process. If trial definitions or outcome classifications change partway through the process, the random experiment itself has changed, and these distributions no longer apply.

In short, these distributions rely on binary outcomes, independent trials, and a constant probability of success. The hypergeometric distribution is the exception to the last two assumptions, and we'll return to that distinction later.

For most of these distributions, I'll use a die-rolling example. Assume we're playing a game where rolling a 6 is advantageous. In this scenario, rolling a 6 is binary because an observation can be either a 6 or not. The probability of rolling a 6 is 1/6, or about 0.1667. We'll use the binomial, geometric, and negative binomial distributions to calculate probabilities for how many 6s we'll roll, when they'll first appear, and the likelihood of observing a certain number of 6s.

I'll show you when to use each distribution and how to interpret the results. I'll cover both how to understand the probability for a specific outcome and the cumulative probability for a range of outcomes.

Binomial Distribution

The binomial distribution is a discrete probability distribution that calculates the likelihood that an event will occur a specific number of times in a set number of opportunities. It is an extension of the Bernoulli distribution that can model only 1 trial.

For example, the binomial probability distribution can answer the following questions. What is the probability of getting:

- Six heads when you toss the coin ten times?
- 12 women in a sample size of 20?
- Three defective items in a batch of 100?
- Two flu infections over 20 years?

Assumptions, Parameters, and Notation

The binomial distribution models the probability of exactly x successes in n trials.

You'll use the binomial distribution when you have Bernoulli Trials. These trials involve random variables that follow the assumptions discussed earlier: binary outcomes, independent trials, and a constant probability of success. Analysts label one outcome as a success (i.e., the event) and the other as a failure.

A Bernoulli trial is a single trial with two possible outcomes. A binomial experiment consists of a fixed number of Bernoulli trials where the probability of success remains constant. The experiment counts the number of successes (x) out of the total number of trials (n).

You can think of the binomial distribution as modeling the number of successes (x) in a sample size of n when the probability of an event is known and constant.

The binomial distribution has two parameters, n and p.

- **n**: the number of trials.
- **p**: the event or success probability.

You denote a binomial distribution as b(n,p).

Alternatively, you can write X~b(n,p), which means that your random variable X follows a binomial probability distribution with n trials and an event probability of p.

For example, we define success as rolling a six on a die, and failure as rolling anything other than a six. The probability of rolling a six is 1/6 = 0.1667.

If rolling sixes is our random variable X, and we roll the die ten times, we can use the following notation for the binomial distribution that models the number of sixes:

X~b(10, 0.1667)

Using and Interpretations

Let's get into some examples because that brings it to life!

I'll start by using statistical software to calculate the binomial probabilities and create distribution plots. This process will help you understand what you can learn from it. Later, I'll walk you through the calculations so you can see how this distribution works.

Suppose you're playing a game where rolling sixes on a die is really good. You want to know the probability of rolling exactly three sixes in ten die rolls. In this example, the number of events is 3 (x), the number of trials is 10 (n), and the probability (p) is 1/6 = 0.1667.

My statistical software tells me that the likelihood is:

Probability Density Function

```
Binomial with n = 10 and p = 0.1667

X  P( X = x )
3     0.155095
```

The binomial distribution calculates a probability of 0.155095 for rolling precisely three sixes in ten rolls. Over the long run, about 15.5% of samples containing 10 die rolls will have exactly three sixes.

That's interesting, but perhaps not so helpful by itself. We're also interested in the chances of rolling other numbers of sixes. Seeing the distribution of probabilities for all possible numbers of sixes is much more helpful. In general, seeing the overall outcome pattern is a key advantage of using probability distributions.

The binomial distribution graph is useful because it displays the probability of all numbers of successes (x's) out of the total number of trials (n). The following graph shows the likelihood of rolling exactly 0 sixes, 1 six, 2 sixes, ..., and up to 10 sixes in ten die rolls. Using this approach, the chart covers the complete range of possible successes up to the total number of trials.

I like these graphs because they emphasize how we're working with a distribution, and it's easy to see which values happen more frequently.

In the chart, each bar represents the probability of rolling a specific number of sixes out of ten die rolls. The graph does not show the chances for seven and higher because the likelihoods of that many sixes in just ten rolls are too low to display on the chart.

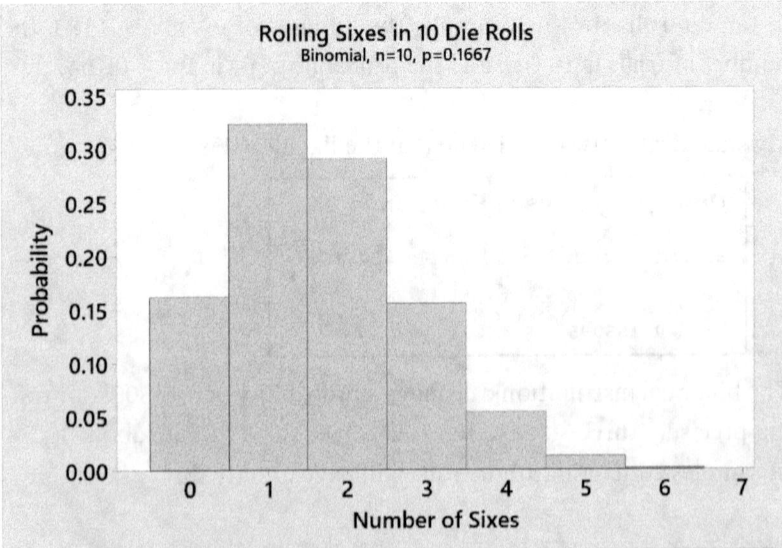

The graph indicates the probability of rolling no sixes is about 16%. The highest chance is rolling one six (32%). Although rolling two sixes occurs almost as frequently. Probabilities drop off quickly starting with three sixes. Additionally, the bar for three sixes matches our earlier result of 0.155095.

The binomial probability distribution is excellent for understanding the likelihood of obtaining an exact number of events (x) within a certain number of trials (n). However, many times you're not interested in just one specific value for a binomial random variable. For example, in the die rolling context, suppose you know from experience that rolling three or more sixes within ten rolls means you're doing well. So, you might want to learn the probability of rolling *at least* three sixes.

Finding probabilities for ranges of values involves using the cumulative distribution function, which keeps a running total of probabilities as the number of events increases. Recall from Chapter 10 that cumulative distributions are great for "at most" questions, and their complements are perfect for "at least" questions. For at least three sixes, we need to use the complement.

The following graph displays the cumulative probability of rolling sixes in ten die rolls.

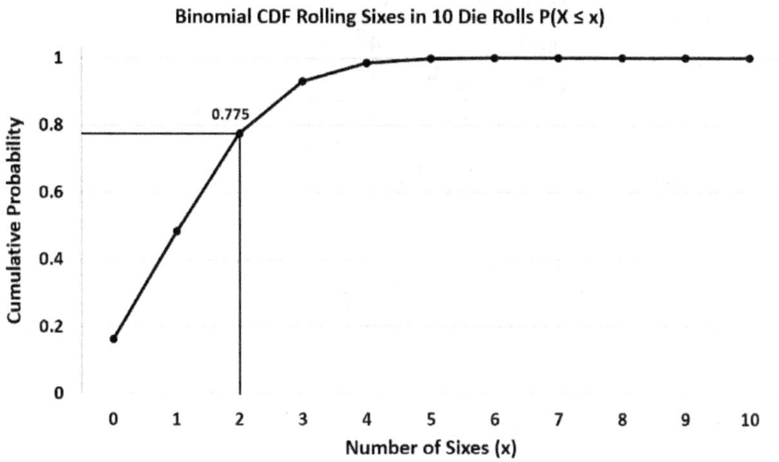

To answer the question of rolling at least three sixes, we start by finding the cumulative probability of rolling two or fewer sixes. Why two? Because rolling at least three sixes includes every outcome *except* those with zero, one, or two sixes. In other words, the probability of rolling at least three sixes is the complement of the cumulative probability of rolling at most two sixes.

The cumulative probability for rolling two sixes is 0.775. Subtract that from 1 to calculate the ≥ 3 sixes: 1 - 0.775 = 0.225. Hence, the likelihood for rolling three or more sixes in ten rolls is 0.225, not quite 1 in 4.

As you can tell, I really like graphing these distributions because a picture is worth 1,000 words. However, you are also likely to see tables that present the same information. The following table displays both the exact and cumulative probabilities for the number of sixes rolled.

Note that the table covers the full range of values from 0 to 10 die rolls, while the graph doesn't display the extremely low probabilities for rolls 7 through 10. Additionally, the cumulative probability for 9 sixes has been rounded up to 1, but it is infinitesimally lower than 1.

x = # of 6s	P(X = x)	P(X ≤ x)
0	0.161441	0.161441
1	0.322959	0.484400
2	0.290733	0.775134
3	0.155095	0.930229
4	0.054296	0.984525
5	0.013034	0.997559
6	0.002173	0.999732
7	0.000248	0.999981
8	1.86E-05	0.999999
9	8.28E-07	1
10	1.66E-08	1

Using a Binomial Distribution Calculator

Next, we'll use my binomial distribution calculator to calculate the binomial probabilities and cumulative probabilities. Find the link in the book's Introduction for all the free resources, including the calculators.

Let's use this calculator to recreate the preceding die examples where we found the probability of rolling exactly three sixes and at least three sixes in 10 die rolls. In the calculator, enter the following and click **Calculate**:

- **Probability (p)**: 0.1667
- **Number of Trials (n)**: 10
- **Number of Events (x)**: 3

The calculator's numeric output displays probabilities for x = 3 (three events), including exactly, less than or equal to, and greater than. The graph shows the probabilities for all possible values of X from 0 to 10, where each bar represents the probability of observing that number of events in 10 trials. The graph highlights the bar corresponding to our x-value. It excludes x-values with probabilities too small to display.

To find the probability for three sixes exactly, look in the numeric output for $P(X = 3)$, which is 0.1551, matching our previous results within rounding.

Next, the calculator also displays $P(X \geq 3) = 0.2249$, which matches the results for at least three sixes in 10 die rolls using the complement of the cumulative distribution.

Now, try one yourself. Imagine you're drawing a random sample of 20 from a population where 10% are statisticians. You're hoping that your study will have 3 or fewer statisticians because they'll gang up and ask too many pesky questions about your study design. What is the likelihood of obtaining ≤ 3 statisticians?

See the correct answer at the end of the binomial distribution section.

Tips for Solving Binomial Probability Problems

Now that you know how to interpret binomial probabilities, you'll need to know how to solve probability problems using them. These problems typically come in several standard flavors, so know what cues to watch for.

Exactly x Successes: P(X = x)

Use this approach when a probability question asks for the likelihood of exactly a specific number of successes in a fixed number of trials. What to look for in the problem statement:

- Words such as *exactly*, *precisely*, or *equal to.*
- A fixed number of trials.
- Two possible outcomes per trial.
- A constant probability of success.

These hints indicate that the problem seeks the binomial probability for a single value.

How to solve it:

1. Identify n (trials), x (successes), and p (probability of success).
2. Calculate $P(X = x)$ using the binomial probability formula or calculator.
3. Interpret the result as the long-run proportion of samples with exactly that many successes.

This type of problem is the most basic binomial probability.

Tip: Use the 'exactly x' method for problems where $x = 0$.

At Most x Successes: $P(X \le x)$

Use this approach when a probability question asks for the probability of observing no more than a certain number of successes.

Look for a problem statement that includes phrases such as *at most*, *no more than*, *X or fewer*, or *less than or equal to.*

These clues indicate that the question is asking for a cumulative probability up to and including X.

How to solve it:

1. Identify n, X, and p.
2. Use a cumulative binomial probability directly, or sum the probabilities for all values from 0 through X.
3. Interpret the result as the long-run proportion of samples with at most x successes.

At Least x Successes: $P(X \geq x)$

Use this approach when a probability question asks for the likelihood of observing x or more successes.

Look for problem statements that include phrases such as *at least*, *x or more*, or *greater than or equal to.*

These clues indicate that the solution requires taking the complement of a cumulative probability.

How to solve it:

1. Identify n, X, and p.
2. Calculate the cumulative probability for $x - 1$ or fewer successes.
3. Subtract that probability from 1 to obtain $P(X \geq x)$.
4. Interpret the result as the long-run proportion of samples with x or more successes.

Note: Step 2 is the tricky part that I called out in Chapter 10. For discrete distributions, shifting by $x - 1$ produces the correct complementary probability for $\geq x$.

Between Two Values

Use this approach for probability questions about successes within a range. Look for phrases like *between*, *from A to B*, *at least*, *at most*, or inequalities such as $P(A \le x \le B)$, $P(A < x \le B)$, $P(A \le x < B)$, or $P(A < x < B)$. These problems define both a lower and an upper boundary.

To find probabilities between two values, use cumulative probabilities and subtraction. The upper boundary gives the total probability up to that point, while the lower boundary determines what to remove.

For discrete distributions, probabilities occur at whole-number values. Adjust boundary values depending on whether they are included.

For the upper boundary, use the largest value included in the range. For the lower boundary, subtract all values below the range, which might require including or excluding the boundary value itself.

How to solve it

1. Identify n, p, and the lower and upper boundary values.
2. Determine whether each boundary is inclusive or exclusive.
3. Find the cumulative probability for the upper boundary using the appropriate value:
   - Inclusive: $X \le B$: use $P(X \le B)$.
   - Exclusive: $X < B$: use $P(X \le B - 1)$.
4. Find the cumulative probability for the upper boundary using the appropriate value:
   - Inclusive: $X \ge A$: use $P(X \le A - 1)$.
   - Exclusive: $X > A$: use $P(X \le A)$.
5. Subtract the lower cumulative probability from the upper cumulative probability.
6. Interpret the result as the long-run proportion of samples with successes in that range.

Tip on Inequalities and Endpoints

When working with ranges, always determine whether the endpoints are inclusive or exclusive. Phrases like *at least* and *at most* include the stated value, while phrases like *greater than* or *less than* exclude it.

A helpful strategy is to first rewrite the problem using an explicit inequality (such as $X \geq A$ or $X < B$). This makes it easier to identify the correct boundary values and avoid off-by-one errors.

These same strategies apply to the geometric, negative binomial, hypergeometric, and Poisson distributions that follow.

Now that you've seen what the binomial distribution describes and how its probabilities behave, we'll briefly look at *how* the binomial probability formula produces these results.

Formula and Manual Calculation Example

Next, I'll walk you through the formulas for finding the probability, expected value, and standard deviation of the binomial distribution and provide worked examples. As I promised in Chapter 9, you'll use combinations again. This counting method is an essential part of the binomial formula.

The binomial distribution formula for probabilities is the following:

$$\mathrm{P(x)} = \mathrm{nCx} \times \mathrm{p}^{x} \times (1 - \mathrm{p})^{n-x}$$

$$nCx = \frac{n!}{x!\,(n-x)!}$$

Where:

- n: number of trials.
- x: number of successes
- p: probability of a success.
- (1 – p): probability of failure.

Use this formula to calculate the probability of observing exactly x successes in n trials.

Binomial Probability = (number of ways to obtain x successes) × (probability of any one such outcome)

nCx counts the number of ways to have exactly x successes among n trials. Because the order of outcomes does not matter for this count, it represents the number of combinations without repetition.

Each of these combinations has the same probability. That probability equals the probability of success raised to the number of successes, multiplied by the probability of failure raised to the number of failures.

The binomial formula multiplies the number of possible combinations by the probability of any one of them to obtain the total probability of observing exactly x successes.

Let's work through an example calculation to show the formula in action!

For this example, we'll use the binomial formula to calculate the probability of rolling exactly three sixes in ten die rolls. Here are the values to enter into the formula:

- $n = 10$
- $x = 3$
- $p = 0.1667$
- $(1-p) = 0.8333$

For the number of combinations, we have:

$$10C3 = \frac{10!}{3!(10-3)!} = 120$$

Now, let's enter our values into the binomial distribution formula.

$$\begin{aligned} P(x) &= nCx \times p^x \times (1-p)^{n-x} \\ &= 120 \times 0.1667^3 \times 0.8333^7 \\ &= 120 \times 0.0046 \times 0.2790 = 0.1540 \end{aligned}$$

This calculation finds that the probability of rolling three 6s in 10 rolls is 0.1540.

If you need to calculate a cumulative probability for a binomial random variable, calculate the likelihood for each individual outcome and then sum them for all outcomes of interest.

For example, to calculate the probability of at least 3 sixes in 10 rolls, you could find the probabilities for 3, 4, up to 10 sixes and sum them. A more efficient approach is to use the complement: subtract $P(X \leq 2)$ from 1, which requires fewer calculations.

Expected Value

The formula for the expected value of the binomial distribution is the following:

$n \times p$

Multiply the number of trials (n) by the success probability (p). This value represents the average or expected number of successes.

For example, if you roll a die ten times, and the probability of rolling a six is 0.1667, plug these into the formula.

$E(X) = 10 \times 0.1667 = 1.667$

The mean for this binomial distribution is 1.667. On average, we'd expect to roll this many sixes in ten rolls. Of course, the actual counts of successes will always be either zero or a positive integer.

Standard Deviation

The binomial distribution formula for the standard deviation is the following:

$$\mathrm{SD(X)} = \sqrt{n \times p \times (1 - p)}$$

As before, n and p are the number of trials and success probability, respectively. (1 – p) is the likelihood of failure.

The standard deviation of the binomial distribution is a function of the number of trials (n) and the probabilities of success (p) and failure (1 – p). For a fixed number of trials, the standard deviation is largest when the probabilities of success and failure are both 0.5. As those probabilities move away from 0.5 in opposite directions, the standard deviation decreases. Additionally, the standard deviation increases as the number of trials (n) increases.

For our die example, we have n = 10 rolls, a success probability of p = 0.1667, and a failure probability of (1 – p) = 0.833. Let's enter these values into the formula.

$10 \times 0.1667 \times 0.8333 = 1.3891$. That's the variance, which uses squared units.

Take the square root of the variance to find the standard deviation.

$SD(X) = \sqrt{1.3891} = 1.1786$

The standard deviation measures the variability of the probabilities around the expected value of the binomial distribution.

Revisiting the 100 Repeated Independent Attempts Example

In Chapter 4, we assessed the probability of at least one success when you have a 1-in-100 chance of success and repeat the attempt 100 times. The answer, 0.634, was surprising because it seems like a success should be nearly guaranteed after so many attempts.

Now that you have more statistical tools at your disposal, we can examine this result more deeply. At its core, the problem follows a binomial distribution because it involves a fixed number of independent attempts with a constant probability of success. Rather than focusing on a single sequence of 100 attempts, the binomial distribution allows us to examine the full range of possible outcomes across many such sequences.

We know that the expected value for a binomial distribution equals:

$E(X) = n \times p$

For this example:
$E(X) = 100 \times 0.01 = 1$

Our intuition was correct. You expect 1 success in every 100 attempts . . . on average. And there's the rub. We must also consider the variability around that average. Even though you expect one success out of 100 attempts, it won't work out that way exactly due to random variability.

Let's calculate the standard deviation to see how much the outcomes will vary around the EV of 1.

$$SD(X) = \sqrt{100 \times 0.01 \times (1 - 0.01)} = 0.995$$

Notice that the standard deviation is nearly as large as the expected value itself. Consequently, outcomes such as 0, 1, or 2 successes are all quite plausible.

Now let's examine the full binomial distribution.

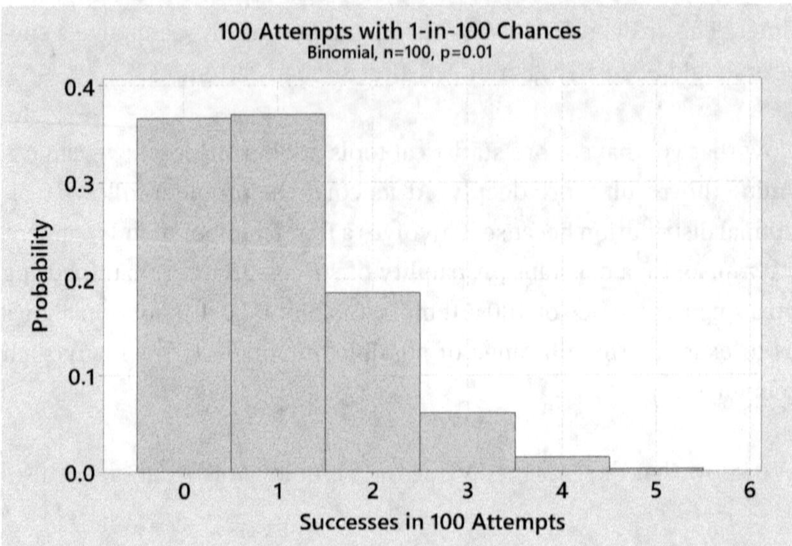

The graph is very revealing. We see that 1 success *is* the single most likely outcome, but it only barely beats out 0 successes (37.0% vs. 36.6%). Even two successes still occur with a respectable probability of 18.5%.

The original example assessed the probability of at least 1 success, which is the complement of 0 successes. The first bar in the distribution represents 0 successes with probability 0.366. Therefore:

P(at least 1 success) = 1 – 0.366 = 0.634

That result matches our previous conclusion. Here, the key insight is that the expected value describes the long-run average across many sets of 100 attempts, not the exact outcome of any single set. Random variability spreads the successes unevenly across different sequences. Some groups of 100 attempts contain no successes, while others contain several.

CALCULATOR SOLUTION: There is an 86.7% chance of having ≤ 3 statisticians in your sample of 20 people. Equivalently, you can say that 86.7% of all random samples of 20 people drawn from that population will have ≤ 3 statisticians.

Geometric Distribution

The geometric distribution calculates the probability that the first success occurs on a specific trial. Use this distribution when you need to understand how many attempts are necessary to produce the first successful outcome.

For example, the geometric distribution can answer the following questions. What is the probability of the first:

- Six in a series of die rolls?
- Year of catching the flu over the years?
- Person to support a law during a repeated sampling for an interview?
- Product to have a defect in a random sample from an assembly line?
- Successful attempt for a project or task?

The geometric distribution is "memoryless." Memoryless is a distribution property indicating that the probability of the next success does not depend on how many trials have already occurred without a success.

For example, when you start observing die rolls looking for the next six, the probability of when the next six appears does not depend on when the last six appeared or when you start watching for sixes.

Assumptions, Parameters, and Notation

The geometric distribution models the probabilities of a random variable that counts the number of trials required to obtain the first success. This distribution uses the same Bernoulli trial framework as the binomial distribution. Hence, it applies only when the process satisfies the assumptions of binary, independent outcomes that have a constant success probability.

The key difference from the binomial distribution lies in what the random variable measures. Instead of fixing the number of trials and counting successes, the geometric distribution fixes the number of successes at 1 and counts the number of trials needed to achieve that first success.

You can think of the geometric distribution as modeling the time until the first success.

The geometric distribution has a single parameter, p.

- p: the probability of success on each trial.

You denote a geometric distribution as g(p).

Alternatively, you can write $X \sim g(p)$, which means that your random variable X follows a geometric probability distribution with a success probability of p.

Using the same die-rolling example as before, suppose success is defined as rolling a six, with probability $p = 1/6 = 0.1667$. If X represents the number of rolls required until the first six appears, then the geometric distribution is written as:

$X \sim g(0.1667)$

Note that there are two forms of the geometric distribution:

- The first success occurring on the xth trial.
- The number of failures before the first success.

Fortunately, these two forms are equivalent if you interpret X correctly. The difference comes down to how you count the trials.

The former definition counts all trials, including the final, successful one. The latter definition counts only the failures before the success. Success on the xth trial is equivalent to having $x - 1$ failures. Statisticians sometimes refer to the first form as the shifted geometric distribution because it moves the final trial over by counting the success.

Suppose I want to calculate the probability of first rolling a six on the third trial. Using the number of failures form, we don't count the successful trial, only the two preceding failures. Hence, if we succeed on the third roll, we have two failures, and the two forms of the geometric distribution give the same probability.

I use the success on the xth trial version throughout this book.

Using and Interpretations

The geometric distribution models the probabilities for the first event occurring during a specific trial when the likelihood of an event is known. Let's bring it to life with an example!

As I did for the binomial distribution, I'll start by using statistical software to calculate a single geometric probability, then move on to distribution plots and manual calculations.

We'll again use rolling a six on a die as our scenario, with a success probability of 1/6 = 0.1667. We're interested in the first six occurring on the third roll.

My software displays the probability as the following:

Probability Density Function

```
Geometric with p = 0.1667

X  P( X = x )
3    0.115755

* NOTE * X = total number of trials.
```

The geometric distribution gives a probability of 0.115755 that the first six will occur on the third roll. Over the long run, about 11.6% of sequences will have the first six occur on the third roll.

Please note the full implications of this scenario. The final probability accounts for the fact that the first two rolls are not sixes, and the third roll is a six. That's the only sequence of events allowing the first success to occur on the third attempt. Consequently, the calculations factor in the probability of the event not occurring on certain trials (failures) and then appearing on a specific attempt (success).

As with the binomial distribution example, let's explore the probability of rolling the first six on other rolls.

The geometric distribution graph displays the probability of rolling the first six in precisely 1, 2, 3, etc. rolls, up to 30.

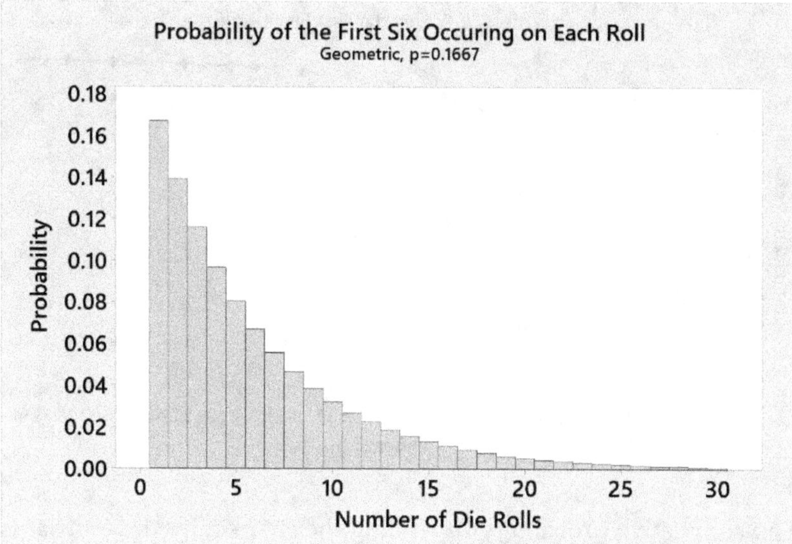

Each bar in the graph shows the probability of rolling the first six on the given trial. For example, the likelihood of rolling the first six on the third roll is 0.115755. This value matches the previous statistical output. The distribution continues past 30 rolls, but those probabilities are infinitesimal.

The geometric distribution helps you understand when an event might first occur.

However, in practice, you might be not interested in a specific trial. For example, you could be interested in the probability of getting your first six within the first six rolls. In that case, you're interested in finding the total probability across multiple trials using the cumulative function.

For our example, we want to learn the cumulative probability that the first 6 appears within the first six rolls (i.e., ≤ 6). The following graph displays the cumulative probability function for rolling the first six.

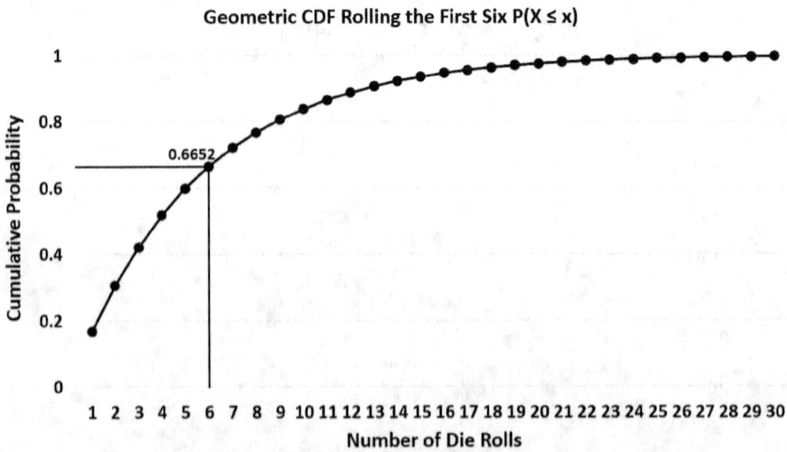

The cumulative probability for rolling the first six by the 6th trial is 0.6652.

Interestingly, you might think you're virtually guaranteed to get a 6 when you roll the die six times. However, the cumulative function indicates you have a two-thirds (66%) chance of getting the first six within the first six rolls.

Using a Geometric Distribution Calculator

Now we'll use my geometric distribution calculator to calculate probabilities and cumulative probabilities. Find the link in the book's Introduction for all the free resources, including the calculators.

For the die example, I calculated the probability of the first success (rolling a six) occurring on the third trial as 0.115755. Let's use the calculator to solve this problem.

In the calculator, enter the following and click **Calculate**:

- **Probability of Success (p)**: 0.1667
- **Number of Trials (x)**: 3
- Be sure the "*Include success trial*" box is checked.

The calculator's numeric output displays probabilities for x = 3 (three trials until the first success), including exactly, less than or equal to, and greater than. The graph shows the probabilities for all possible values of die rolls (X), where each bar represents the probability that the first success occurs on that number of trials. The graph highlights the bar corresponding to our x-value. It excludes x-values with probabilities too small to display.

To find the probability for three sixes exactly, look in the numeric output for P(X = 3), which is 0.1158, matching our previous results within rounding.

Now, try one yourself. Suppose you have a team working on a project, and you give them four attempts to complete the project successfully due to time constraints. It's a very complex task, and they have only a 30% chance of correctly completing the project on each attempt. If they fail, they'll need to try again. What is the cumulative probability of succeeding on attempts 1 – 4?

The answer is at the end of the geometric distribution section!

Formula and Manual Calculation Example

Now, I'll walk you through the formulas for the geometric distribution with the worked examples.

The geometric distribution formula is the following:

$$P(x) = (1 - p)^{x-1} \times p$$

Where:

- x is the number of trials.
- p is the probability of a success for each trial.

For the geometric distribution, the random variable counts the trial on which the first success occurs. For a given value of x, there is only one possible sequence of outcomes: x – 1 failures followed by a success.

Because this sequence is fixed, there are no combinations to count. That's why the geometric distribution formula does not include a binomial coefficient. Instead, the formula takes the probability of failure (1 – p) and raises it to the power of the number of failures (x – 1). This finds the likelihood of failures before the trial of interest. The formula then multiplies that by the probability of success, p, on the xth trial.

That gives you the probability of the first success happening on the xth trial.

Let's work through an example calculation to bring the formula to life!

We'll use the geometric distribution formula to calculate the probability of rolling the first six on the third roll. That's the example we used before, and now we'll calculate it by hand. Here are the values to enter into the formula:

- $x = 3$
- $p = 0.1667$

$$\begin{aligned} P(x) &= (1-p)^{x-1}p \\ &= 0.8333^2 \times 0.1667 \\ &= 0.6944 \times 0.1667 = 0.1158 \end{aligned}$$

This calculation confirms the previous statistical software results within rounding error.

If you need to find the cumulative geometric probability for a range of trials, calculate the probability for each attempt and then sum the probabilities.

Expected Value

For the geometric distribution, the expected value represents the average number of trials required to obtain the first success.

The formula for the geometric distribution EV is the following:

$$E(X) = \frac{1}{p}$$

This formula depends only on the probability of success, p.

For example, suppose success is rolling a six, with probability p = 0.1667. Entering this value into the formula gives:

$$E(X) = \frac{1}{0.1667} = 6$$

The expected value is 6. On average, we would expect the first six to occur on the sixth roll.

Standard Deviation

The standard deviation for the geometric distribution measures how much the number of trials needed to obtain the first success varies across repeated sequences.

The formula for the standard deviation of a geometric distribution is the following:

$$SD(X) = \sqrt{\frac{1-p}{p^2}}$$

As before, p is the probability of success on each trial, and 1 - p is the probability of failure.

Using our die example with p = 0.1667 and 1 - p = 0.8333, substitute these values into the formula:

$$SD(X) = \sqrt{\frac{0.8333}{(0.1667)^2}} = 5.476$$

This produces a standard deviation of approximately 5.476.

The standard deviation describes the amount of variability in the number of trials required to obtain the first success. Because waiting times for the first success can vary substantially, the geometric distribution often has a relatively large standard deviation compared to its expected value. Here, the standard deviation of 5.476 is nearly equal to the expected value of 6.

CALCULATOR SOLUTION: Finally, the solution to the problem in the geometric distribution calculator section is that the cumulative probability of completing the project successfully across attempts 1–4 is 0.7599.

Negative Binomial Distribution

The negative binomial distribution calculates the probability that a specified number of events occurs on a particular trial. When you provide an event probability and the number of successes (r), this distribution calculates the likelihood of observing the rth success on the xth trial. Statisticians also refer to this distribution as the Pascal distribution.

The negative binomial distribution is a generalization of the geometric distribution. The geometric distribution models the number of trials required to obtain the first success, while the negative binomial extends this idea to the number of trials required to obtain the rth success. In this sense, the geometric distribution is a special case of the negative binomial distribution when r = 1.

For example, the negative binomial distribution can answer the following questions. What is the probability of the following:

- Rolling the 5th six on the 20th roll of a die?
- Getting the 10th defective item on the 1000th item inspected?
- Selecting the 10th woman as the 15th participant?

The key distinction lies in what the random variable measures. Instead of fixing the number of trials and counting successes (as in the binomial distribution), or stopping after the first success (as in the geometric distribution), the negative binomial distribution fixes the number of successes and counts how many trials are required to achieve them.

Assumptions, Parameters, and Notation

Like the binomial and geometric distributions, the underlying process must have independent binary outcomes with a constant success probability.

You can think of the negative binomial distribution as modeling the waiting time until the rth success occurs.

The negative binomial distribution has two parameters, r and p.

- r: the number of successes of interest.
- p: the probability of success on each trial.

You denote a negative binomial distribution as nb(r,p).

Alternatively, you can write X ~ nb(r,p), which means that your random variable X follows a negative binomial probability distribution with r successes and a success probability of p.

For example, rolling a six on a die is our random variable X, we specify 5 successes, and the probability is 0.1667. The negative binomial distribution notation for this scenario is the following:

X~nb(5, 0.1667)

Using and Interpretations

The negative binomial distribution models the probabilities for the rth success occurring on the xth trial when you know the event probability. Let's bring it to life with an example!

I'll use the same sequence of explaining an individual probability, distribution, cumulative function, and formulas as before.

Suppose you want to find the probability of getting the 5th six on the 20th roll. In this example, the number of successes is 5 (r), the number of trials is 20 (x), and the success probability (p) is 1/6 = 0.1667.

The software indicates that the likelihood is:

Probability Density Function

```
Negative binomial with p = 0.1667 and r = 5

 X  P( X = x )
20    0.0323655

* NOTE * X = total number of trials.
```

The negative binomial distribution calculates a probability of 0.0323655 for rolling the 5th six on the 20th roll. In the long run, about 3.2% of sequences will have the 5th six occur on the 20th roll.

As before, we'd also like to know the chances for rolling the 5th six on other rolls by graphing the distribution.

For example, the following distribution graph displays the probability of rolling the fifth six on the 6th, 7th, 8th, etc., die rolls. Each bar indicates the likelihood of rolling the 5th six on the specified number of die rolls. The fifth roll is theoretically your first chance for getting the 5th six, but the probability is too low to display. Instead, the negative binomial distribution graph starts with the sixth roll. The graph stops at 81 rolls because the probability of obtaining the 5th six after that is too low to display.

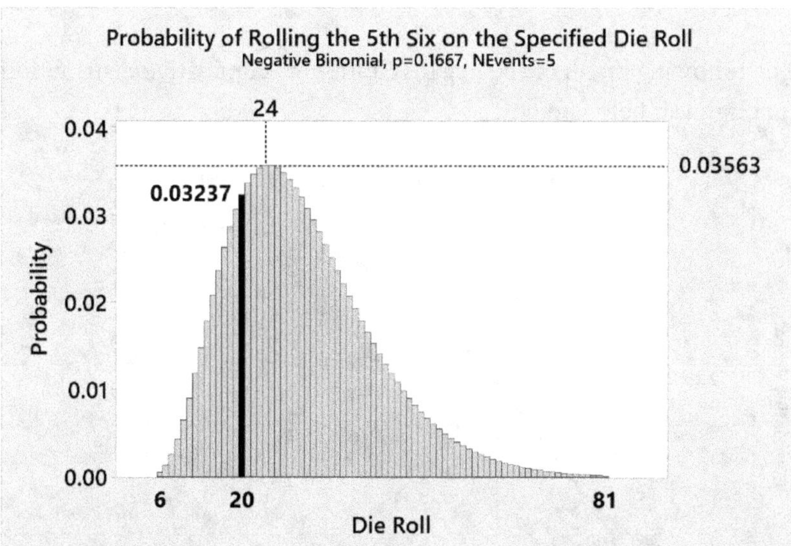

On the graph, I've highlighted the bar corresponding to the previous statistical output for the probability of rolling the 5th 6 on the 20th roll, which equals 0.03237.

More generally, the chart indicates that the maximum likelihood (0.03563) of rolling the fifth six occurs on the 24th roll, which is the tallest bar. Before 24 rolls, your probability of throwing the 5th 6 increases for each successive roll. After 24 rolls, the likelihood for each roll decreases. On the declining portion of the curve, you've had so many rolls that you've probably already rolled five 6s.

The negative binomial distribution helps you understand the probability of the rth success occurring on the xth trial. However, you're frequently not interested in the chances for only one particular attempt. As with the other distributions, you might want to learn the cumulative probability of the rth success occurring over a range of trials.

For example, you might be interested in answering the question: How many times do I need to roll the die to have a 50% cumulative probability of getting five sixes?

The following graph of the negative binomial cumulative distribution function can help you out!

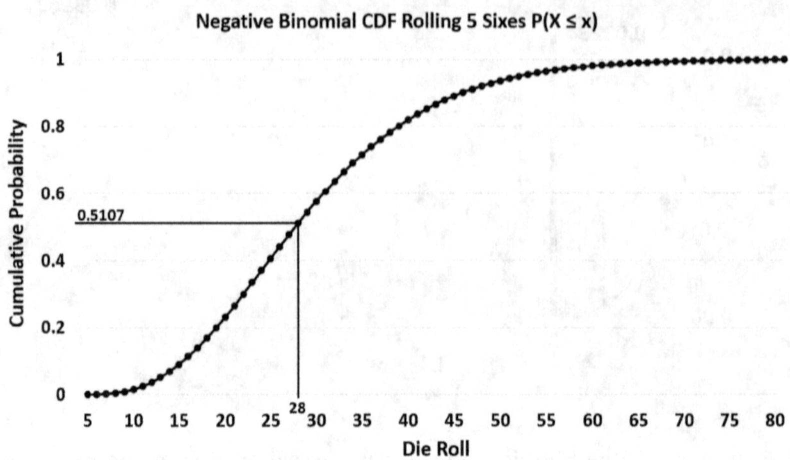

In the CDF graph, look for the die roll that corresponds to approximately a 0.50 probability of five successes. The cumulative probability of obtaining the 5th six in the first 28 rolls is 0.5107.

Using a Negative Binomial Distribution Calculator

Let's use my negative binomial distribution calculator to calculate probabilities and cumulative probabilities. Find the link in the book's Introduction for all the free resources, including the calculators.

First, we'll use this calculator to solve the previous die example where we found the probability of getting exactly five sixes in 20 rolls. In the calculator, enter the following and click **Calculate**:

- **Event Probability (p)**: 0.1667
- **Number of Events (r)**: 5
- **Total Number of Trials (X)**: 20

The calculator's numeric output displays probabilities for X = 20 (the number of trials required to reach 5 events), including exactly, less than or equal to, and greater than. The graph shows the probabilities for all values of die rolls (X), where each bar represents the probability that the 5th event occurs on that trial. The graph highlights the bar corresponding to our X-value. It excludes X-values with probabilities too small to display.

To find the probability of exactly 5 sixes in 20 trials, look in the numeric output for P(X = 20). The calculator displays a probability of 0.0324, matching our previous results to within rounding error.

Now, try one yourself. Imagine you're drawing a random sample of 13 individuals for jury duty. Assume that females comprise 50% of the population. What is the probability of selecting the 7th woman on the 13th selection?

Find the answer at the end of the negative binomial section.

Now, let's proceed to the formula for those who want to calculate the probabilities manually.

Formula and Manual Calculation Example

Let's walk through the formulas for the negative binomial distribution with the worked examples.

The negative binomial distribution formula is the following:

$$\mathrm{P(x)} = \mathrm{C(x-1, r-1)} \times \mathrm{p}^r \times (1-\mathrm{p})^{x-r}$$

$$\mathrm{C(x-1, r-1)} = \frac{x-1!}{r-1!\left((x-1)-(r-1)\right)!}$$

Where:

- x = number of trials.
- r = the number of successes
- p = the probability of a success.

C (x – 1, r – 1) is the binomial coefficient that counts how many different sequences of outcomes result in the rth success occurring on the xth trial. Because the final trial must be a success, this coefficient counts the number of ways to arrange the remaining r – 1 successes among the first x – 1 trials. Technically, it represents the number of combinations without repetition for choosing the locations of those successes.

The binomial coefficient for the negative binomial distribution is smaller than it is for the binomial because there are fewer possible combinations for r successes in x trials when the final attempt *must* be a success. Conversely, the binomial distribution does not have that constraint, allowing for more combinations.

The negative binomial distribution formula multiplies this number of possible sequences by the probability of any one such sequence. That probability equals the probability of success raised to the number of successes, multiplied by the probability of failure raised to the number of failures. Together, these components give the probability that the rth success occurs exactly on the xth trial.

Working through a calculation will bring the formula to life!

We'll use the negative binomial distribution formula to calculate the probability of rolling the 5th six on the 20th die roll. Enter these values into the formula:

- x = 20
- r = 5
- p = 0.1667

For the number of combinations, we have:

$$C(20-1,5-1) = \frac{19!}{4!\,(19-4)!} = 3876$$

Now, let's enter our values into the negative binomial distribution formula.

$$\begin{aligned} P(x) &= C(x-1, r-1) \times p^r \times (1-p)^{x-r} \\ &= 3876 \times 0.1667^5 \times 0.8333^{15} \\ &= 3876 \times 0.000129 \times 0.064867 = 0.032434 \end{aligned}$$

This hand calculation verifies the statistical software solutions within rounding error.

If you need to find the cumulative negative binomial probability for a range of trials, calculate the probability for each attempt and then sum the probabilities for all trials of interest.

For example, to calculate the likelihood of getting the 5th six sometime within 20 rolls, calculate the probability for the 5th six on the 5th, 6th, 7th, …, 20th rolls. Then sum that set of probabilities.

Expected Value

The expected value of the negative binomial distribution is the average number of trials required to obtain r successes.

The formula for the expected value of a negative binomial distribution is:

$$E(X) = \frac{r}{p}$$

This depends on the number of successes you want to achieve (r) and the probability of success on each trial (p).

Using our die example, suppose success is rolling a six with probability p = 0.1667, and we want r = 5 successes. Entering these values into the formula gives:

$$E(X) = \frac{5}{0.1667} = 30$$

The expected value is approximately 30. On average, we would expect it to take about 30 rolls to obtain the 5th six.

Standard Deviation

The formula for the standard deviation of a negative binomial distribution is:

$$SD(X) = \sqrt{\frac{r \times (1-p)}{p^2}}$$

As before, p is the probability of success and 1 – p is the probability of failure. The standard deviation increases when successes are rarer (smaller p), and it also increases as the number of required successes (r) increases.

For our example with r=5, p = 0.1667, and 1 – p = 0.8333, substitute into the formula:

$$SD(X) = \sqrt{\frac{5 \times 0.8333}{(0.1667)^2}} = 12.24$$

This gives a standard deviation of approximately 12.24. The standard deviation describes how much the number of trials required to obtain the 5th success tends to vary around the mean.

CALCULATOR SOLUTION: The probability of picking the 7th woman on the 13th selection is 0.1128.

Hypergeometric Distribution

The hypergeometric distribution calculates the likelihood that an event occurs k times in x trials when sampling from a small population without replacement.

Like the binomial distribution, the hypergeometric distribution calculates the probability of a specified number of events in a given number of trials. However, unlike the binomial distribution, it does not assume that the likelihood of an event's occurrence is constant. In fact, the hypergeometric distribution assumes that the probability changes because you are drawing from a small population without replacement.

For instance, when you draw an ace from a deck of cards, the probability decreases for drawing another ace on the next draw because the deck has fewer aces.

The hypergeometric distribution can answer the following questions. What is the probability of getting:

- Two red candies when we draw five candies from a jar containing five red candies and 10 white candies.
- Drawing five cards of the same suit from a regular deck of cards.
- 8 women on a jury of 13 people when drawing randomly from a jury pool of 50 people evenly split between men and women?

As the population size increases, the hypergeometric distribution more closely approximates the binomial distribution.

Assumptions, Parameters, and Notation

The hypergeometric distribution models the probability of obtaining a given number of successes when sampling from a small population without replacement.

As with the binomial distribution, each observation has two possible outcomes, one labeled a success and the other a failure. However, the hypergeometric distribution differs in a key way: the observations are not independent because each draw changes the composition of the remaining population. As a result, the probability of success changes from one draw to the next.

The hypergeometric distribution has three parameters:

- **N**: the population size.
- **K**: the number of successes in the population.
- **n**: the sample size drawn from the population.

You denote a hypergeometric distribution as h(N, K, n).

Alternatively, you can write $X \sim h(N, K, n)$, which means that your random variable X follows a hypergeometric probability distribution with population size N, K successes in the population, and a sample size of n.

For example, suppose during jury selection the court has a pool of 50 juror candidates, of which 25 are women, and you randomly draw 13 people without replacement for jury selection. The following is the hypergeometric notation for this scenario:

$X \sim h(50, 25, 13)$

Using and Interpretations

The hypergeometric distribution models the probabilities for exactly k events occurring in n selections from a finite population. Let's look at an example to bring it to life!

Again, we'll start by calculating a single probability and move on to the distribution from there.

Suppose we're interested in the possible outcomes for a jury selection. We want to know the probability of drawing 8 women for a jury of 13 when there are 25 female and 25 male candidates. For this example, assume the jurors are randomly selected from the candidate pool.

We'll need the following information to solve this problem:

- Total population size is 50 candidates (N).
- Number of events (Female) in the population (all candidates) is 25 (K)
- Jury size is 13 (n).
- Outcome of interest is selecting 8 women (k).

The hypergeometric distribution accounts for how the probabilities change with each selection. As we select men and women from the candidate pool, the remaining pool changes because there are no replacements. Each woman we choose reduces the number of women in the candidate pool, thus lowering the likelihood that the following selection will be a woman. Conversely, selecting a man increases the chances that the next juror will be a woman.

My statistical software tells me that the likelihood is:

Probability Density Function

```
Hypergeometric with N = 50, K = 25, and n = 13

X  P( X = x )
8    0.161934
```

The hypergeometric probability distribution calculates a probability of 0.161934 of selecting 8 women in 13 draws. Over the long run, approximately 16.2% of jury selection sequences of 13 people from populations with this composition will include 8 women.

That's interesting, but perhaps limited by itself. We're also interested in the chances of selecting other numbers of female jurors. Let's look at the full distribution to see which values occur more frequently.

The hypergeometric distribution graph is helpful because it shows the probability of obtaining different numbers of successes (k) out of the total number of trials (n). The following chart shows the probability distribution of selecting exactly 0, 1, 2, ..., and up to 13 women in 13 selections. With this approach, the hypergeometric distribution graph covers the complete range of possible successes up to the total number of trials.

The graph does not show the chances for fewer than 2 or more than 11 because those likelihoods are too low to display on the chart.

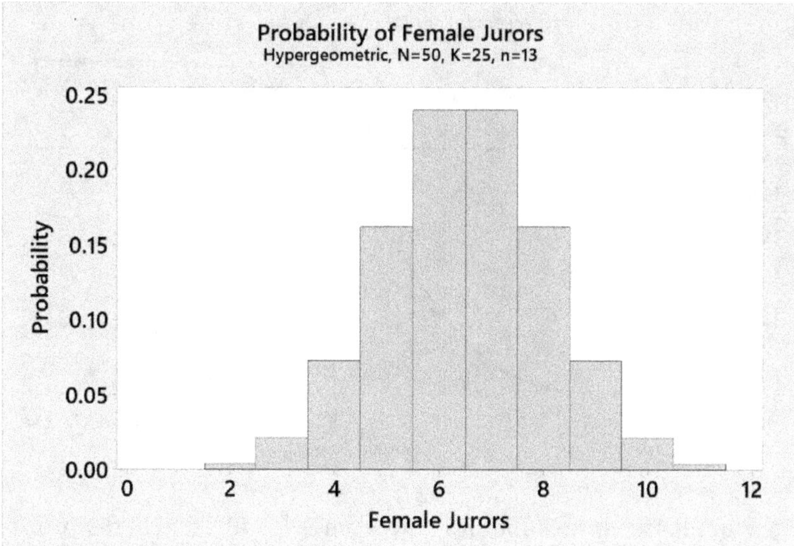

In the chart, each bar represents the probability of selecting a specific number of women during the 13 selections. The bar for 8 corresponds with the probability (0.161934) shown in the previous output. At a glance, we can see that selecting 6 or 7 female jurors is the most likely outcome, with both having equal probabilities of approximately 0.24.

As with the other binary distributions, you might not be interested in just one specific number of outcomes but a range. For instance, you might want to learn the probability of selecting *at least* eight women. Again, we turn to the hypergeometric cumulative distribution function, and in this case, we need its complement to answer the "at least" question.

On the following CDF plot, we need to find the cumulative probability of ≤ 7 to find the complementary probability for ≥ 8.

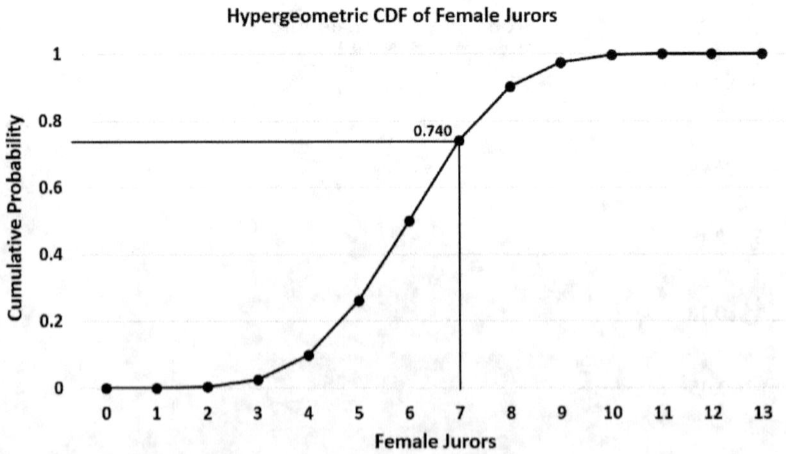

The cumulative probability for 7 women is 0.740. Hence, the probability for randomly choosing eight or more women in 13 selections is 1 – 0.740 = 0.260, approximately 1 in 4.

Using a Hypergeometric Distribution Calculator

Let's turn to my hypergeometric distribution calculator to calculate probabilities and cumulative probabilities. Find the link in the book's Introduction for all the free resources, including the calculators.

We'll use this calculator to recreate the jury selection example from earlier where we found the probability of drawing 8 women for a jury of 13 when there are 25 female and 25 male candidates.

In the calculator, enter the following and click **Calculate**:

- **Population size (N)**: 50
- **Number of Events in Population (K)**: 25
- **Sample Size (n)**: 13
- **Number of Events in Sample (k)**: 8

The graph shows the probabilities for all possible values of k from 0 to 13, where each bar represents the probability of selecting that number of women in a sample of 13 from a population of 50 containing 25 women. The graph highlights the bar corresponding to our k-value. It excludes k-values with probabilities too small to display.

To find the probability for selecting exactly 8 women, look in the numeric output for $P(X = 8)$, which is 0.1619, matching our previous results within rounding.

For the probability of at least 8 women, find $P(X \geq 8)$. The calculator shows a probability of 0.2601. This result matches our graphical example using the complement of the cumulative distribution.

Now, try one yourself. Imagine you're drawing from a jar that contains 5 red candies and 10 non-red candies. What is the probability that you'll draw two red candies with five random draws from the jar?

See the correct answer at the end of the manual calculation section because we'll solve it there! Now, onto the formula for those who want to learn how it works.

Formula and Manual Calculation Example

The hypergeometric distribution formula is the following:

$$P(X = k) = \frac{({}_{K}C_{k}) \times ({}_{N-K}C_{n-k})}{{}_{N}C_{n}}$$

Where:

- N is the population size.
- n is the sample size.
- K is the number of events or successes in the population.
- k is the number of events in the sample.

Use the hypergeometric distribution formula to calculate the probability of obtaining exactly k successes in a sample of size n drawn from a finite population without replacement. Notice how the formula incorporates both the characteristics of the population (N and K) and the characteristics of the sample (n and k).

Note that n cannot exceed N, and k cannot exceed either K or n. The sample cannot be larger than the population, and the number of successes in the sample cannot exceed either the sample size or the successes in the population.

The hypergeometric distribution is based on combinations without repetition because the order of selection does not matter and it involves sampling without replacement. For example, $_KC_k$ represents the number of ways to choose k successes from the K successes available in the population.

Given a population of size N containing K successes, the formula counts the number of possible samples of size n that contain exactly k successes and n – k failures. It does this by multiplying the number of ways to choose k successes from the population by the number of ways to choose the remaining n – k failures from the non-successes. This product is then divided by the total number of possible samples of size n that can be drawn from the population.

The resulting probability represents the likelihood that a randomly selected sample of size n contains exactly k successes.

Let's work through an example calculation to bring the formula to life!

We'll use the hypergeometric distribution formula to calculate the probability of drawing red candies from a jar. The jar contains 5 red candies and 10 non-red candies, for a total of 15. We'll randomly draw five candies from the jar.

Let's calculate our chances of getting two red candies in our five draws!

We'll enter the following values in the hypergeometric distribution formula:

- N = 15 total candies in the jar.
- n = 5 draws for the sample.
- K = 5 red candies in the jar.
- k = 2 red candies in the sample.

For these calculations, I'll use my combinations calculator to obtain the number of combinations for each term in the equation.

$$P(X = k) = \frac{({}_{K}C_{k})*({}_{N-K}C_{n-k})}{{}_{N}C_{n}}$$

$$P(X = 2) = \frac{({}_{5}C_{2})*({}_{15-5}C_{5-2})}{{}_{15}C_{5}} = \frac{({}_{5}C_{2})*({}_{10}C_{3})}{{}_{15}C_{5}}$$

$$P(X = 2) = \frac{10 * 120}{3003} = 0.3996$$

The probability of drawing precisely two red candies in our five random draws from the jar is 0.3996. This result also answers the hypergeometric calculator problem I gave you earlier!

Expected Value

The expected value of the hypergeometric distribution is its mean. For the hypergeometric distribution, the expected value represents the average number of successes you would expect to observe in the sample when repeatedly drawing samples of the same size from the population.

The formula for the hypergeometric distribution EV is the following:

$$E(X) = n \times \frac{K}{N}$$

Here, N is the population size, K is the number of successes in the population, and n is the sample size. The ratio K/N represents the proportion of successes in the population.

In our candy example, the jar contains 15 candies total, 5 of which are red. We randomly draw 5 candies from the jar. Substituting these values into the formula gives:

$$E(X) = 5 \times \frac{5}{15} = 1.667$$

The expected value is 1.667. On average, we would expect to draw about 1.667 red candies when selecting five candies at random from the jar. Of course, the actual number of red candies must be an integer, but the expected value represents the long-run average over many repeated samples.

Standard Deviation

The formula for the standard deviation of a hypergeometric distribution is:

$$SD(X) = \sqrt{n \times \frac{K}{N} \times \left(1 - \frac{K}{N}\right) \times \frac{N-n}{N-1}}$$

The first three terms of this expression resemble the standard deviation of a binomial distribution, but it uses K/N to represent P. The last term, N – n / N -1, accounts for sampling without replacement from a finite population.

Using the values from our candy example, we substitute:

- N = 15
- K = 5
- n = 5

$$SD(X) = \sqrt{5 \times \frac{5}{15} \times \left(1 - \frac{5}{15}\right) \times \frac{10}{14}} = 0.891$$

Evaluating this expression gives a standard deviation of approximately 0.891. This value describes how much the number of red candies selected typically varies around the expected value of 1.67 when repeatedly drawing five candies from the jar without replacement.

The binomial, geometric, negative binomial, and hypergeometric distributions describe the probabilities of the number of events and when they occur. This information can be invaluable for planning purposes. I kept the die and candy examples intentionally simple so they are easy to understand. However, for a science-based, real-world example, read about the effectiveness of flu shots in the next section.

Modeling Flu Outcomes

The following example uses flu infection data to show how these binary distributions can be applied in a realistic setting to reason about probabilities.

Flu shot benefits are like investing year after year. The cumulative effect magnifies the differences over time. I'm going to model flu outcomes over decades in the same way financial planners illustrate the differences between courses of action.

Investment return rates fluctuate over the years, just as influenza rates do. Financial planners use reasonable long-term averages to estimate different outcomes. Similarly, I'll use the average infection rate for vaccinated and unvaccinated individuals. This approach produces comparative results for annual flu vaccination versus no flu vaccination over decades.

To model this statistically, I'll use probability distribution plots. Because the data are binary (infected or not infected), I'll use the binomial and geometric distributions. For the graphs, the assumptions are that the average infection rate for the:

- Unvaccinated is 7.0% annually.
- Vaccinated is 1.9% annually.

I based both of these average infection rates on multiple randomized controlled studies of flu shots. See the references at the end of the book for a listing.

We'll learn what effect that 5.1% absolute difference has over the course of decades. I'll compare two different scenarios: getting the flu shot every year versus never getting the influenza vaccination. Using our estimates, I'll answer two questions:

- How long until my first case of the flu on average?
- How many times will I get the flu?

In both plots, the left panel displays the no-flu-shot scenario, while the right panel shows the annual influenza vaccine scenario.

How long until my first case of the flu on average?

I'll use the geometric distribution to model the probability of catching the flu for the first time. Each bar in the graph indicates the likelihood of first getting the flu in a specific year. The risk for any given year is small. More importantly, I've shaded the graphs to indicate the number of years until the cumulative probability reaches 50%. This shading is a different way to show cumulative probabilities over a range.

The chart stops at 65 years because flu shot effectiveness decreases around that age.

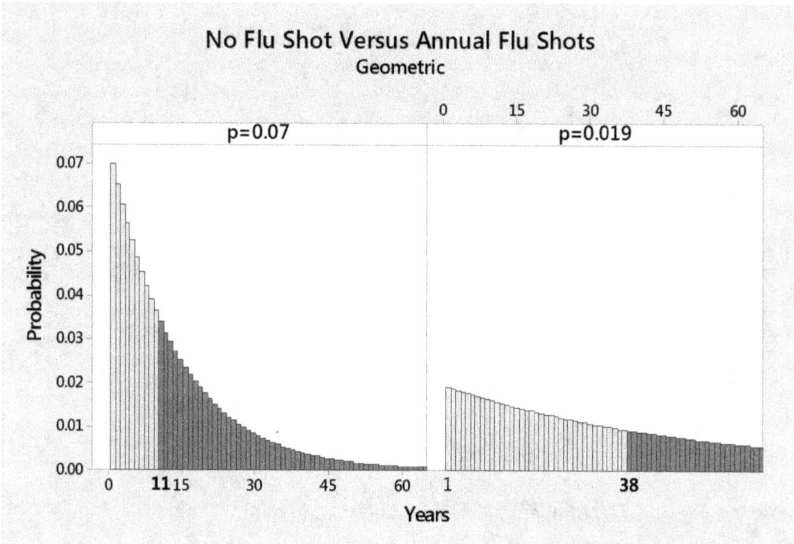

The chart shows how your probability of first contracting the flu is much higher early on when you aren't vaccinated (left) than when you are regularly vaccinated.

The left panel indicates that without flu shots, you have a 50/50 chance of getting the flu in 11 years. On the other hand, the right panel displays the probabilities for annual vaccination. You don't reach a 50% cumulative probability until 38 years. Furthermore, when you never get the flu shot, you have only a 6.8% chance of not catching the flu in 38 years (not shown)!

How often will I catch the flu?

Let's statistically model the number of times you can expect to catch the flu in 20 years. For this graph, I'll use the binomial distribution. Each bar indicates the likelihood of catching the flu the specified number of times. I've shaded the bars to represent the cumulative probability of catching the flu at least twice in 20 years.

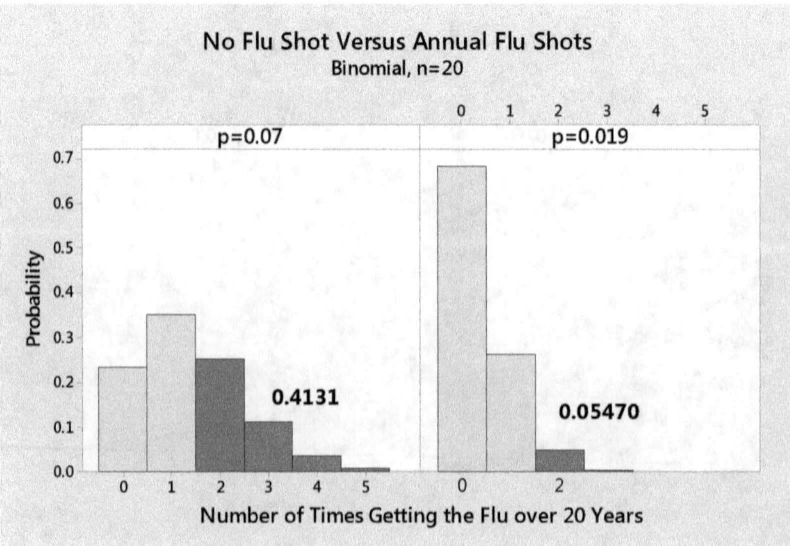

A significant difference jumps out at you, which demonstrates the power of graphing probability distributions! The tallest bar on the graph is the one in the right panel that represents zero cases of the flu in 20 years when you get flu shots. When you vaccinate annually, you have a 68% chance of not catching the flu over the next 20 years! Conversely, if you don't vaccinate, you have only a 23% of escaping the flu entirely.

In the left panel, the distribution extends much farther than in the right panel. Without vaccinations, you have a 41% chance of getting the flu at least twice in 20 years compared to 5% with annual vaccinations. Some unlucky unvaccinated folks will get the flu four or five times during that period!

Review and Next Steps

In this chapter, you learned how different probability distributions arise from binary random experiments. Each distribution answers a different type of question and, in some cases, the process structure is different. The key skill is recognizing which distribution answers your question and fits the structure of the random experiment.

The following summarizes when to use each distribution and the key features to look for in a probability question.

Binomial Distribution: Fixed Number of Trials

Use the binomial distribution when you are counting the number of successes in a fixed number of trials. They all apply only to binary outcomes.

Look for problems that include:

- A fixed number of trials (for example, “in 10 rolls” or “out of 20 inspections”).
- Independent trials.
- A constant probability of success.
- Questions about *how many successes occur.*

If the question specifies a set number of trials and asks how many successes occur within them, the binomial distribution is the right tool.

Geometric Distribution: Waiting for the First Success

Use the geometric distribution when you are interested in the time to the first success.

Look for problems that include:

- Questions phrased as “How many trials until the first success?”
- Independent trials.
- A constant probability of success.

If the process continues until the first success occurs and then stops, the geometric distribution applies.

Negative Binomial Distribution: Waiting for the rth Success

Use the negative binomial distribution when you are counting how many trials are needed to achieve a specified number of successes.

Look for problems that include:

- Questions such as "How many trials until the 5th success?"
- A target number of successes greater than one.
- Independent trials.
- A constant probability of success.

The negative binomial distribution generalizes the geometric distribution by extending it from the first success to the rth success.

Hypergeometric Distribution: Sampling Without Replacement

Use the hypergeometric distribution when you are sampling from a finite population without replacement.

Look for problems that include:

- Questions about how many successes appear in the sample.
- A known population size.
- A known number of successes in the population.
- Sampling without replacement.
- Dependent sampling.
- A changing probability of success.

If selecting one item affects the probabilities of future selections, the hypergeometric distribution is appropriate.

In the next chapter, you'll learn about another discrete distribution that uses a different kind of structure altogether. Instead of counting successes or waiting for events to occur, you'll model counts over time or space using the Poisson distribution. This shift introduces a rate-based way of modeling randomness, focusing on how often events occur over time or space rather than on proportions from repeated trials.

Formula Quick Reference

- **Binomial:** nCx × p^x × (1 – p)^(n – x)
  - **EV:** n × p
  - **SD:** sqrt(n × p × (1 – p))
- **Geometric:** (1 – p)^(x – 1) × p
  - **EV:** 1 / p
  - **SD:** sqrt((1 – p) / p^2)
- **Negative Binomial:** (x – 1)C(r – 1) × p^r × (1 – p)^(x – r)
  - **EV:** r / p
  - **SD:** sqrt(r × (1 – p) / p^2)
- **Hypergeometric:** [(K C x) × ((N – K) C (n – x))] / (N C n)
  - **EV:** n × (K / N)
  - **SD:** sqrt[n × (K/N) × (1 – K/N) × ((N – n)/(N – 1))]

Chapter 11 Practice Problems

You may use any method to solve these problems, including manual calculations, online calculators, or statistical software. Focus on identifying the correct distribution and approach. If you are preparing for an exam, use the calculation method that matches your testing conditions.

Question 1: A researcher conducts independent trials where each trial has the same probability of success.

For each scenario, identify the appropriate probability distribution. Do not perform any calculations.

- **A.** The probability that the first success occurs on the 6th trial.
- **B.** The probability that exactly 6 successes occur in 20 trials.
- **C.** The probability that the 6th success occurs on the 20th trial.

Question 2: A box contains 50 items, 10 of which are defective. A sample of 5 items is selected.

For each scenario, identify the appropriate probability distribution. Do not perform any calculations.

- **A.** Each item is selected and then returned to the box before the next selection.
- **B.** Items are selected one after another without being returned to the box.

Question 3: A customer support center finds that 30% of incoming calls require escalation to a supervisor. Calls are handled independently.

What is the probability that the first call requiring escalation occurs on the 4th call?

Question 4: A quality inspector checks items coming off a production line. Each item has a 5% chance of being defective, and inspections are independent.

What is the probability that at most 2 defective items are found in the first 20 inspections?

Question 5: A warehouse contains 40 boxes, 12 of which contain fragile items. A worker randomly selects 8 boxes without replacement.

What is the probability that the worker selects exactly 3 boxes containing fragile items?

Question 6: A basketball player makes 80% of free throws. Assume each attempt is independent.

What is the probability that the player makes at least 9 out of 12 free throws?

Question 7: A researcher conducts a series of independent trials where each trial has a 20% chance of success.

What is the probability that the 3rd success occurs on the 10th trial?

Question 8: A shipment contains 25 components, of which 7 are faulty. A technician randomly selects 6 components without replacement.

What is the probability that the technician selects between 1 and 3 faulty components, inclusive?

CHAPTER 12

Counting on the Poisson Distribution

In Chapter 11, you modeled randomness using binary outcomes and success probabilities tied to individual trials. In this chapter, the focus shifts to using event rates over time or space rather than success probabilities during trials.

The Poisson distribution is a discrete probability distribution that describes the probability of counts of events occurring within a fixed interval, such as a period of time, a region of space, or a unit of exposure. This interval, called the observation unit, defines the context over which events are counted. It is named for Siméon Denis Poisson. The name *Poisson* is pronounced *pwah-SON*, which reflects its French origin.

In the Poisson distribution, count data represent the number of events. Common examples of situations modeled by the Poisson distribution include counts of the following:

- Births in a hospital per day.
- Meteors seen per hour.

- Cigarettes smoked per day.
- Customer arrivals at a service center per hour.
- Defects per manufactured item.
- Emails received in a fixed time period.

In each case, the focus is not on whether an event occurs on a particular trial, but on how many events occur within a specified interval.

One of the earliest and most influential empirical applications of the Poisson distribution was by Ladislaus Bortkiewicz, a Russian economist, who analyzed the annual number of Prussian army officer deaths caused by horse kicks from 1875 to 1894. His study showed that this distribution can model real-world counts.

The Poisson distribution was developed to model count data directly. Count data are discrete, non-negative integers (0, 1, 2, 3, etc.) and often skewed, so they are not well described by continuous models such as the normal distribution. Analysts often use the Poisson distribution to model rare events, but it is not limited to them.

The Poisson distribution works by focusing on the average rate of occurrence rather than tracking individual trials. Instead, it describes the probability of observing different numbers of events even when the number of possible opportunities is large or not directly observed. You just need to know the average number of events per interval, which is typically estimated from observed counts.

The Poisson and binomial distributions are similar because they both model the occurrence of events. However, while the number of trials (n) in a binomial distribution defines the upper bound for the number of events (x), the Poisson distribution places no upper bound on the number of events. Hence, the number of possible events in a Poisson distribution ranges from 0 to infinity.

For example, while the number of meteors observed per hour might fall within a typical range, the Poisson distribution does not impose an upper limit. However, the binomial distribution for the number of heads during 10 coin tosses places a hard limit on the number of heads. Obviously, the number of heads cannot exceed the number of coin tosses.

Counts that are extremely large relative to the average rate are theoretically possible but so unlikely that their probabilities are virtually zero. For example, when the average count is 2.24, observing 20 or 30 in one interval is extraordinarily unlikely.

Another practical difference between the Poisson distribution and the binary distributions from the previous chapter is how analysts typically determine their parameters. With binomial-type models, you often derive the parameters from the process itself rather than collecting data.

In contrast, researchers generally use Poisson models after recording event counts repeatedly and estimating the rate from those observations. This shift places greater emphasis on using observed data to estimate distribution parameters, a theme that continues in later chapters.

Assumptions, Parameters, and Notation

The Poisson distribution is based on a random experiment that counts the number of times an event occurs within a fixed interval. Although this structure differs from the trial-based experiments in Chapter 11, it still relies on clearly defined assumptions.

To model data using a Poisson distribution, the following conditions must hold:

- Events occur independently of one another.
- Events occur at a constant average rate over the interval.
- Two events cannot occur at exactly the same instant.
- The interval of interest is fixed in advance.

The Poisson distribution uses a single parameter, lambda (λ), which is the mean number of occurrences during an observation unit. A common way to estimate lambda is to calculate the sample mean count per observation unit (i.e., the mean rate of occurrence).

For example, if $\lambda = 2.24$, then on average, you would expect about 2.24 events to occur in each interval. Some intervals will have fewer events, some will have more, but over many repetitions, the average count will approach 2.24.

You can denote a Poisson distribution as Poisson(λ). Alternatively, you can write $X \sim$ Poisson(λ), which means that the random variable X follows a Poisson distribution with mean rate λ.

For a Poisson distribution, the random variable X always represents a count of events, and its possible values are 0, 1, 2, 3, and so on, with no upper bound. The distribution assigns probabilities to each of these counts based on the assumed event rate and the length of the interval being studied.

Using and Interpretation

The Poisson distribution models the probabilities for exactly x events occurring within a fixed interval when events occur at a constant average rate. Let's work through an example to see how this distribution works in practice.

I'll use the same approach as in the previous chapter. I'll start by interpreting a single value and then move on to the full distribution and cumulative probabilities. This approach helps illustrate what the Poisson distribution tells us about the overall pattern of outcomes. Later,

I'll walk through the calculations so you can see how the formula works.

Suppose a safety inspector needs to monitor the number of falls per day at a large nursing home. He randomly selects 50 days to observe and records the number of falls that occur each day. Each observation is a count, such as 0 falls, 1 fall, 2 falls, and so on. Based on these observations, the average rate of falls is estimated to be $\lambda = 2.24$ falls per day.

In this example, the random variable X represents the daily number of falls.

Probability Density Function

```
Poisson with mean = 2.24

X  P( X = x )
3    0.199422
```

My statistical software tells me that the probability of observing exactly three falls in a day is 0.199422. In the long run, you'd expect about one-fifth of the days to have three falls when the average rate is 2.24 falls per day.

That result is informative, but looking at just one outcome doesn't tell the whole story. Probability distributions are especially valuable because they reveal the overall pattern of outcomes, not just individual probabilities.

The following Poisson distribution graph displays the probability of observing various counts of falls in a single day. While the number of falls can be any non-negative integer in theory, the probabilities for very large counts are extremely small and not shown in the graph.

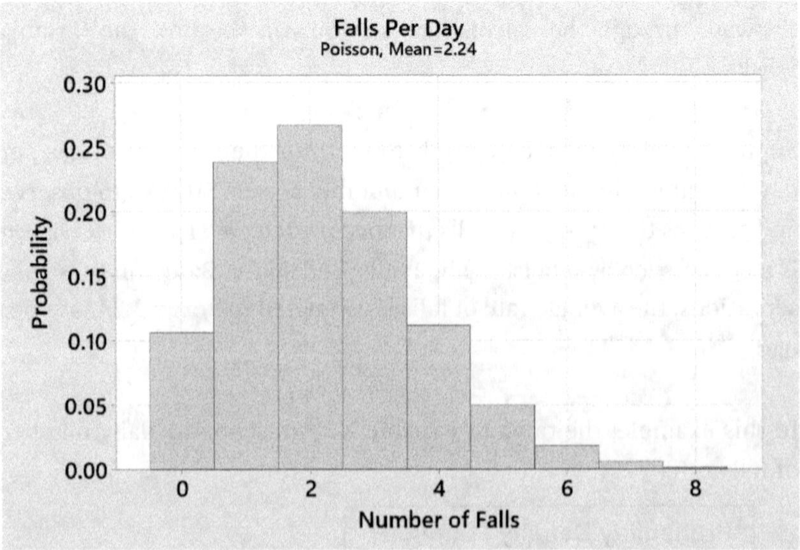

It's easy to see which counts are most common and how quickly probabilities decrease as the number of events moves away from the average rate.

In the graph, each bar represents the probability of observing a specific number of falls in one day. The tallest bar is for two falls per day, which is the closest value to the average rate of 2.24. In this example, counts increase to the average and then decline.

Many Poisson distributions are right-skewed, as in this example. In practice, this skew means most days have few events, but some days can have much higher counts. These high-count days might seem like outliers, but they are an expected part of the process and don't necessarily require a special cause explanation.

The Poisson distribution is especially useful for understanding the likelihood of observing exact counts of events within a fixed interval. However, as with the binary distributions, many practical questions involve a range of values rather than a single value.

For example, the safety inspector might be concerned about days with three or more falls. In that case, the question becomes: what is the probability of observing at least three falls in a day?

Recall that cumulative distributions are helpful for answering "at most" questions, and their complements are useful for answering "at least" questions. To find the probability of at least three falls, we begin by finding the cumulative probability of two or fewer falls.

The following cumulative probability graph shows the probability of observing 0, 1, 2, and so on in a day. Because "at least 3" includes 3, 4, 5, and so on, its complement is "at most 2." So we find $P(X \leq 2)$ and subtract it from 1, which corresponds to the probability of observing three or more falls.

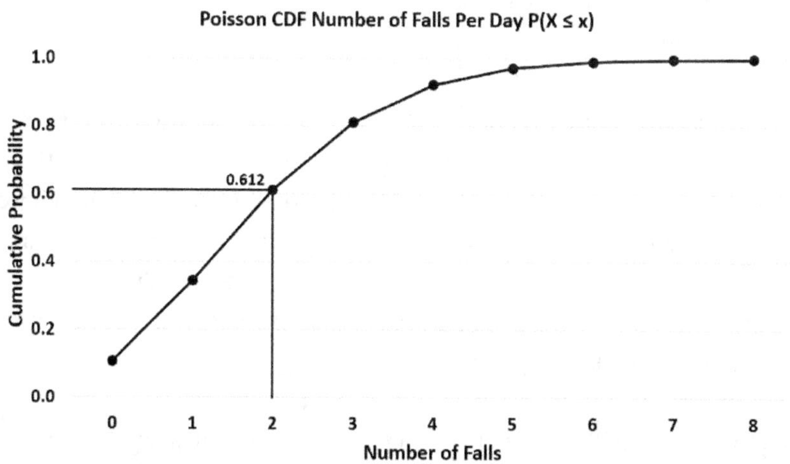

Hence, 1 - 0.612 = 0.388. The probability of observing at least 3 falls in a day is 0.388. In the long run, 38.8% of days will have at least 3 falls.

In general, the same problem-solving strategies introduced for the binomial distribution in Chapter 11 also apply to Poisson probability questions. Identifying whether a question asks for an exact count probability, a cumulative probability, or a complement remains the key first step. Then determine whether the bounds are inclusive or exclusive.

Using a Poisson Distribution Calculator

Next, we'll use my Poisson distribution calculator to find Poisson probabilities and cumulative probabilities. Find the link in the book's Introduction for all the free resources, including the calculators.

Let's use this calculator to recreate the nursing home example from before. Recall that the estimated average rate of falls is $\lambda = 2.24$ falls per day and we're interest in the probability of at least 3 falls per day. In the calculator, enter the following and click **Calculate:**

- **Average Rate (λ)**: 2.24
- **Number of Events (x)**: 3

The calculator's numeric output displays probabilities for x = 3 (three falls per day), including exactly, less than or equal to, and greater than. The graph shows the probabilities for all possible values of X, where each bar represents the probability of observing that number of falls in a day when the average rate is 2.24. The graph highlights the bar corresponding to our x-value. It excludes x-values with probabilities too small to display.

To find the probability of observing exactly 3 falls in a day, look in the numeric output for $P(X = 3)$, which is 0.1994, matching our previous results within rounding.

For the probability of at least 3 falls, find $P(X \geq 3)$. The calculator shows a probability of 0.3880. This result matches the result obtained previously using the cumulative distribution and its complement.

Now, try one yourself.

Suppose a large volume call center receives an average of 1.5 customer complaints per hour. What is the probability that the call center receives at least two complaints in an hour?

The answer is at the end of the Standard Deviation section.

Changing the Length of the Interval

In Poisson models, the event rate λ is always tied to a specific interval. In the nursing home example, λ = 2.24 represents the average number of falls per day. If the length of the interval changes, the value of λ must change as well.

For example, over a two-day period, the expected number of falls would be twice as large (4.48), whereas converting to a half-day interval halves λ (1.12). The Poisson distribution would still apply, but it would use a different rate that reflects the longer or shorter interval.

Before using the Poisson distribution, the interval of interest must be clearly defined and applied consistently.

Formula and Manual Calculations

Next, I'll walk you through the formulas for finding the probability, expected value, and standard deviation of the Poisson distribution and provide worked examples.

The Poisson distribution formula for probabilities is the following:

$$P(x) = \frac{e^{-\lambda} \times \lambda^{x}}{x!}$$

Where:

- λ (lambda) is the average number of events per interval.
- x is the number of events of interest.
- e is the mathematical constant approximately equal to 2.71828.

Use this formula to calculate the probability of observing exactly x events within a fixed interval.

Unlike the binomial probability formula, the Poisson probability formula does not include a combination term. That's because Poisson models do not consist of a fixed number of trials with different possible arrangements of successes and failures. Instead, the formula directly models the likelihood of observing a given count based on the specified event rate.

The term $e^{-\lambda}$ adjusts the overall scale of the probabilities so that they sum to 1, while λ^x reflects how the likelihood changes as the number of observed events increases. Dividing by x! keeps the probabilities from increasing too quickly as the number of events grows.

Together, these terms create a balance. As x increases, λ^x pushes the probability upward at first. However, x! grows much faster for larger values of x, which eventually pulls the probability back down. This push-and-pull produces the familiar rise to a peak followed by a decline. The most likely value typically occurs near λ.

Let's work through an example calculation to bring the formula to life. Recall the nursing home example from the previous section, where the estimated average rate of falls is $\lambda = 2.24$ falls per day. Suppose we want to calculate the probability of observing exactly two falls in a day.

Here are the values to enter into the formula:

- $\lambda = 2.24$
- $x = 2$

$$P(2) = \frac{e^{-2.24}(2.24)^2}{2!} = 0.267$$

This result indicates that, in the long run, you'd expect about 26.7% of days to have exactly two falls when the average rate is 2.24 falls per day.

If you need a cumulative probability for a Poisson random variable, calculate the probabilities for the relevant individual counts and add them together.

As with the binomial distribution, use cumulative probabilities to answer "at most" questions and their complements to answer "at least" questions.

Unlike the binomial distribution, the Poisson distribution has no upper bound. Consequently, you must use complements to get an exact solution rather than the alternative method of summing all larger values for "at least" questions.

Expected Value

The expected value of the Poisson distribution is its mean rate of occurrence (λ).

Hence, the formula for the expected value is very simple:

$$E(X) = \lambda$$

For the nursing home example:

$E(X) = 2.24$

This means that, on average, you would expect about 2.24 falls per day over a long period of observation.

Standard Deviation

The standard deviation of the Poisson distribution is given by:

$$SD(X) = \sqrt{\lambda}$$

Using $\lambda = 2.24$, just take the square root.

$$SD(X) = \sqrt{2.24} = 1.497$$

The standard deviation describes how much the daily number of falls tends to vary around the average rate. While some days will have fewer falls and others more, most daily counts will fall within a few standard deviations of the mean.

One distinctive characteristic of the Poisson distribution is that its expected value and variance (i.e., the standard deviation squared) are both equal to λ. In practice, real-world count data sometimes show more or less variability than the Poisson model allows. Consequently, the Poisson distribution is not appropriate for all count data.

Calculator Answer: The probability of receiving at least two complaints in an hour is 0.4422. Over the long run, about 44.22% of hourly intervals will have at least two complaints. Use the Poisson distribution calculator with the following values and look at the $P(X \geq 2)$ result:

- $\lambda = 1.5$
- $x = 2$

Review and Next Steps

In this chapter, you learned how the Poisson distribution models randomness in a fundamentally different way from the binary distributions in the previous chapter. Rather than counting successes across a fixed number of trials, the Poisson distribution models event counts over a fixed interval, using an average event rate to describe long-run behavior.

You saw how the Poisson distribution naturally applies to count data, such as accidents, arrivals, defects, and other events that occur independently over time or space. You also learned how Poisson probabilities describe the likelihood of observing exact counts, how distribution plots reveal the overall pattern of outcomes, and how cumulative probabilities and complements help answer practical "at most" and "at least" questions.

Use the Poisson distribution when a probability question includes clues such as:

- Counting how many times an event occurs in a fixed period of time or space.
- Describing events with an average rate (for example, "per hour," "per day," or "per mile").
- Working with only non-negative integer outcomes such as 0, 1, 2, 3, and so on.
- Modeling counts of events over a fixed interval without a fixed number of trials.

If a problem focuses on how often events occur, rather than whether a specific trial succeeds or fails, the Poisson distribution is often the appropriate tool.

Chapters 11 and 12 together introduced two major ways of modeling randomness for discrete data:

- **Binary, trial-based models**, where probabilities are tied to individual trials and success/failure outcomes
- **Rate-based count models**, where randomness is described by how frequently events occur over an interval

In the next chapter, you'll take another step forward by moving from discrete outcomes to continuous random variables. Continuous distributions model measurements rather than counts, and probabilities are described using smooth curves rather than bars.

Formula Quick Reference

- Poisson: (e^(−λ) · λ^x) / x!
  - EV: λ
  - SD: sqrt(λ)

Chapter 12 Practice Problems

For questions 4 and 5, you may use any method to solve these problems, including manual calculations, online calculators, or statistical software. Focus on identifying the correct approach. If you are preparing for an exam, use the calculation method that matches your testing conditions.

Question 1: For each scenario, determine whether the binomial distribution or the Poisson distribution is more appropriate.

A. A quality inspector selects 15 items from a production line and records how many are defective.
B. A website records the number of visitors who arrive during randomly selected 10-minute intervals throughout the day.
C. A survey records how many of 50 randomly selected voters support a candidate.
D. A hospital tracks how many patients arrive at the emergency department during each hour.

Question 2: For each scenario, determine whether the Poisson distribution is appropriate. If not, briefly explain why.

A. Counting the number of cars passing a checkpoint per hour during normal traffic conditions.
B. Counting the number of customers entering a store per hour during a flash sale where traffic increases rapidly over time.
C. Counting the number of defects per sheet of material when defects tend to cluster in certain areas.
D. Counting the number of emails received per day when emails arrive independently at a roughly steady rate.

Question 3: A call center records the number of complaints received each hour. Answer the following:

A. In this context, what does λ represent?
B. Does λ mean the call center will receive exactly that number of complaints every hour?
C. If $\lambda = 3$, would it be unusual to observe 0 complaints in an hour? Explain briefly your reasoning.

Question 4: A hospital infection-control team tracks the number of needle-stick injuries among staff members each week. Based on past records, the average number of needle-stick injuries is 0.8 per week.

What is the probability that the hospital has no needle-stick injuries in a given week?

Question 5: A city traffic engineer monitors the number of traffic accidents at a busy intersection each month. Historical data indicate that the average number of accidents is 2.6 per month.

What is the probability that there are at least 3 accidents in a given month?

CHAPTER 13

Continuous Distributions and the Normal Curve

So far, you have worked with probability models that describe discrete outcomes. In those settings, probabilities attach directly to individual values. A probability mass function assigns a probability to each possible outcome, and those probabilities add up to one.

In this chapter, we shift to a different kind of randomness. Many real-world quantities do not come in neat, countable steps. Measurements such as time, height, weight, and temperature vary continuously. For these situations, probability no longer exists at single points. Instead, it spreads smoothly across a range of values.

To model this type of randomness, we use probability density functions, or PDFs. A PDF does not assign probabilities to specific values. Instead, it describes how densely probability is distributed across the number line. Probabilities come from the area under the curve over an interval, not from the height of the curve at a single point. This change has important implications for how probabilities are interpreted and calculated, and understanding that shift is a central goal of this chapter.

We will focus primarily on the normal distribution, the most widely used continuous probability model. The normal distribution plays a central role in probability and statistics because it appears naturally in many random processes, describes measurement error well, and serves as a reference point for comparing and standardizing values. Its symmetry and well-understood shape also make it an ideal starting point for learning how continuous models work.

Rather than emphasizing manual calculations, this chapter focuses on interpretation. You will see how probabilities arise from areas under the normal curve, how standardization simplifies comparisons, and how to use tables and software. Along the way, we will connect these ideas to earlier concepts such as expected value, spread, and long-run behavior.

Near the end of the chapter, I briefly introduce several other common continuous distributions. These models exist because not all random processes follow a normal shape. The main goal is to recognize when a particular model is appropriate and what kinds of patterns each distribution is designed to capture.

By the end of this chapter, you should understand how probability density functions differ fundamentally from the discrete models you encountered earlier and how to use the normal distribution.

Probabilities for Continuous Distributions

You know that you have a continuous distribution if the variable can assume an infinite number of values between any two values. Continuous variables are often measurements on a scale, such as height, weight, and temperature.

Unlike discrete probability distributions where each particular value has a non-zero likelihood, specific values in continuous distributions have a zero probability. For example, the likelihood of measuring a temperature that is exactly 32 degrees is zero.

Why? Consider that the current temperature can be an infinite number of other temperatures that are infinitesimally higher or lower than 32. Statisticians say that an individual value has an infinitesimally small probability that is equivalent to zero.

Hence, probabilities for continuous distributions are calculated for ranges of values rather than single points. A probability indicates the likelihood that a value will fall within an interval. This property is simple to demonstrate using probability distribution plots.

A probability density function (PDF) describes how probability is distributed over the possible values of a continuous random variable. On a graph, a PDF is the curve that shows how density changes across values. The density at a specific value is the height of the curve at that point. Unlike discrete probabilities, the probability density at a specific value is not the probability that the variable equals that value.

Instead, the density tells you how much probability is packed per unit around that value. Taller parts of the curve indicate where values are more concentrated, meaning values near that location are more likely to occur. Shorter parts of the curve indicate where values are less common.

In practice, you usually will not use probability density values directly. The height of the curve shapes the distribution and allows software to calculate probabilities, but it is not itself a probability. Instead, you will focus on probabilities for intervals, which come from the area under the curve.

To summarize, the height of the curve shows how densely values cluster around a point, while probability comes from the total area under the curve over an interval. Because the probability of any exact value is zero, you must always calculate probabilities over ranges of values.

For example, on the following graph, the probability density at 100, or any other value, tells you how tightly probability is concentrated around that IQ score. This density is just one of an infinite number of points along the PDF curve, and the probability of having a score of exactly 100 is zero. To get an actual probability, you need to look at a range, such as IQ scores between 120 and 140.

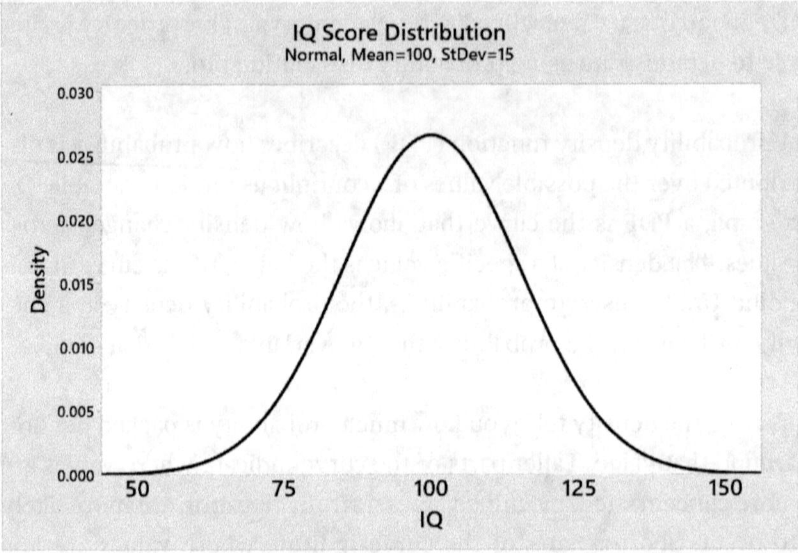

The curve shows how values are distributed across possible IQ scores. The curve peaks at 100, meaning values near 100 are the most common. As you move away from 100 in either direction, the curve becomes lower, indicating that values farther from the mean are less common. For instance, values near 50 or 150 are extremely rare. The steepness of the curve near the center shows that values change rapidly in density around the mean, while the flatter tails show that changes in density are more gradual for extreme values.

However, the graph hasn't given us any probabilities yet.

Finding probabilities on a probability density plot involves finding areas under the curve. For starters, the total area under the curve equals 1, representing the total probability of 1 over the full range of possible values in the distribution. In other words, the likelihood of a value falling anywhere in the complete distribution curve is 1. Finding the probability for a smaller range of interest involves finding the portion of the area under the curve that corresponds to that range.

As with discrete distributions, we often calculate probabilities for common types of ranges, such as less than a value, greater than a value, between two values, and more extreme values. For continuous distributions, these probabilities correspond to areas under the curve for those ranges and are typically found using cumulative distribution functions (CDFs).

For example, the following graph displays the distribution of IQ scores. Suppose we need to find the probability that the next IQ score we measure will fall between 120 and 140.

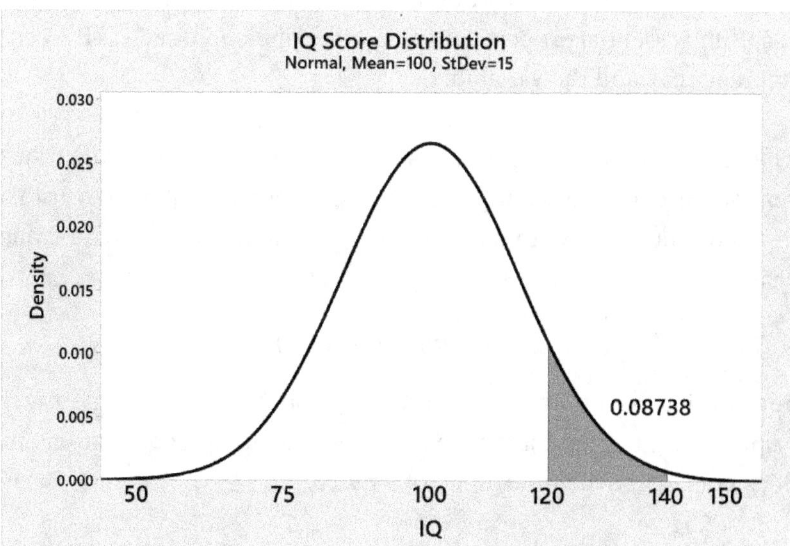

The graph shades that area and indicates that the proportion of the area under the curve between 120 and 140 is 0.08738. Consequently, the probability of the next IQ value falling within that range is 0.08738.

We will see how to calculate these probabilities later in the chapter. For now, focus on how probabilities correspond to areas.

Finally, you can't have an area under the curve with only a single value, which is another way to explain why the probability equals zero for an individual value. Consider that the probability for a PDF in statistics equals an area. For a non-zero area, you must have both a non-zero height and a non-zero width because Height × Width = Area. In this context, the height is the curve's height on the graph, while the width relates to the range of values. When you have a single value, you have zero width, which produces zero area. Hence, a zero chance!

Just as there are different types of discrete distributions for different kinds of discrete data, there are different distributions for continuous data. Each probability distribution has parameters that define the shape and essential properties of an entire population, such as the central tendency and the variability.

The most well-known continuous distribution is the normal distribution because it models a wide variety of phenomena. We'll cover this distribution and how to use it to find probabilities for most of this chapter.

A Note on Inequalities for Continuous Variables

For continuous random variables, the probability of a single exact value is zero. Because of that, including or excluding a boundary point does not change the probability of a range.

As a result:

- $P(X \leq a) = P(X < a)$
- $P(X \geq a) = P(X > a)$

In practice, this means that "less than" and "less than or equal to" represent the same probability for continuous distributions. The same is true for "greater than" and "greater than or equal to."

This happens because probabilities for continuous variables are based on areas under a curve, not on individual points. A single boundary value has no area, so adding or removing it does not affect the total probability.

For discrete distributions, this distinction matters because individual values can have positive probability. For continuous distributions, it does not.

Normal Distribution

The normal distribution is the most important probability distribution in statistics because it fits many natural phenomena. For example, heights, blood pressure, measurement error, and IQ scores follow the normal distribution. It is also known as the Gaussian distribution and the bell curve.

The normal distribution is symmetric, with most observations clustering around the central peak, and the probability of values farther from the mean tapering off equally in both directions. Extreme values in both tails of the distribution are similarly unlikely.

We'll start with the general properties of the normal distribution and its parameters. Then, we'll move on to using it to find probabilities.

Height data are normally distributed. The distribution in this example models real data I collected from 14-year-old girls during a study.

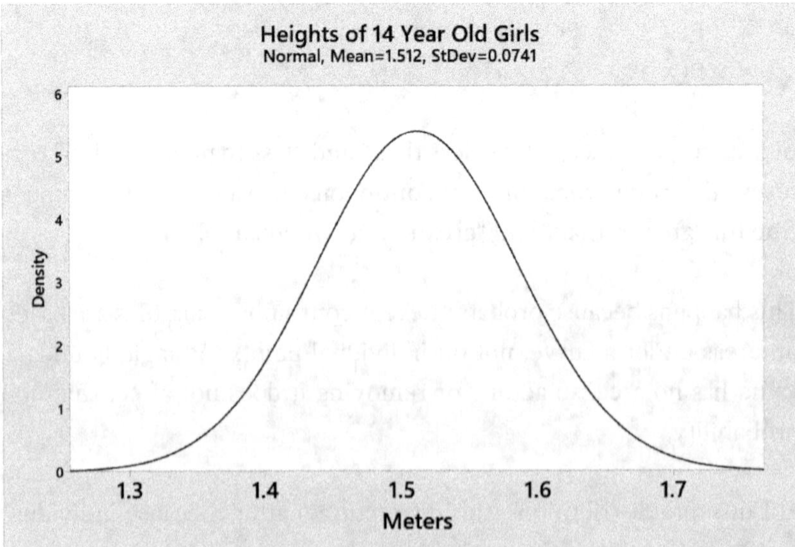

As you can see, the distribution of heights follows the typical pattern for all normal distributions. Most girls are close to the average (1.512 meters). Small differences between an individual's height and the mean occur more frequently than substantial deviations from the mean. The standard deviation is 0.0741m, indicating the typical distance individual girls tend to fall from the mean height.

Again, the distribution is symmetric. The number of girls shorter than average equals the number of girls taller than average. In both tails of the distribution, extremely short girls occur as infrequently as extremely tall girls.

Parameters

As with the discrete parametric models, the parameters for the normal distribution define its shape and probabilities entirely. The normal distribution has two parameters, the mean and standard deviation. The normal distribution does not have just one form. Instead, the shape changes based on the parameter values, as shown in the following graphs.

The mean, μ, is the central tendency of the distribution. In normal distributions, it defines the peak's location and is the expected value. Most values cluster around the mean. On a graph, changing the mean shifts the entire curve left or right on the X-axis.

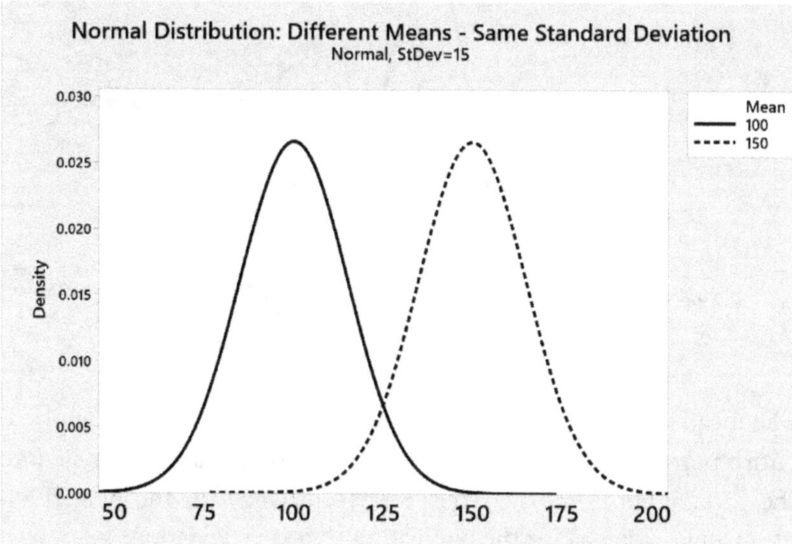

The standard deviation, σ, is a measure of variability. It defines the width of the normal distribution. The standard deviation indicates how far the values tend to fall from the mean. It represents the typical distance between the observations and the mean.

On a graph, changing the standard deviation either tightens or spreads out the width of the distribution along the X-axis. Larger standard deviations produce distributions that are more spread out.

For narrow distributions, values are more likely to fall close to the mean. As the spread increases, values are more likely to fall farther from the mean.

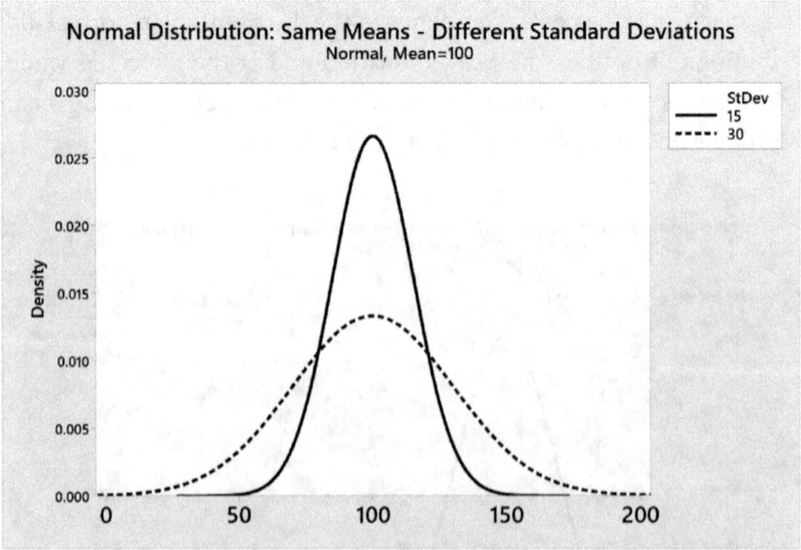

The mean and standard deviation are parameter values that apply to entire populations. For the normal distribution, statisticians denote the parameters using the Greek symbol μ (mu) for the population mean and σ (sigma) for the population standard deviation.

Use the same notation style as we did for discrete distributions to denote a normal distribution. $X \sim N(\mu, \sigma)$ refers to a distribution that follows a normal distribution with a population mean of μ and a standard deviation of σ. For example, the distribution of IQ scores is denoted as $X \sim N(100, 15)$.

In practice, population parameters are usually unknowable because it's generally impossible to measure an entire population. However, you can use random samples to estimate these parameters.

In probability problems involving the normal distribution, you are typically told to assume that the variable follows a normal distribution and are given the mean and standard deviation. These two parameters fully specify the distribution and provide all the information needed to find probabilities.

Unlike the discrete distributions, we do not calculate the mean and standard deviation directly from the probability density function for the normal distribution. Instead, these values are treated as given parameters or are estimated from data. In practice, you will usually be provided with μ and σ when working with normal distributions.

Regardless of the parameter values, all normal distributions share the following properties:

- They're all symmetric. The normal distribution cannot model skewed distributions.
- The mean, median, and mode are all equal.
- Half of the population is less than the mean, and half is greater than the mean.

Formula

The normal distribution formula gives the probability density for a given value of a continuous random variable (X).

The normal distribution equation is the following:

$$f(x) = \frac{1}{\sigma\sqrt{2\pi}} e^{-\frac{(x-\mu)^2}{2\sigma^2}}$$

Keep in mind that this formula does not give the probability for a specific value. Remember, the probability for any specific value in a PDF is always zero. Instead, the formula returns the probability density, or height of the curve, at any given value of x. However, the total area under the curve equals 1, and you can find the probability of a range of values by calculating the area under the curve for that range.

For continuous distributions, probabilities are obtained by accumulating probability density across a range of values. The density function tells us the *rate* at which the probability increases as we move along the number line. Conceptually, to find the total probability for an interval, we add up these tiny contributions across the range. Mathematically, this accumulation is done using integration, which computes the area under the curve.

Note: You do not need to perform integration by hand in this book.

In practice, you won't calculate these probabilities manually because the normal distribution equation and the integration process are complex. For exact solutions, you'll generally use statistical software. My website has free statistical calculators to solve normal distribution problems.

Additionally, statisticians have devised several simpler ways to obtain probabilities from the normal distribution using the empirical rule, z-scores, and tables. Stay tuned for more on these methods!

Normal Cumulative Distribution Function

In the previous sections, you explored the normal distribution using probability density function (PDF) graphs. Those graphs showed how probability density is distributed across values. In this section, we shift focus to how probability accumulates across values. To do that, we again return to cumulative distribution functions (CDFs).

You already encountered these functions when working with discrete probability distributions. In the discrete case, the CDF sums the probabilities of all outcomes up to a given value. The same idea applies here, but instead of adding individual probabilities, the CDF accumulates probability continuously across a range of values. For continuous distributions, the CDF represents the total probability that a random variable is less than or equal to a given value.

Graphically, this probability corresponds to the total area under the PDF curve to the left of that value. The CDF always increases from 0 to 1 as x moves from negative infinity to positive infinity. Additionally, these cumulative probabilities are equivalent to percentiles. A cumulative probability of 0.80 is equivalent to the 80th percentile.

Understanding cumulative probability will help you interpret a cumulative z-table, which lists these accumulated probabilities for the standard normal distribution. Later in this chapter, you'll use this table to solve a variety of probability problems.

For example, consider the height of an adult male in the United States. We can use the cumulative distribution function to find the probability that a person is 6 feet tall or less.

Adult male heights in the U.S. follow a normal distribution with a mean of 69.2 inches and a standard deviation of 2.66 inches. Consequently, we'll need to use a normal CDF with these parameters to answer our question. Because we're working in inches, I'll enter 72 inches for 6 feet.

The typical CDF statistical output from your software or online calculator will look like the following:

Cumulative Distribution Function

```
Normal with mean = 69.2 and standard deviation = 2.66

 x   P( X ≤ x )
72    0.853745
```

The probability that a randomly selected adult male in the U.S. will be 6 feet tall or shorter is 0.853745. Equivalently, you can say that a 6' tall adult male is at the 85.4th percentile.

Comparing Distributions

Cumulative distribution functions are useful for comparing two distributions. By comparing the CDFs of two random variables, we can determine whether one distribution is more likely to be less than a given value than the other.

Imagine we're a clothing manufacturer and want to compare the prevalence of 6' tall men to women. We'll use the normal CDF to find the probability that an adult woman will be 6' tall or less. Women's heights follow a normal distribution with a mean of 64.3 inches and a standard deviation of 2.58 inches.

Cumulative Distribution Function

```
Normal with mean = 64.3 and standard deviation = 2.58

 x   P( X ≤ x )
72    0.998580
```

The statistical output for the normal CDF indicates that women have a probability of 0.99858 for being ≤ 6'. That's equivalent to the 99.9th percentile.

85.4% of men and 99.9% of women are shorter than 6'. To compare how often heights above 6 feet occur in each population, we can compare the probabilities in the upper tails. By dividing the complement probabilities (1 - p) for the greater than probabilities, we find that the probability that a man is taller than 6 feet is about 103 times the probability that a woman is taller than 6 feet.

$$\frac{(1-0.853745)}{(1-0.99858)} = \frac{0.146255}{0.00142} = 102.99$$

As a clothing manufacturer, knowing that is helpful. A woman more than 6 feet tall is a rarity!

Graphing Normal CDFs

I always think graphs bring statistical concepts to life. So, let's graph a cumulative distribution function to see it. We'll return to the normal CDF for men's heights.

On a cumulative distribution function plot, the horizontal axis displays the x values, while the vertical axis displays cumulative probabilities or percentiles. The curve represents pairs of x-values and their corresponding cumulative probabilities. For normal CDFs, the function sums from negative infinity up to the value of x, which is $(-\infty, x]$ in interval notation. Continuous variables produce a smooth curve, as the following graph shows, while discrete variables produce the stepped function you saw before.

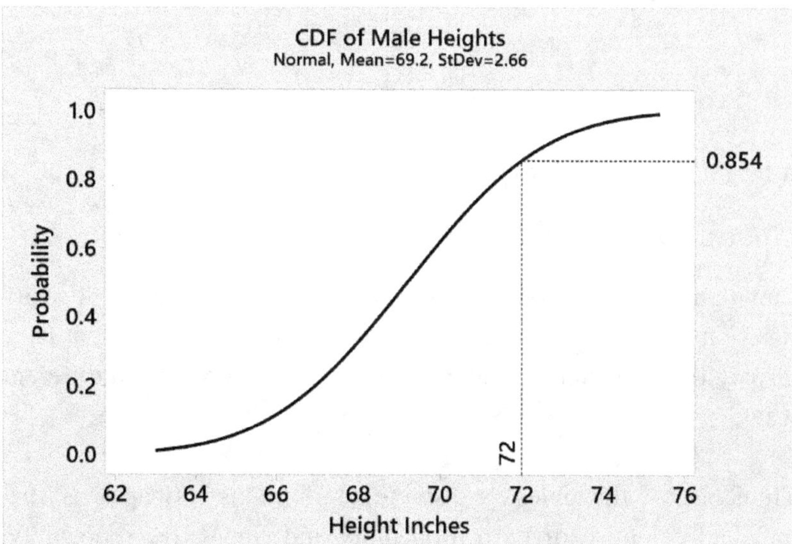

On the CDF graph for men's heights, I've added a reference line at 6' (i.e., 72") to show the corresponding probability of 0.854, matching the earlier answer within rounding. Using these graphs, you can easily find probabilities and percentiles for other values. For instance, 70 inches (5' 10") is around the 60th percentile.

For comparison, see the women's chart. While the graph ends at 72 inches, the distribution actually extends to infinity in both directions.

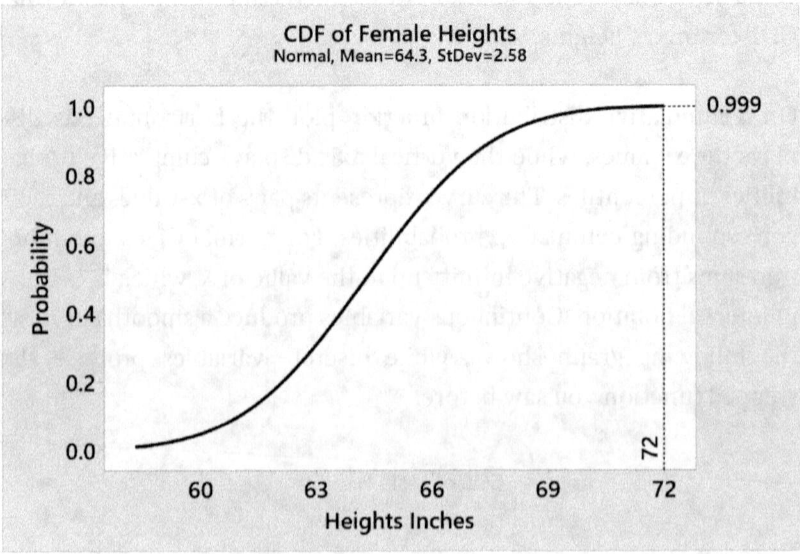

A height of 6 feet is in the tail of the distribution.

CDF vs. PDF

Cumulative distribution functions (CDFs) and probability density functions (PDFs) both describe a random variable's distribution. They display the same underlying probability information but in different ways.

Think back to the tables for discrete distributions in Chapter 10 that included columns for both probability and cumulative probability. They are two different sides of the same coin. We have a similar idea here with PDFs and CDFs. In simple terms, the PDF shows the shape of the distribution, while the CDF shows how probability accumulates as the value of the random variable increases.

Let's explore this briefly!

PDFs can find cumulative probabilities by calculating the area under the curve up to a particular value. The following PDF shows the probability for the shaded area representing male heights up to 6' (72").

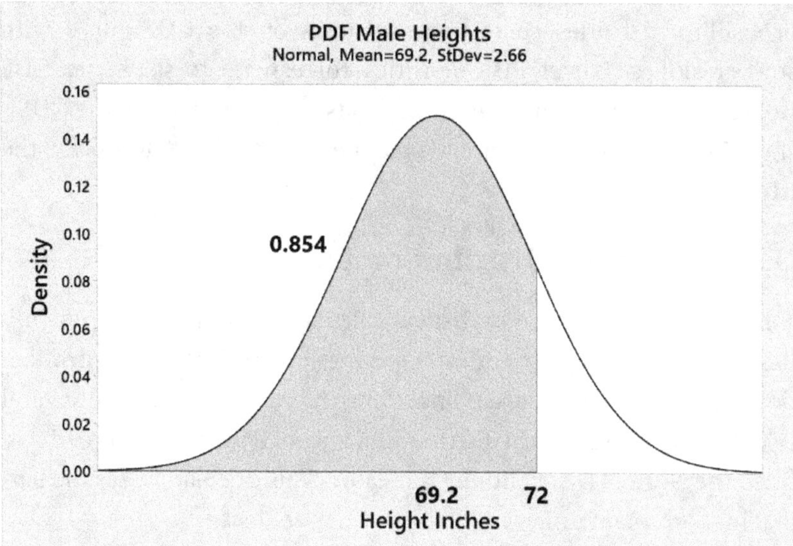

It finds the same probability as the CDF, showing how they present the same underlying information in different formats.

Now, imagine that you started with the shaded area to the left side of the PDF and systematically move it to the right while recording the cumulative probabilities. That process produces the CDF!

The PDF gives the probability density, indicating how concentrated values are around each point. In comparison, the cumulative distribution function accumulates probability up to each value by adding the area under the curve from left to right.

In this manner, the probability density on a PDF is the rate of change for the CDF. This relationship mirrors what you saw with discrete distributions in Chapter 10. In those tables, the cumulative probability column was a running total of the probability column, just as the CDF accumulates probability density from the PDF.

Consequently, the ranges where the PDF curve has relatively high probability densities correspond to areas on the CDF curve with steeper slopes. Lower PDF densities correspond to shallower CDF slopes. As the PDF's curve approaches its peak at the mean, the CDF's slope increases to its maximum steepness. After the PDF's peak, the CDF slope flattens.

Using a Normal Distribution Calculator

Next, let's use one of my statistical calculators to calculate probabilities for the normal distribution. Find the link in the book's Introduction for all the free resources, including the calculators. I have several calculators for the normal distribution, but we'll use the one creatively titled the Normal Distribution Calculator. This calculator uses the natural measurement units, which is very convenient.

Assume that a pizza restaurant's delivery times follow a normal distribution with a mean of 30 minutes and a standard deviation of 5 minutes. Our pizza took 38 minutes to be delivered. We want to determine the percentile for our delivery time. In other words, what is the probability of having a pizza delivered in 38 minutes or less?

In the calculator, enter the following values and click **Calculate**:

- **Observed Value (x)**: 38
- **Population Mean (μ)**: 30
- **Standard Deviation (σ)**: 5

The calculator displays probabilities for ranges involving our value of x = 38, including less than, greater than, between, and more extreme. Each probability has a corresponding normal distribution graph that shades the appropriate area(s) to help identify your probability of interest.

To find the probability for a delivery time of less than 38 minutes, look in the output for $P(X < x)$. The probability of a pizza being delivered in 38 minutes or less is 0.94520. Alternatively, our delivery time was approximately the 95th percentile.

Now, try one yourself. Suppose you have an apple that weighs 110 grams. Assume that apple weights follow a normal distribution with a mean of 100 grams and a standard deviation of 15.

What is the probability that the next randomly selected apple will be greater than this weight?

See the correct answer in the z-table section, where we solve it using z-scores.

This calculator also finds probabilities for greater-than, in-between, and more-extreme-than type problems. We'll solve all these types of problems in the z-table section.

Next, we transition to several methods for using the normal distribution to find probabilities without a computer.

The Empirical Rule

The empirical rule is our first way to quickly estimate probabilities from the normal distribution. You just need to know the mean and standard deviation, and then use simple addition and subtraction.

When a variable follows a normal distribution, use the empirical rule to determine the probability that the next observed value will fall within a specified number of standard deviations from the mean. The empirical rule specifies the proportion of values that fall within 1, 2, and 3 standard deviations of the mean.

Mean ± standard deviations	Probability
1	68%
2	95%
3	99.7%

For example, in a normal distribution, 68% of the observations fall within ±1 standard deviation of the mean. Equivalently, the probability that a randomly selected observation will fall in that range is 0.68.

To apply the empirical rule, multiply the standard deviation by the number of standard deviations you want to examine, then add and subtract that amount from the mean to find the range of values.

Let's return to our pizza delivery scenario where delivery times follow a normal distribution with a mean of 30 minutes and a standard deviation of 5 minutes.

Using the empirical rule, we can determine that 68% of the delivery times are between 25-35 minutes (30 ± 5), 95% are between 20-40 minutes (30 ± 2×5), and 99.7% are between 15-45 minutes (30 ±3×5). The chart illustrates this property graphically.

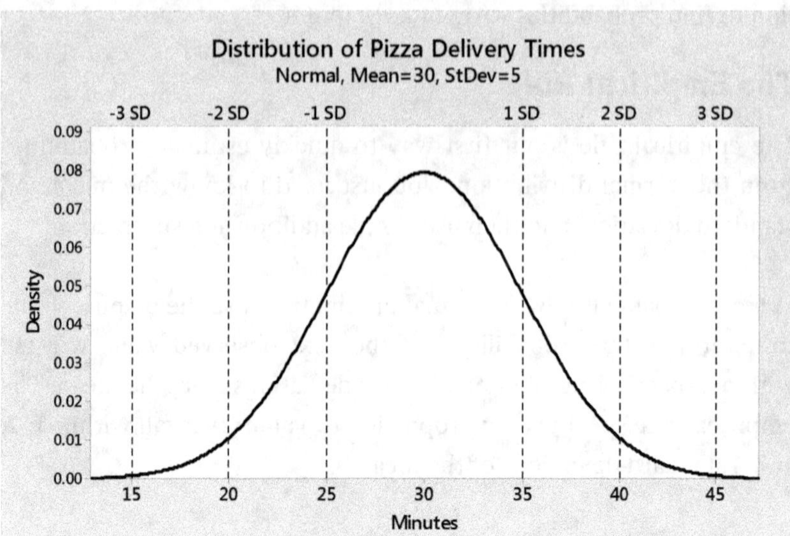

You can extend the empirical rule by breaking the ranges into smaller parts and taking advantage of the fact that the normal distribution is symmetric. For example, about 68% of values fall within one standard deviation of the mean, so about 34% fall between the mean and one standard deviation above it or below it. Consequently, about 84% of values fall below the mean + 1SD because this range includes the 50% below the mean and the 34% between the mean and 1SD above it.

These probabilities can be very helpful for quick, back-of-the-envelope calculations. However, you'll often want to find probabilities for ranges that you specify, rather than only ranges based on standard deviations. You can still do this without using computers, but you will first need to learn about the standard normal distribution and z-scores.

Standard Normal Distribution

Normal distributions can have many different properties. They all share the same bell-shaped form, but they differ in where they are centered and how spread out they are. Some normal distributions are tall and narrow, while others are wider and flatter. These differences are determined by the mean and the standard deviation.

The standard normal distribution is a special version of the normal distribution that removes differences in location and spread. It is centered at zero and has a standard deviation of one. By converting values from any normal distribution into this common scale, we can work with a single standardized curve rather than many different ones. This distribution is also known as the Z-distribution, and we denote the standard normal variable as $Z \sim N(0, 1)$. Values in this distribution are called z-scores.

When you know a variable follows a normal distribution, you only need three pieces of information to find probabilities: the value(s) of interest, the mean, and the standard deviation. Standardization uses this information to translate the problem into the standard normal setting, where you can easily find probabilities in a statistical table.

In effect, the standard normal distribution provides a common reference that allows you to find probabilities from different normal distributions.

Z-scores

The standard normal distribution provides a common reference for working with normal probability models, but using it requires a way to translate values from the original scale to the standardized scale. That translation is done using z-scores. A z-score expresses how far a value lies from the mean, measured in units of standard deviation. Statisticians refer to this conversion process as standardization because it places your raw data on a standardized scale. Similarly, z-scores are also known as standard scores.

After you calculate the z-score for a raw value, you can find probabilities directly from the standard normal distribution.

Use the following formula to convert raw measurements into z-scores:

$$Z = \frac{(x - \mu)}{\sigma}$$

X represents the raw value of the measurement of interest. Mu (μ) and sigma (σ) represent the parameters for the population from which the observation was drawn.

To calculate the standard score for an observation:

1. Take the raw measurement.
2. Subtract the mean.
3. Divide by the standard deviation.

A z-score represents the number of standard deviations above or below the mean that a specific observation falls. A positive score indicates the raw value is above the mean. Conversely, a negative score represents a value below the average. The mean has a z-score of 0, which appears at the peak of the following graph.

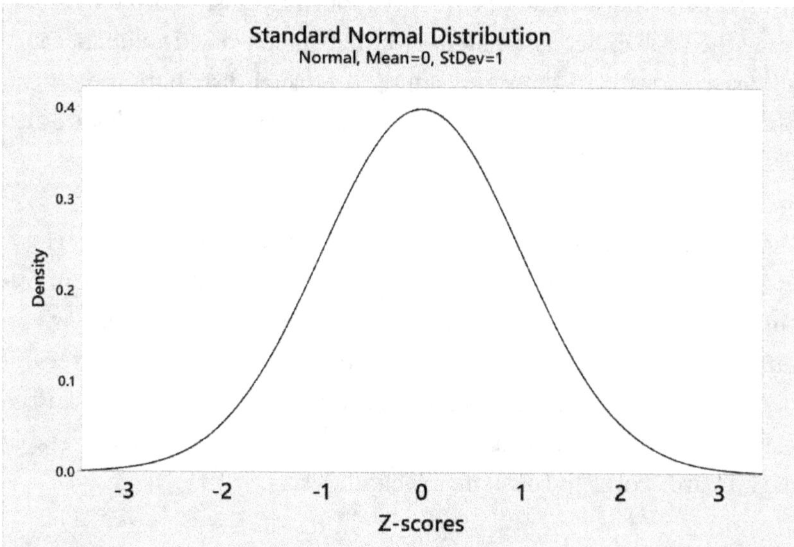

For example, a standard score of 1.5 indicates that the observation is 1.5 standard deviations above the mean. By locating that value on the standard normal distribution graph, you know it is above average and values in that region are somewhat unusual.

Now, let's calculate an example z-score. Suppose we have an apple that weighs 110 grams. We also know that this type of apple has weights that follow a normal distribution with a mean of 100 and a standard deviation of 15.

Z = (110-100) / 15 = 0.667

By placing this value in the standard normal distribution curve, you can see that this value is above the mean but isn't particularly rare. Additionally, you can use z-scores to find probabilities by using a statistical table.

Using a Z-table

After converting a value to a z-score, your next task is often to determine the probability associated with that standardized value. Because all z-scores refer to the same standard normal distribution, you can obtain probabilities from a single reference table. These tables are also known as the standard normal table.

Typically, I use statistical software to find areas under the curve. However, when you're working with the normal distribution, you can calculate areas by looking up z-scores in a table. Because there are an infinite number of different normal distributions, publishers can't print a table for each distribution. However, you can transform the values from any normal distribution into z-scores, and then use a table of standard scores to calculate probabilities.

Z-tables come in various formats, but I will stick with the most common. The z-table I use for the rest of the book reports cumulative probabilities to the left of a given z-score, $P(Z \leq z)$. In other words, they provide the area under the curve to the left of a z-score. This area represents the probability that z-scores will fall within a region of the standard normal distribution. However, keep in mind that z-tables are accurate only for populations that follow a normal distribution.

Note: The complete z-table is available in Appendix A.

As you should remember from the discrete distributions, cumulative probabilities are a great way to find probabilities for ranges of values. Additionally, all probabilities for continuous distributions are for ranges. Hence, you're often using cumulative probabilities for continuous variables.

You can use these z-tables to find probabilities corresponding to ranges of z-scores. Alternatively, you can find raw data values that correspond to given probabilities and their corresponding z-scores.

I'll go over the methods for solving all these types of common probability problems but let's first look at the anatomy of a z-table! In the examples that follow, I'll show you how to use the table and graph what we're doing so you can see it visually.

A z-table is often divided into two sections, negative and positive z-scores. Negative z-scores are below the mean, while positive z-scores are above the mean. Row and column headers define the z-score, while table cells represent the area.

In the z-score table, its components represent the following:

- Row headings define the z-score to the tenth's place.
- Column headings add the z-scores' hundredth's place.
- Cells within the table represent the area under the standard normal curve to the left of the z-score.

For example, to find the area for z = -2.23, look at the row (-2.2) and column (0.03) intersection, as shown in the truncated z-table.

Z	**0.00**	**0.01**	**0.02**	**0.03**	**0.04**	**0.05**
-2.5	0.00621	0.00604	0.00587	0.00570	0.00554	0.00539
-2.4	0.00820	0.00798	0.00776	0.00755	0.00734	0.00714
-2.3	0.01072	0.01044	0.01017	0.00990	0.00964	0.00939
-2.2	0.01390	0.01355	0.01321	0.01287	0.01255	0.01222
-2.1	0.01786	0.01743	0.01700	0.01659	0.01618	0.01578
-2.0	0.02275	0.02222	0.02169	0.02118	0.02068	0.02018

The area to the left of z = -2.23 is 0.01287. The probability of z-values falling below this value in a standard normal distribution is 0.01287 or 1.287%.

Suppose you could graph the entire z-table by taking all the z-scores and plotting their cumulative probabilities. Here's what that CDF graph looks like, along with the value from the table.

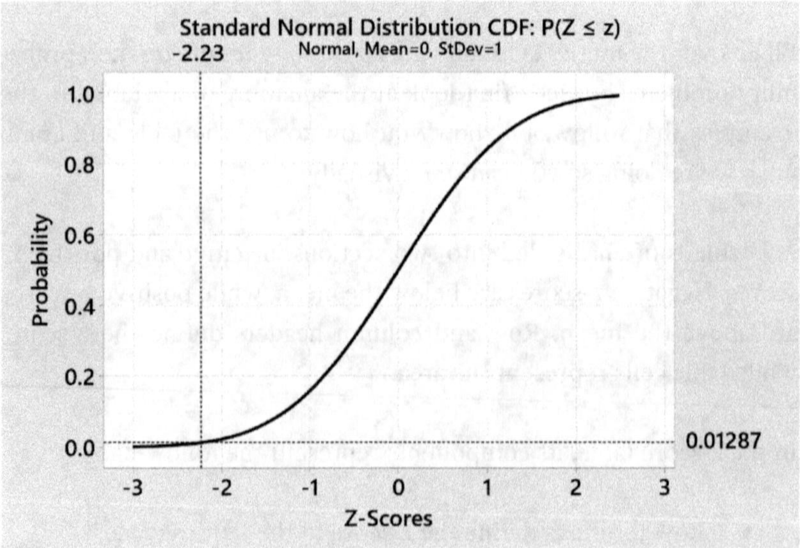

Here's what it looks like using a PDF graph with the shaded area for the probability. Two different ways to show the same thing.

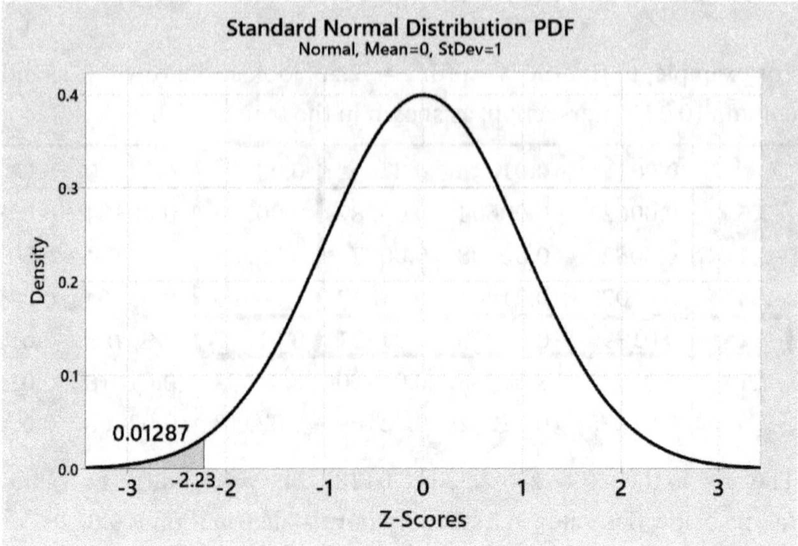

Going forward, I will illustrate the examples using PDF plots with shading.

Solving Probability Problems with Z-tables

In this section, you will learn how to use a z-table to solve common probability questions involving the normal distribution.

By looking up a z-score in the table, you can find probabilities that the next observed value will be greater or less than a particular value, fall between two values, or be more extreme than two values. Additionally, you can solve reverse probability problems where the problem provides a probability, and you need to use a z-table to find the raw values that correspond to it.

Before using a z-table, it often helps to sketch a quick diagram of the normal curve and mark the values involved in the problem. The diagram does not need to be precise or to scale. Its purpose is simply to give you a visual picture of where the values lie relative to the mean. Start by marking the mean at the center, then place the value or values of interest to the left or right based on whether they fall below or above the mean.

This simple step helps you identify whether the problem involves a left-tail, right-tail, between-values, or outside-values probability. A rough sketch also helps you see which probabilities to include or subtract when using the z-table, reducing errors and making the overall process clearer.

When comparing z-table results to a statistical calculator, you will often notice small differences. Z-tables typically round z-scores to two decimal places, while calculators use more precise values.

Reminder: For continuous distributions, ≤ and < are equivalent, and so are ≥ and >.

Less Than or Greater Than

The most straightforward problem is finding a probability below a z-score because the z-table shows that without requiring further calculations. You can also use this type of result as a percentile. Statisticians also refer to these less-than questions as left-tail probabilities because they include everything to the left of (i.e., less than) a cutoff value.

We denote this type of problem as $P(Z \leq z)$, where z is the cutoff z-score. Keep in mind that z represents a standardized value, not the raw data value. For many probability problems, you first need to calculate the z-score from the information the problem provides.

Suppose you have an apple that weighs 110 grams. Assume that apple weights follow a normal distribution with a mean of 100 grams and a standard deviation of 15.

The z-score for our apple is (110-100) / 15 = 0.67.

Let's use that z-score to find the probability that the next apple we weigh will be less than this value.

This truncated z-table shows the area for our z-score.

Z	0.05	0.06	0.07	0.08	0.09
0.0	0.51994	0.52392	0.52790	0.53188	0.53586
0.1	0.55962	0.56356	0.56749	0.57142	0.57535
0.2	0.59871	0.60257	0.60642	0.61026	0.61409
0.3	0.63683	0.64058	0.64431	0.64803	0.65173
0.4	0.67364	0.67724	0.68082	0.68439	0.68793
0.5	0.70884	0.71226	0.71566	0.71904	0.72240
0.6	0.74215	0.74537	0.74857	0.75175	0.75490
0.7	0.77337	0.77637	0.77935	0.78230	0.78524
0.8	0.80234	0.80511	0.80785	0.81057	0.81327
0.9	0.82894	0.83147	0.83398	0.83646	0.83891

By looking at the intersection for 0.6 and 0.07, the z-table indicates that the area under the curve for z = 0.67 is 0.74857. This value indicates that 74.857% of all apple weights will be lower than our apple weight of 110 grams. Alternatively, you can say that our apple is at the 74.857th percentile.

The following chart illustrates this solution.

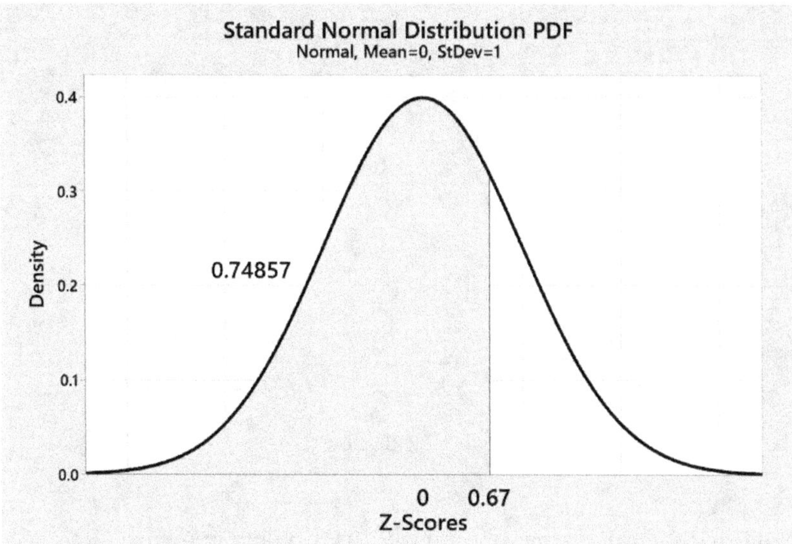

Finding Areas Above a Z-score

To find areas above a z-score using a z-table, understand that the total area under the curve equals 1. Consequently, to find the area above a z-score, use the complement probability. Find the area below the z-score in the z-table and subtract it from 1.

Area above the z-score = 1 – area below the z-score

These problems are also known as right-tail probabilities because they include all the area to the right of the value. We denote them as $P(Z \geq z)$.

For the apple example, the z-table indicated that the area below z = 0.67 was 0.74857. Consequently, the area above that z-score is 1 – 0.74857 = 0.25143. In other words, the probability that the next apple will weigh more than the current apple is 0.25143.

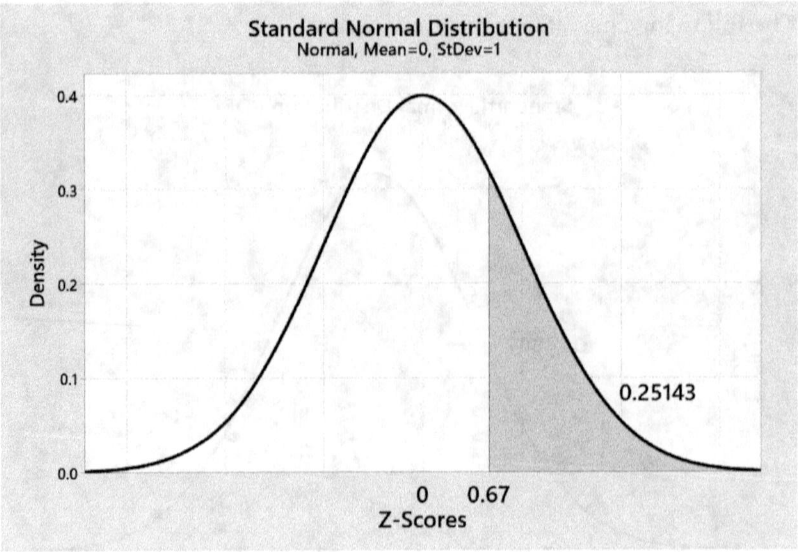

Note that the calculator solution for this problem in the previous section is slightly different. It produces $P(X \geq 110) = 0.25249$. Statistical calculators are more accurate than the z-table method because statistical tables often use fewer decimal places for z-scores.

Between Two Values

Some problems ask you to find the area between two z-scores. For example, the wording might ask you to find the probability of values falling within a specified number of standard deviations from the mean. The number of standard deviations becomes your positive and negative z-scores. Alternatively, you might need to calculate the z-scores from raw data values that the problem provides.

We denote this type of problem as $P(z_1 \leq Z \leq z_2)$, where z_1 and z_2 are the standardized cutoff values. This probability represents the proportion of values that fall between the two z-scores on the standard normal distribution.

The following process uses a z-table to find the probability for the range between any two z-scores:

1. Find the area below the higher z-score.
2. Find the area below the lower z-score
3. Take the larger area and subtract the smaller area from it.

Area for range between z-scores = Larger area – Smaller area

Subtracting out the smaller area removes that region from the larger total, leaving only your middle range of interest.

For example, the empirical rule describes the probabilities for z-score ranges of ±1, ±2, and ±3 standard deviations from the mean. However, what is the probability of observations falling within ± 1.5 standard deviations from the mean?

The following two truncated z-tables show the areas for z-scores of 1.5 and -1.5.

Z	0.00	0.01	0.02	0.03	0.04
1.2	0.88493	0.88686	0.88877	0.89065	0.89251
1.3	0.90320	0.90490	0.90658	0.90824	0.90988
1.4	0.91924	0.92073	0.92220	0.92364	0.92507
1.5	0.93319	0.93448	0.93574	0.93699	0.93822
1.6	0.94520	0.94630	0.94738	0.94845	0.94950
1.7	0.95543	0.95637	0.95728	0.95818	0.95907

Z	0.00	0.01	0.02	0.03	0.04
-1.7	0.04457	0.04363	0.04272	0.04182	0.04093
-1.6	0.05480	0.05370	0.05262	0.05155	0.05050
-1.5	0.06681	0.06552	0.06426	0.06301	0.06178
-1.4	0.08076	0.07927	0.07780	0.07636	0.07493
-1.3	0.09680	0.09510	0.09342	0.09176	0.09012

The z-tables indicate that the areas for the higher and lower z-scores are 0.93319 and 0.06681, respectively.

Therefore, the area between these two z-scores is 0.93319 – 0.06681 = 0.86638.

We can conclude that 86.638% of values will fall within ± 1.5 standard deviations of the mean in a normally distributed population. The chart illustrates this answer.

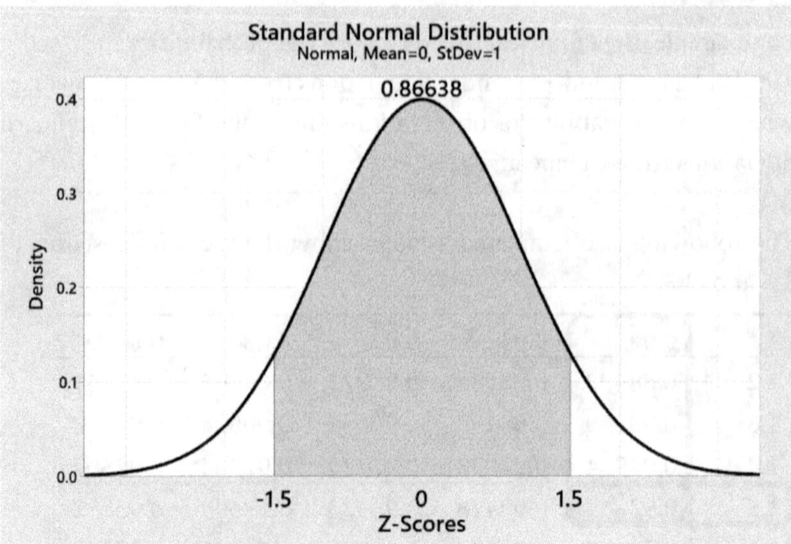

Outside Two Values

Problems might ask you to use a z-table to find the probability for distribution regions that are more extreme than the given z-scores. In this context, more extreme refers to values that are further away from the mean than the two z-scores. These are two-tailed probabilities because they involve both tails of the distribution.

We denote this type of problem as $P(Z \leq z_1 \text{ or } Z \geq z_2)$, where z_1 and z_2 are the standardized cutoff values on opposite sides of the mean. This probability represents the proportion of values that fall outside the two z-scores, in the distribution's more extreme regions.

To calculate this type of information using a z-table, do the following:

1. Find the area greater than the higher z-score using the method described earlier.
2. Find the area less than the lower z-score.
3. Add the two areas for the total.

For example, a problem might ask you to find the area of z-scores that are more extreme than ± 2.5 standard deviations from the mean.

There's a shortcut for this method when you have the ± values of the same z-score, as we do in this example. The shortcut is that the standard normal distribution is symmetrical. Simply double the area of the negative z-score to find the total area in both tails.

In our example, the area less than z = -2.5 equals the area greater than z = 2.5, thanks to the symmetry.

Z	0.00	0.01	0.02
-2.7	0.00347	0.00336	0.00326
-2.6	0.00466	0.00453	0.00440
-2.5	0.00621	0.00604	0.00587
-2.4	0.00820	0.00798	0.00776

The z-table indicates that the area to the left of z = -2.5 is 0.00621. Consequently, the area to the right of z = 2.5 is also 0.00621.

Sum these two areas to find the total area for these regions: 0.00621 + 0.00621 = 0.01242.

We can conclude that 1.242% of observations will fall more than ±2.5 standard deviations from the mean in a normally distributed population.

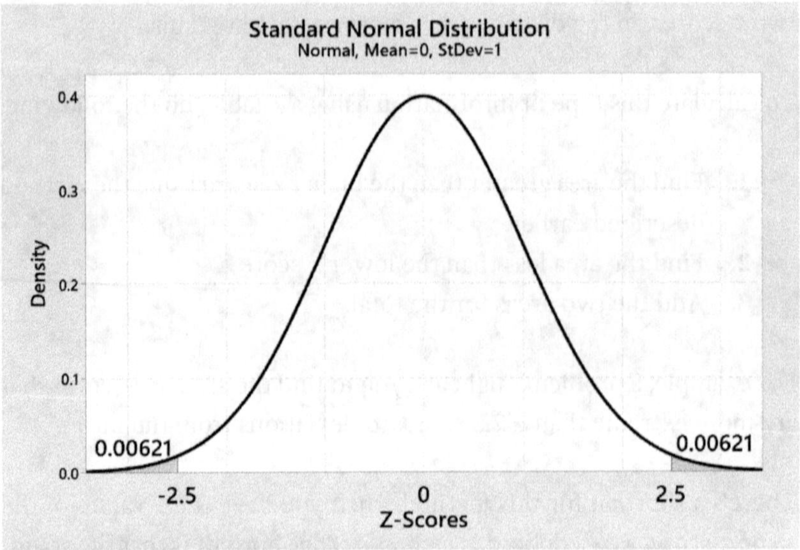

Finding Values for Given Probabilities

In the previous sections, you used z-scores and z-tables to find probabilities associated with known values. In those problems, you started with raw data values, converted them to z-scores, and then used the standard normal distribution table to find probabilities. In this section, you reverse that process. Instead of starting with a value and finding a probability, you start with a probability and you determine the values in natural units that correspond with it.

These problems focus on identifying cutoff values that capture a specified proportion of the distribution. For example, you might need to find the value that marks the top 5% of outcomes, the value below which 90% of outcomes fall, or the range of values that contains the middle portion of the distribution. Solving these problems still relies on z-scores and the standard normal distribution, but now you will use the z-table to locate probabilities first and then work backward to find the associated values on the original scale.

In the sections that follow, learn how to find cutoff values for single-tail probabilities, central ranges between two cutoffs, and more extreme regions outside two cutoffs.

Less Than Cutoff Values

This type of single-tailed probability question asks you to find the value on the original scale that corresponds to a specified probability in one tail of a specified normal distribution.

Let's go back to our apple distribution. Find the apple weight that is greater than 95% (i.e., the 95th percentile) of all apple weights in a normal distribution with a mean of 100 and standard deviation of 15.

To solve a less than a cutoff value question like this one, use the following procedure:

1. Find the given probability in the cumulative z-table.
2. Note the corresponding z-score.
3. Convert the z-score to the original units.

Tip: If the given probability is greater than 0.5, the z-score will be positive. If the probability is less than 0.5, the z-score will be negative. If the probability is 0.5, the z-score is 0.

The true z-score usually falls between two table entries, so choosing the closest value produces the most accurate estimate.

Returning to our apple problem, we'll start by finding the z-score for 0.95. Because the probability is greater than 0.5, we'll look in the

Positive Z-score Table. As the following truncated table indicates, the z-score for 0.95 falls between 1.64 and 1.65.

Z	0.02	0.03	0.04	0.05	0.06
1.4	0.92220	0.92364	0.92507	0.92647	0.92785
1.5	0.93574	0.93699	0.93822	0.93943	0.94062
1.6	0.94738	0.94845	0.94950	0.95053	0.95154
1.7	0.95728	0.95818	0.95907	0.95994	0.96080
1.8	0.96562	0.96638	0.96712	0.96784	0.96856
1.9	0.97257	0.97320	0.97381	0.97441	0.97500

Because the probability for 1.64 (0.94950) is slightly closer to our target of 0.95, we'll use that z-score.

Now, we need to convert z = 1.64 to the original units. The formula for doing that is the reverse of the earlier formula for converting a data value to a z-score.

$$x = \mu + z\sigma$$

Where:

- x = value in the original units
- μ = mean of the distribution
- z = z-score
- σ = standard deviation

So, plugging in the μ and σ values from the problem and the z that we found in the table:

$x = 100 + 15 \times 1.64 = 124.6$

The upper cutoff value for this problem is 124.6 grams. Specifically, there is a 0.95 probability that the next apple randomly drawn from this population will weigh less than 124.6 grams. Equivalently, an apple that weighs 124.6 grams is at the 95th percentile.

The following graph illustrates the solution. Notice how the x-axis has been converted back to the natural units of grams.

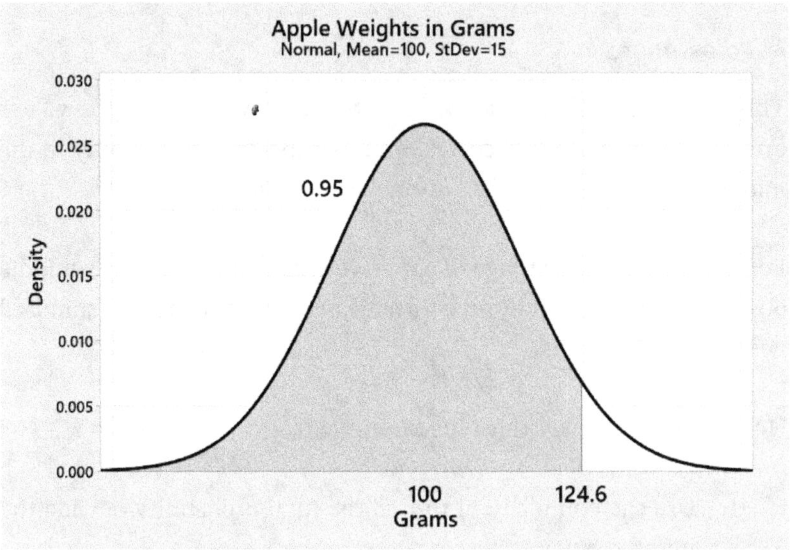

Greater Than Cutoff Values

For greater-than cutoff problems, you can use the same overall process as the less-than approach with one small adjustment. Because the z-table reports cumulative probabilities to the left, first convert the greater-than probability into an equivalent less-than probability using complements.

Subtract the given probability from 1 to find the corresponding left-tail probability. Then follow the same steps used in the less-than procedure: locate the probability in the z-table, find the corresponding z-score, and convert that z-score back to the original scale using the mean and standard deviation.

For example, suppose we are asked to find the apple weight for which there is only a 0.05 probability that the next apple will be greater than it. First subtract the probability from 1: 1 – 0.05 = 0.95. Then find this probability in the z-table and convert the corresponding z-score to the original scale, just as we did in the previous less-than example. We find that 5% of observations exceed 124.6 grams, which is consistent with 95% falling below that value in the previous example.

Middle Ranges

This type of middle probability question asks you to find the values on the original scale that correspond to a specified probability in the middle portion of a specified normal distribution.

For example, find the range of apple weights that contains the middle 60% of a normal distribution with a mean of 100 grams and a standard deviation of 15 grams.

Here's the process for this type of problem:

1. Use the complement to find the total probability outside the middle region.
2. Divide the complement by 2 to find the excluded probability in each tail.
3. Find the z-score (which will be negative) in the cumulative table that corresponds to step #2.
4. Symmetry dictates that the other z-score is the positive value.
5. Convert both z-scores to the original units.

Tip: The symmetry of the normal distribution helps simplify this type of problem.

Because we need to find the middle 60% of apple weights, we start by subtracting 0.6 from 1: 1 - 0.6 = 0.4. That complement finds the proportion outside the middle range. Then we divide that in half to find how much is excluded from each tail: 0.4/2 = 0.2.

The value of 0.2 defines the lower bound of the range because it excludes the lowest 20%. So, we need to locate that value in the table.

Z	0.01	0.02	0.03	0.04	0.05
-0.9	0.18141	0.17879	0.17619	0.17361	0.17106
-0.8	0.20897	0.20611	0.20327	0.20045	0.19766
-0.7	0.23885	0.23576	0.23270	0.22965	0.22663
-0.6	0.27093	0.26763	0.26435	0.26109	0.25785
-0.5	0.30503	0.30153	0.29806	0.29460	0.29116

The value that is closest to our target probability of 0.2 is z = -0.84.

Symmetry tells us the upper bound for the range must be z = +0.84, which excludes the highest 20%. That leaves us with the middle 60% falling between the z-score interval [-0.84, 0.84].

Next, we need to convert these two z-scores to the original units:

- $x = 100 + 15 \times -0.84 = 87.4$
- $x = 100 + 15 \times 0.84 = 112.6$

Apple weights have a 0.6 probability of falling within the range of 87.4 to 112.6 grams. The following graph illustrates these results.

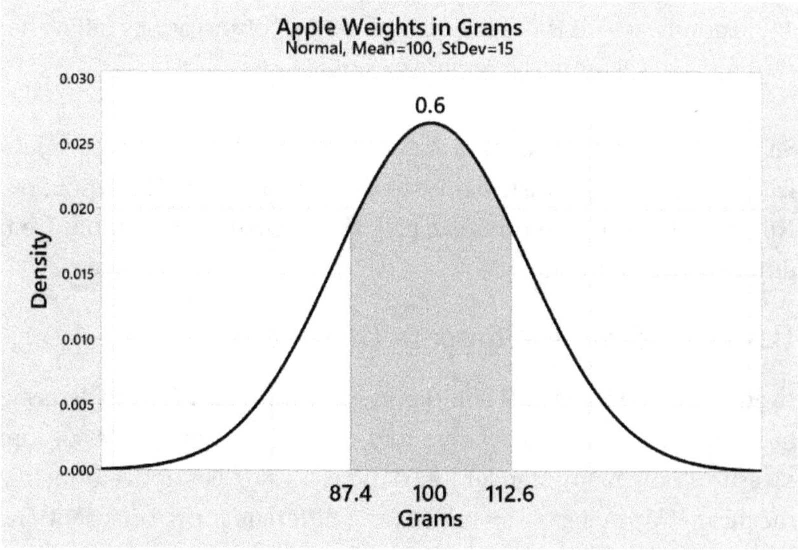

Outside Ranges

This type of outside probability question asks you to find the values on the original scale that correspond to a specified probability in the outer portions of a specified normal distribution.

This process is nearly identical to the middle-range procedure. The only difference is that the problem already gives you the outside probability, so you do not need to use complements. You simply divide the outside probability equally between the two tails.

For example, suppose we are asked to find the apple weights that define the outer 40% of the distribution. Because we already know that 40% of observations fall outside the middle range, we don't use complements. Instead, divide this probability evenly between the two tails: 0.40 ÷ 2 = 0.20. Then find this probability in the z-table and convert the corresponding z-scores to the original scale, just as we did in the previous middle-range example.

We find that 20% of observations fall below 87.4 grams and 20% above 112.6 grams, which is consistent with 60% of observations falling between those values in the previous example.

So far, we've used z-scores to find probabilities by converting values from a single normal distribution to the standard normal distribution. Now you'll see how z-scores can help you compare observations from different distributions.

Using Z-scores to Compare Distributions

In addition to finding probabilities, z-scores provide a way to describe how unusual a value is relative to a normal distribution. A z-score measures how many standard deviations a value lies above or below the mean. By placing observations from different normal distributions onto the same scale, z-scores allow you to compare their relative positions within their own distributions.

This process, called standardization, converts observations from populations with different means and standard deviations to a common scale. After standardizing the data, you can place the observations within the standard normal distribution to compare where each falls relative to its own distribution. In other words, standardization allows you to compare apples to oranges. Isn't statistics great!

Suppose we literally want to compare apples to oranges. Specifically, let's compare their weights. Imagine we have an apple weighing 110 grams and an orange weighing 100 grams.

If we compare the raw values, it's easy to see that the apple weighs more than the orange. However, let's compare their standard scores. To do this, we'll need to know the properties of the weight distributions for apples and oranges. Assume that the weights of apples and oranges follow a normal distribution with the following parameter values:

	Apples	**Oranges**
Mean weight grams	100	140
Standard deviation	15	25

Now let's calculate the z-scores:

- Apple = (110-100) / 15 = 0.667
- Orange = (100-140) / 25 = -1.6

The z-score for the apple (0.667) is positive, which means that our apple weighs more than the average apple. It's not an extreme value by any means, but it is above average for apples. On the other hand, the orange has a fairly negative z-score (-1.6). It's pretty far below the mean weight for oranges. I've placed these z-scores in the standard normal distribution.

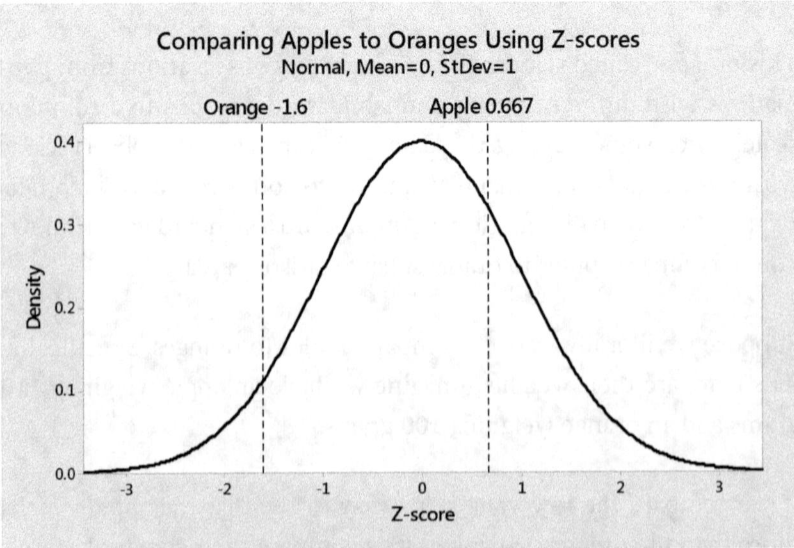

While our apple weighs more than our orange, we are comparing a somewhat heavier than average apple to a downright puny orange! Using z-scores, we've learned how each fruit fits within its own distribution and how they compare to each other.

Normal Approximations to Discrete Distributions

In earlier chapters, you saw how the formulas for discrete probability distributions such as the binomial and Poisson can sometimes produce extremely large numbers, creating computational challenges. For these distributions, large numbers arise when the number of trials, event counts, or rate parameters become large.

In binomial formulas, factorial and combination terms can become extremely large even when calculating a single probability. Similarly, Poisson probabilities can involve very large factorial values. These calculations were difficult to perform by hand, making normal approximations useful for simplifying the math.

In this context, an approximation means using one probability distribution to estimate probabilities that technically come from another distribution. Approximations can work because different probability distributions have very similar shapes under particular conditions. When the normal distribution approximates a discrete distribution, the probability calculated using the normal distribution will be very close to the probability from the discrete distribution.

Historically, statisticians used approximations to simplify these calculations. Before computers and statistical software were widely available, calculating exact probabilities involving large values could be slow and impractical. Approximations provided a faster way to obtain probability estimates. Indeed, some hypothesis tests even used approximations. However, these results were still approximations rather than exact answers.

Today, computers and statistical software can quickly calculate exact binomial and Poisson probabilities, even for very large numbers. Because of this, you rarely need to rely on normal approximations in practical work.

Previously, textbooks emphasized these approximations because they made difficult probability calculations easier to perform by hand. However, many textbooks continue to teach approximations because they help illustrate how different probability distributions become similar under certain conditions.

For this reason, I'm just going to explain the conditions that produce these approximations for the binomial and Poisson distributions and show you several graphs. My goal here is to teach you the concept so you know what it means if you hear about approximations.

Binomial Approximation

Normal distributions can adequately approximate the binomial distribution when the probability of success is not too close to 0 or 1, and the number of trials is reasonably large. The normal approximation uses $n \times p$ and the square root of $n \times p \times (1-p)$ for its mean and standard deviation, respectively. Where:

- n = number of trials
- p = event probability

In the graph, notice how the two distributions have almost the same shapes and produce similar probability densities for all values of x.

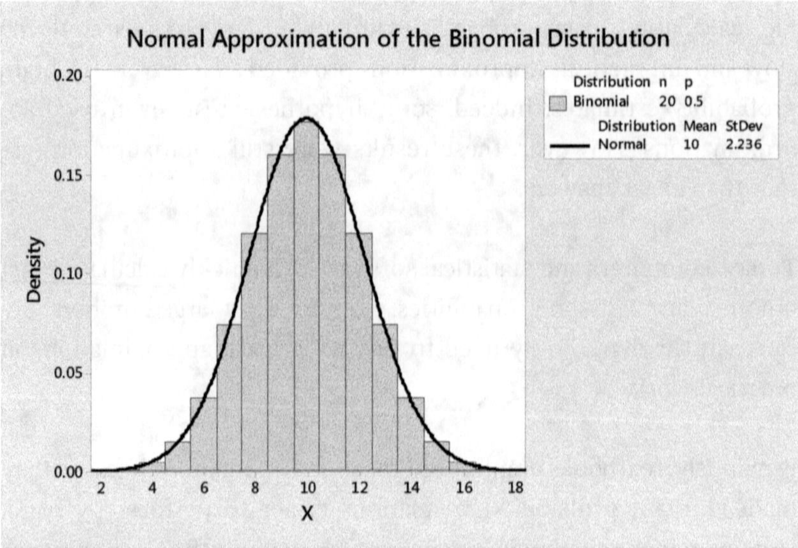

However, as p gets closer to 0 or 1, and n decreases, the approximation deteriorates because the binomial distribution gets skewed (asymmetrical), which the normal distribution can't model.

Poisson Approximation

As the average rate (λ) in a Poisson distribution increases, the distribution becomes less skewed and increasingly approximates the normal distribution, as shown in the graph. The shape becomes more symmetric and begins to resemble the familiar bell-shaped curve of the normal distribution.

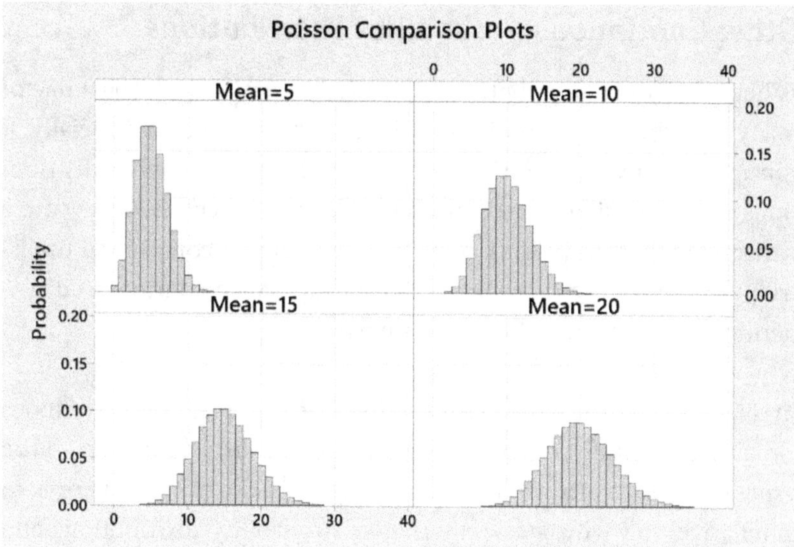

The normal distribution can adequately approximate the Poisson distribution when the mean is ~20 or more. The normal approximation uses the lambda and the square root of lambda for its mean and standard deviation, respectively. The following probability plot shows a normal distribution that closely follows a Poisson distribution with a lambda of 16.

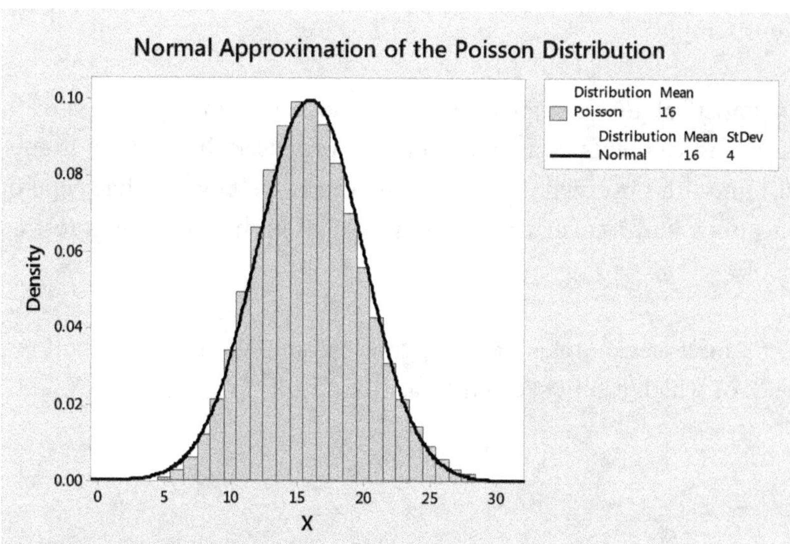

Other Continuous Probability Distributions

While the normal distribution is essential in statistics, it is just one of many probability distributions, and it does not fit all data. Crucially, it can model only symmetrical distributions. Fitting skewed distributions requires using other probability density functions. There are a variety of other probability density functions that correspond to distributions of different shapes and properties. Most continuous distributions have between one and three parameters.

In this section, I'll show you a few other standard PDFs for continuous variables. Generally, if you're using these other distributions, you'll be using statistical software to find probabilities. So, my goal here is to familiarize you with a few alternative probability distributions and when you might use them.

Before using a PDF to find a probability, you must identify a distribution and parameter values that reasonably describe the population you are studying. To accomplish that, you'll typically gather a random sample from that population and use a combination of parameter estimation techniques, subject-area knowledge, and hypothesis tests to identify the distribution and parameters that most likely produced your sample.

In statistical analysis, analysts sometimes use formal goodness-of-fit tests and QQ plots to assess whether a proposed distribution is consistent with observed data. Those methods are beyond the scope of this book and I cover them in more detail in my hypothesis testing book.

Let's look at examples of the lognormal and Weibull distributions, both of which can fit skewed data.

Lognormal PDF

The lognormal distribution models right-skewed data that arise from multiplicative growth processes or when values cannot fall below zero.

The following graph illustrates the probability density function for a lognormal distribution that models the distribution of body fat percentages for teenage girls. The data are from a study I performed.

The body fat percentages came from the same individuals as the heights earlier in this chapter. However, while their heights are normally distributed, their weights are not. Instead, the weights follow the lognormal distribution.

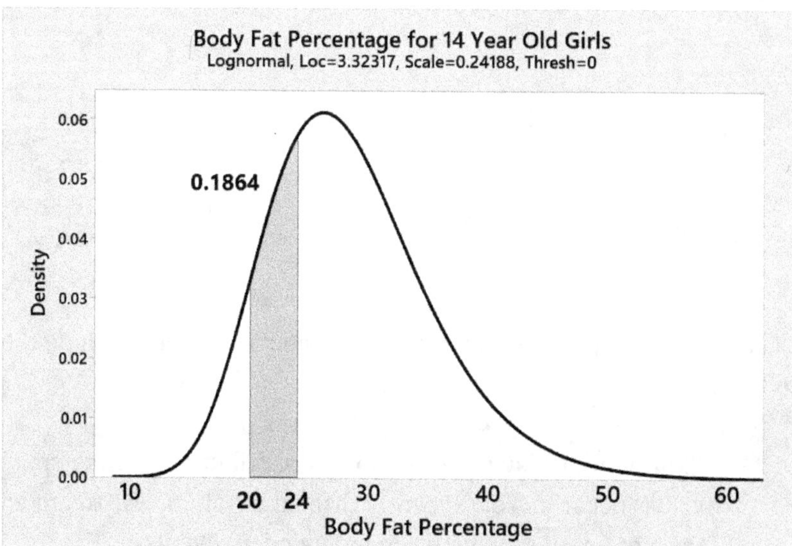

Notice how the graph shows a right-skewed distribution. Additionally, the probability distribution plot indicates that 18.64% of the population will fall between 20% and 24%. The highest probability densities occur near the peak at 26%.

Weibull PDF

The Weibull distribution is a remarkably versatile distribution that analysts use in many settings. It can model skewed distributions in both directions and approximate the normal distribution.

The example displays a left-skewed distribution.

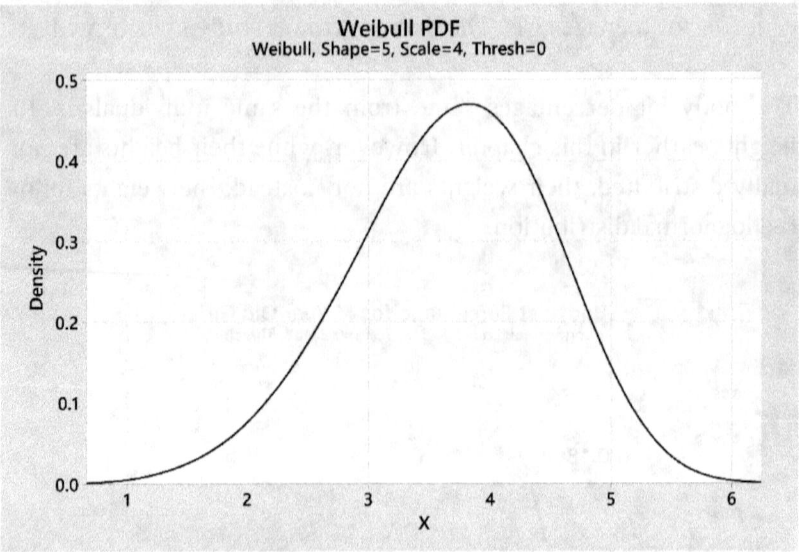

There are a variety of PDFs for other distributions. They include the following:

- **Exponential distribution:** Models distributions where small values occur more frequently than large values. Use to model the amount of time between independent events.
- **Gamma distribution:** Models right-skewed distributions. Use to model the time until the kth event.
- **Uniform distribution:** Models symmetric distributions where all equal-length ranges have the same probability.
- **Beta distribution:** Models distributions with all values falling within a finite range.

Putting Continuous Distributions in Perspective

Different random processes produce different distribution shapes. Some processes produce symmetric patterns that the normal distribution models well, while others produce skewed or bounded patterns that require other probability density functions. Using subject-area knowledge to recognize these patterns helps you choose a distribution that reasonably describes the process you are studying.

The goal is not to memorize many probability distributions or their formulas. Instead, focus on recognizing how data behave and understanding which types of distributions tend to model those patterns. In practice, statistical software will perform most probability calculations for you. Your role is to select an appropriate model and interpret the results correctly.

Review and Next Steps

In this chapter, we moved from discrete probability models to continuous probability distributions. Use continuous distributions when a variable can take any value within an interval. Instead of assigning probabilities to individual values, probability density functions (PDFs) describe how probability is distributed across values. For continuous variables, individual values have zero probability, and probabilities are determined by finding areas under the curve over intervals.

Using the normal distribution as the primary example, you saw how the mean and standard deviation fully determine the distribution's shape and the probabilities it generates. The mean defines the center of the distribution, while the standard deviation controls its spread. The normal distribution is appropriate for symmetric, bell-shaped data and serves as a useful model for many real-world variables.

You then explored several practical tools for working with the normal distribution. You used the empirical rule to make quick probability estimates, z-scores to standardize values, and cumulative z-tables to find probabilities and cutoff values. You examined cumulative distribution functions to understand accumulated probability and saw how PDFs, CDFs, tables, and software all describe the same underlying information from different perspectives. Finally, you saw how the normal distribution relates to other distributions, including when it can approximate discrete distributions and when other continuous distributions are more appropriate.

In the next chapter, we shift focus from modeling individual observations to modeling the behavior of statistics calculated from samples. You will learn how probability distributions describe sample means and other sample-based quantities, why these sampling distributions often have predictable shapes, and how the normal distribution plays a central role in this process. These ideas form the foundation for understanding statistical inference, including why methods such as confidence intervals and hypothesis tests work.

Cumulative distribution functions (CDFs)

- Use CDFs to find the probability that a value is less than a given value.
- Use the CDF complement for greater than problems.
- Find the probability between two values by subtracting the smaller cumulative probability from the larger.

Empirical Rule

- Use the empirical rule to quickly estimate probabilities for normally distributed data.
- Look for wording like "within one, two, or three standard deviations of the mean."
- Apply it by multiplying the standard deviation (σ) by the number of standard deviations and then adding and subtracting the result from the mean (μ) to find the range.

Z-scores

- Use z-scores when you want to convert values to a common scale.
- A z-score tells you how many standard deviations a value is from the mean.
- Use to compare the relative standing of values from different normal distributions.
- Use with a z-table to find probabilities.
- Data value → z-score: $z = (x - \mu) / \sigma$.
- Z-score → data value: $x = \mu + z\sigma$.

Cumulative z-tables

- Cumulative z-tables list left-tail probabilities for the standard normal distribution.
- Requires normal assumption, parameter values, and either x-values or probabilities.
- When given x-values, convert to z-scores and use a table to find probabilities for less than, greater than, between, and more extreme than cases.
- When given probabilities, use a table to find z- values that correspond to probability cutoffs or ranges. Then, covert z to data values.

Practice Problems Chapter 13

Note: Appendix A contains the cumulative z-table.

Question 1: Suppose the time required to complete a manufacturing process follows a normal distribution with a mean of 50 minutes and a standard deviation of 5 minutes. What is the probability that a randomly selected process takes 50 minutes?

A. 0.50
B. 0.34
C. 0.16
D. 0
E. None of the above

Question 2: For a continuous random variable X, what does the cumulative distribution function (CDF) at x represent?

A. The height of the probability density function at x
B. The probability that X equals x
C. The probability that X is less than or equal to x
D. The standard deviation of X

Question3: For a normal distribution, suppose $P(X < \mu - a) = 0.16$, where a is a positive constant. What is $P(X < \mu + a)$?

A. 0.16
B. 0.34
C. 0.50
D. 0.84

Question 4: Student A scored 85 on an exam where the mean was 80 and the standard deviation was 5. Student B scored 90 on a different exam where the mean was 85 and the standard deviation was 10. Which student performed better relative to the other students in the same course?

- **A.** Student A
- **B.** Student B
- **C.** They performed equally well relative to their groups
- **D.** There is not enough information to tell

Question 5: For a continuous random variable X, which statement is true?

- **A.** $P(X \leq x) > P(X < x)$
- **B.** $P(X \leq x) = P(X < x)$
- **C.** $P(X \leq x) < P(X < x)$
- **D.** Cannot be determined

Question 6: In a normal distribution, what does the peak of the curve represent?

- **A.** The value with the highest probability
- **B.** The value with the highest probability density
- **C.** The cumulative probability at that value
- **D.** The standard deviation

Question 7: The weights of adult golden retrievers are approximately normally distributed with a mean of 70 pounds and a standard deviation of 5 pounds.

What proportion of dogs weigh between 65 and 75 pounds?

Question 8: The lifetimes of a certain brand of lightbulbs are normally distributed with a mean of 800 hours and a standard deviation of 100 hours.

What is the probability that a randomly selected lightbulb lasts between 750 and 920 hours?

Question 9: The resting heart rates of adults are normally distributed with a mean of 70 beats per minute and a standard deviation of 8 beats per minute.

What is the probability that a randomly selected adult has a resting heart rate more than 10 beats per minute away from the mean?

Question 10: The scores on a standardized test are normally distributed with a mean of 500 and a standard deviation of 100.

What score corresponds to the 90th percentile?

CHAPTER 14

From Random Variables to Random Statistics

Up to this point, we have modeled randomness in individual outcomes.

- When you rolled a die, the random variable described one result.
- The binomial distribution finds the probability of one specific count of successes.
- The Poisson distribution gives the probability of one particular number of events.
- The normal distribution calculates the probability that a single observation falls within a particular range.

Even when a probability distribution displayed many possible values, each probability calculation focused on a single value of a random variable at a time. Those outcomes might represent counts across multiple trials, but they still describe a single realized result from a random process.

Now we shift perspective.

Instead of modeling one outcome from a process, we will model how repeatedly drawing random samples from the same population produces statistics that vary across samples. We will then calculate probabilities using that behavior.

Imagine there is a fixed population with stable characteristics. For example:

- The true average height of adult women in a particular city.
- The true proportion of voters who support a policy.
- The true average lifetime of a brand of lightbulb.

These population characteristics do not change when you draw a random sample. They are fixed, even if we do not know their exact values.

Now, suppose we draw a random sample from the population of adult women, measure their height, and calculate the mean height, a sample statistic. If we draw another random sample from the same population, we will get a different mean height. And so on.

In this process, the following is true:

1. The population stays fixed.
2. The random sample changes.
3. The sample statistic changes.

That means the statistic is random.

When we repeatedly draw random samples from the same population, the statistics we calculate vary from sample to sample in a predictable pattern. Most sample statistics cluster near the population's true value. Some fall farther away. Extreme sample values are rare. Because the statistic varies across repeated samples, it has its own probability distribution. Therefore, you can treat sample statistics as random variables and calculate probabilities for them.

In earlier chapters, probability distributions described individual outcomes from a process. In this chapter, probability distributions describe statistics calculated from random samples of *multiple* outcomes drawn from a population with fixed properties. Statisticians refer to this type of distribution as a sampling distribution.

In the sections that follow, you will learn how to model probability distributions of sample means, sample totals, and sample proportions. By the end of this chapter, you'll be able to solve the following types of probability problems:

- What is the probability that a random sample of 5 people will have an average IQ of 110 or higher?
- What is the probability that a sample of 30 apples will have a total weight less than 2.5 kg?
- What is the probability that a random sample of 200 lightbulbs will have fewer than 5% defective bulbs?

Before we solve problems like these, we need to build the necessary foundation.

Towards that end, I'll first clarify the distinction between population parameters and sample statistics. Then we will explore sampling distributions in greater depth and examine how their shape and variability behave across repeated samples and different sample sizes. Finally, we examine the central limit theorem, which explains why normal distributions play a central role in modeling sample means, totals, and proportions.

After developing these ideas, we can solve probability problems involving sample statistics.

Parameters vs. Statistics

Before we study how those statistics behave, let's clarify an important distinction by defining population parameters and sample statistics more precisely.

Parameters are numbers that describe the properties of entire populations. Statistics are numbers that describe the properties of samples drawn from those populations.

For example, the average income for the United States is a population parameter. Conversely, the average income for a random sample drawn from the U.S. is a sample statistic. Both values represent the mean income, but one describes the entire population while the other describes only a sample.

There is a helpful mnemonic device for remembering which term applies to which group:

Parameter = Population
Statistic = Sample

A population is the entire group of people, objects, animals, transactions, or events that you are studying. A sample is a portion of that population selected for analysis.

A population parameter is fixed for the defined population, even if we don't know its value. When we repeatedly draw random samples from the same population with fixed properties, the statistics we calculate will vary from sample to sample because each random sample contains different observations.

In short, the statistic changes, but the parameter does not.

Parameters and statistics use numbers to summarize the properties of a population or sample. The type of summary depends on the kind of characteristic being measured.

When you measure a characteristic using a continuous scale, you can calculate summary values such as:

- Mean
- Total (the sum of all observations)
- Proportion

When the characteristic is categorical, the parameter or statistic is often a proportion, such as the proportion of people who agree with a particular law.

Although parameters and statistics represent the same types of summary values, statisticians denote them differently. Typically, Greek letters represent population parameters, and Latin letters represent sample statistics.

Summary Value	**Parameter**	**Statistic**
Mean	μ or Mu	$\overline{x}$ or x-bar
Standard deviation	σ or Sigma	s
Proportion	p	$\hat{p}$ or p-hat

These symbols help distinguish fixed population quantities from random sample-based quantities.

In the following examples, notice how the same subject and summary value can be either a parameter or a statistic. The difference depends on whether the value summarizes an entire population or a sample drawn from that population.

Parameter	Statistic
Mean weight of all German Shepherd dogs.	Mean weight of a random sample of 200 German Shepherds.
Mean weight of a particular variety of apple.	Mean weight of a random sample of 50 apples.
Standard deviation of all transaction times in a particular bank.	Standard deviation of a random sample of 500 transaction times at that bank.
Proportion of all people who prefer Coke over Pepsi.	Proportion of a random sample of 100 people who prefer Coke over Pepsi.

In each pair, the population quantity is fixed. The sample quantity would change if a different random sample were selected.

Using Parameters in Probability Modeling

In probability modeling, a population is a fixed, well-defined group with stable and known parameters. It might include all adult residents of a city, all lightbulbs produced under a particular manufacturing process, or all registered voters in a state at a specific time. For the purposes of this chapter, we treat the population's parameters as fixed and known. The population does not change from sample to sample. Only the random samples change.

When you solve probability problems involving sample statistics, the problem typically provides the population parameters and a sample size. Your task is to find probabilities associated with specific sample statistics.

To do that, you must first understand how those statistics behave across repeated sampling. In the next section, we begin that process by studying sampling distributions.

What is a Sampling Distribution?

A sampling distribution is the probability distribution of a statistic calculated from many random samples of a fixed size drawn from the same population.

The sampling distribution describes how that statistic behaves across repeated sampling. It tells us which values are common, which are rare, and how far the statistic typically falls from the population parameter. A sampling distribution is not the distribution of the data. Instead, it is the distribution of a statistic across many samples.

Because a sampling distribution is a probability distribution, we can calculate probabilities for sample statistics using the same methods as we used for random variables. If a probability problem mentions "the average of a sample" or "the proportion in a sample," you are working with a sampling distribution.

Sampling distributions exist for many statistics. While the sampling distribution of the mean is the most common example, we can also define sampling distributions for the sample total and sample proportion. In this section, I focus on the mean, and we'll extend the ideas to other statistics later.

In the following sections, I'll show you sampling distributions for random variables that follow normal and nonnormal distributions, and I'll demonstrate how they change with sample size.

Simple Apple Example

Let's start with a simple example and move on from there.

Imagine you draw a random sample of 10 apples from an apple population. You calculate the mean weight of that sample as 103 grams. That is one sample mean from one random sample.

If you were to draw another random sample of 10 apples from the same population, you would obtain a different mean. A third sample would produce yet another mean. Each sample differs because random selection produces different combinations of apples.

With this in mind, you decide to collect 50 random samples of the same apple population. Each sample contains 10 apples, and you calculate the mean for each sample.

At this point, you have 50 sample means. When you plot these sample means in a histogram, you create an approximation of the sampling distribution of the mean.

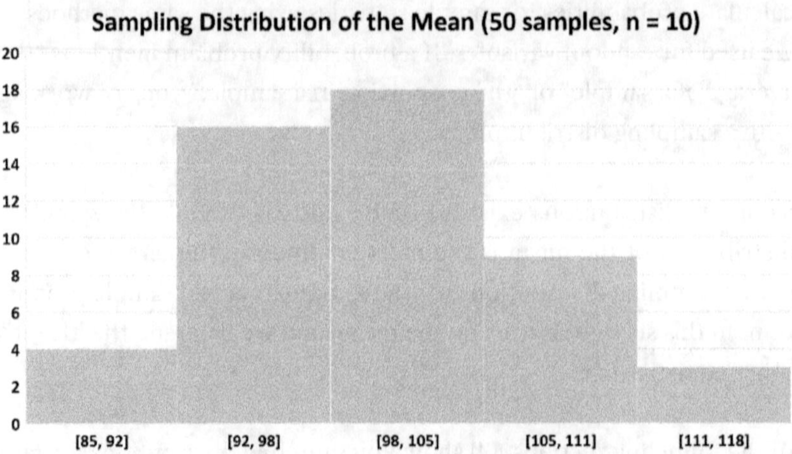

This histogram shows how the sample mean varies across repeated random samples drawn from the same population.

In this simulation, our initial sample mean of 103 falls near the center of the sampling distribution. Means in this central range occur most frequently—18 of the 50 samples (36%) fall within the middle bar. Other samples from the same population produce higher and lower means, but extreme values are less common. None of our 50 sample means fall outside the range of 85–118. That suggests obtaining a mean outside that range is unlikely.

For this simulation, we know the population parameters because I specified that the apple population follows a normal distribution with a mean (μ) of 100 grams and a standard deviation (σ) of 15 grams. These are the fixed parameters of the population from which we are repeatedly sampling.

Notice that the histogram of sample means centers on the population mean of 100. Values become less frequent as they move farther away from that central value. The distribution is also roughly symmetric.

Like probability distributions, sampling distributions have a center, spread, and shape. The center describes the typical value of the statistic, which in this case is the sample mean. The spread describes how much the statistic varies across samples. The shape shows how values are distributed around the center, including whether the distribution is symmetric or skewed. In this example, the sampling distribution is centered near 100, most sample means fall between about 85 and 118, and the distribution appears roughly symmetric.

This distribution is not perfectly smooth because 50 samples is a relatively small number for illustrating a probability distribution. As you'll see shortly, increasing the number of samples produces a smoother representation of the underlying probability distribution.

I used Excel to create this example. I had it randomly draw 50 samples of size 10 from a population with $\mu = 100$ and $\sigma = 15$. If you want to simulate 50 random samples yourself, use the link for my website's resource page in the book's introduction to download the Excel file.

Let's now examine the central tendency and variability of sampling distributions more formally.

Sampling Distributions of the Mean for Normal Distributions

As you saw in the apple example, sampling distributions have their own shapes, central tendencies, and variabilities. Let's start exploring these properties for cases where the parent distribution is normal.

In this context, the parent distribution refers to the probability distribution of individual observations in the population. It describes how single measurements or outcomes behave, including their center and variability. In contrast, sampling distributions describe the probability distribution of statistics for multiple observations (i.e., samples) across repeated samples. Sampling distributions also have their own centers and variability.

Here, I link the properties of a normal parent population to its sampling distributions because the former logically affects the latter.

When the parent population follows a normal distribution, its sampling distributions of the mean have the following properties:

$\bar{X} \sim N(\mu, \sigma / \sqrt{n})$

That reads as:

- The sample mean ($\bar{X}$)
- Is distributed normally (N)
- With mean μ
- And standard deviation $\sigma / \sqrt{n}$.

The following table compares the parent distribution's parameters to those of the sampling distribution.

	Mean	Standard Deviation
Parent Distribution	μ	σ
Sampling Distribution	μ	$\sigma / \sqrt{n}$

Notice how the mean of the parent population is also the central value for the sampling distribution.

However, the variabilities are different. The variability for the parent distribution is a fixed value (σ), while for a sampling distribution, it's related to σ but also depends on the sample size (n). From the formula, we know the sampling distribution always has less variability than the parent distribution when $n > 1$.

Additionally, each sample size (each value of n) produces a probability distribution with a unique spread. Further, the spread decreases as the sample size increases because n is in the denominator.

Statisticians refer to the standard deviation for a sampling distribution as the standard error. Because we're assessing the mean, the distribution's variability is the standard error of the mean. The standard error describes how much the sample mean varies from sample to sample.

In summary, for normally distributed parent populations, the sampling distribution of the mean:

- Centers on the population mean μ.
- Has variability determined by $\sigma / \sqrt{n}$.
- Each sample size produces a unique sampling distribution.
- The distribution narrows as the sample size increases.

Returning to Our Apple Simulation

Let's return to our apple example. We know what probability theory and its sampling distribution equations say. Now, let's see how this works using random sampling to see how reality compares. We'll also get to see nice graphs!

Recall that we specified that the apple population follows a normal distribution with a mean of 100 and a standard deviation of 15. We used a sample size of 10, which shows up in the standard error of the mean. Therefore, we'd expect its sampling distribution to center on $\mu = 100$ and that the standard error will be $15 / \sqrt{10} = 4.743$.

We'll rerun our previous apple sampling simulation, but on a massive scale. This time I'll draw 500,000 samples instead of just 50. To perform this simulation, I'll use the Statistics101 simulation software. Again, see the resources webpage for this book to try it yourself!

This simulation follows the same process as the Excel version. It draws random samples from a population with a mean of 100 and a standard deviation of 15. Each sample has a size of 10. It calculates the sample means and plots them using a histogram. This setup is basically our previous simulation on steroids!

I also have the simulation software calculate the mean and standard error of the sample means, which should be close to the theoretical values of 100 and 4.743. Please remember that these statistics are the mean and standard error of the sample means, not the individual observations.

Because this simulation uses so many repeated samples, it produces a smooth histogram that reveals the underlying probability distribution more clearly. This graph displays the distribution of sample means rather than individual observations. It's a histogram like the one in Excel, but with many more samples and very tiny bars!

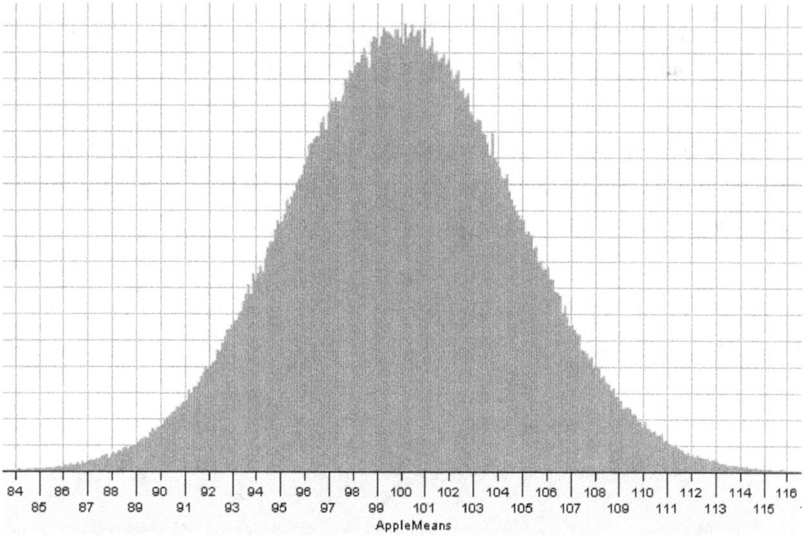


```
Mean: 100.00246487432325
StandardError: 4.744701089684036
```

This sampling distribution clearly follows a normal distribution. Additionally, the calculated mean of the samples and the standard error of the mean almost precisely match the theoretical values. Consequently, statistical equations can find the sampling distribution without drawing all those samples!

Standard Errors and Sample Sizes

As mentioned earlier, the standard error of the sampling distribution depends on the sample size. The formula for the standard error of the mean is: $\sigma / \sqrt{n}$

Notice how the formula is a ratio with the square root of the sample size in the denominator?

This fact causes the denominator's value to increase as the sample size increases. In turn, a larger denominator reduces the standard error. Consequently, sampling distributions based on larger sample sizes should have smaller standard errors, leading the distributions to cluster more tightly around the central population parameter.

For instance, quadrupling the sample size halves the standard error because √4 = 2 is in the denominator. We'll quadruple the sample size in the following simulation to see what happens!

Let's return to the apple example to see this in action. This time, we'll collect:

- 500,000 samples with n = 10
- 500,000 samples with n = 40

Theoretically, increasing the sample size from 10 to 40 should reduce the standard error from 4.743 to 2.372.

Let's run a simulation!

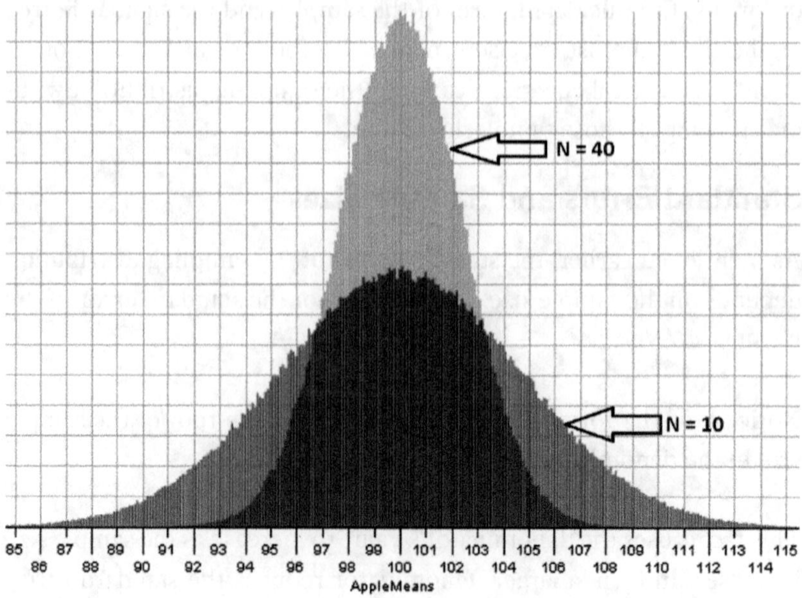

```
Mean10: 100.00070541860869
StandardError10: 4.750226312091001
Mean40: 99.99784699275877
StandardError40: 2.3723878186728786
```

When we run the simulation, both sampling distributions center on the population mean of 100. However, the distribution for n = 40 is noticeably narrower than the distribution for n = 10.

Again, the observed means and standard errors closely match the theoretical values.

What are the practical implications of this difference? A tighter sampling distribution indicates that the sample means cluster more closely to the population mean. Notice how the wider n = 10 spread has more sample means farther away from the population mean (100).

This situation explains a familiar statistical principle: larger samples are more likely to produce sample estimates that are closer to the correct population value than smaller samples.

I'm sure you already knew this old statistical adage, but now you see why that's the case! The mathematics of the sampling distribution shows that sampling variability shrinks as the sample size increases.

So far, you've learned that when the parent population follows a normal distribution, probability theory tells us that the sampling distribution of the mean also follows a normal distribution, specifically $\bar{X} \sim N(\mu, \sigma / \sqrt{n})$. Our large-scale simulations closely matched those theoretical results, confirming that the formulas describe how sample means behave across repeated sampling.

Because theory and simulation agree, we do not need to physically draw hundreds of thousands of samples to understand a sampling distribution. Mathematics allows us to describe its shape, center, and variability.

But a natural question arises: what if the parent population is not normal? What if the original data are skewed?

For these cases, the central limit theorem becomes our friend!

Central Limit Theorem

Population distributions are not always normal. They might be skewed, irregular, or uniform. Yet when we examine sampling distributions in practice, we often observe something surprising: the distribution of sample means still looks approximately normal.

Why does that happen?

The central limit theorem (CLT) provides the answer.

The central limit theorem is one of the most powerful ideas in probability theory. It states that, for sufficiently large sample sizes, the sampling distribution of the sample mean will approximate a normal distribution regardless of the parent population's shape (with a few exceptions).

In short, even when individual observations are not normally distributed, sample averages tend to follow a normal distribution.

That definition contains several important ideas: sampling distributions, sample size, approximation, and the relationship between the population distribution and the distribution of sample means.

In this section, we will carefully unpack those ideas. You'll see why the CLT works, how sample size affects the result, and why the normal distribution plays such a central role in modeling sample statistics.

Distribution of the Variable in the Population

Part of the definition for the central limit theorem relates to "the variable's distribution in the population." This part is easy! In a population, the parent distribution of individual outcomes can follow different probability distributions. These distributions can range from normal to left-skewed, right-skewed, and uniform, among others.

Normal

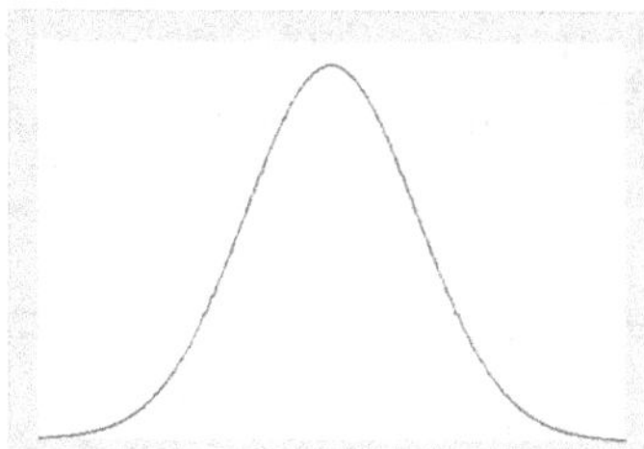

Right-Skewed

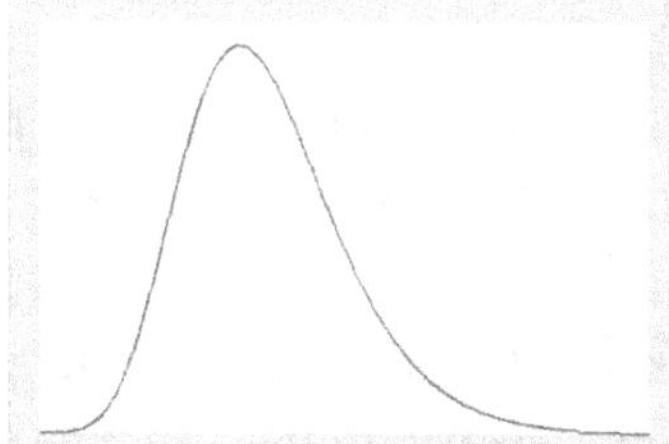

Left-Skewed

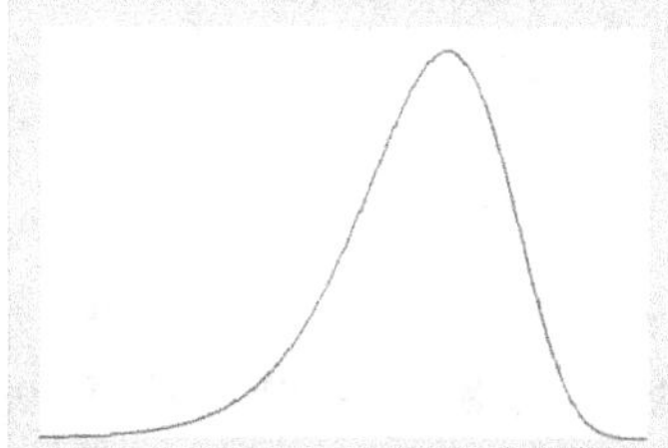

Uniform

This part of the definition refers to the probability distribution of individual outcomes in the population from which you draw a random sample.

The central limit theorem applies to almost all types of probability distributions, but there are exceptions. For example, the population must have a finite variance. That restriction rules out the Cauchy distribution because it has infinite variance.

Additionally, the central limit theorem applies to independent, identically distributed variables. In other words, the value of one observation does not depend on the value of another observation. And, the distribution of that variable must remain constant across all measurements.

Linking the Parent and Sampling Distributions

The central limit theorem links the following two distributions:

- The distribution of individual outcomes in the population.
- The sampling distribution of the mean.

Specifically, the CLT states that regardless of the population distribution's shape, the sampling distribution of the mean tends to approximate the normal distribution.

Previously, you saw how the binomial and Poisson distributions can approximate the normal distribution under particular conditions. The same idea holds for sampling distributions arising from nonnormal parent distributions.

According to the central limit theorem, the probability distribution for individual outcomes can look like the following:

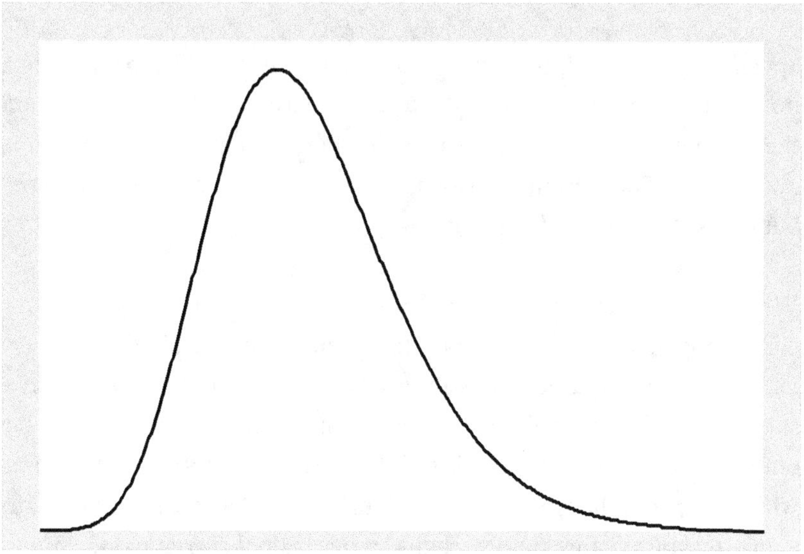

But the sampling distribution can appear like this:

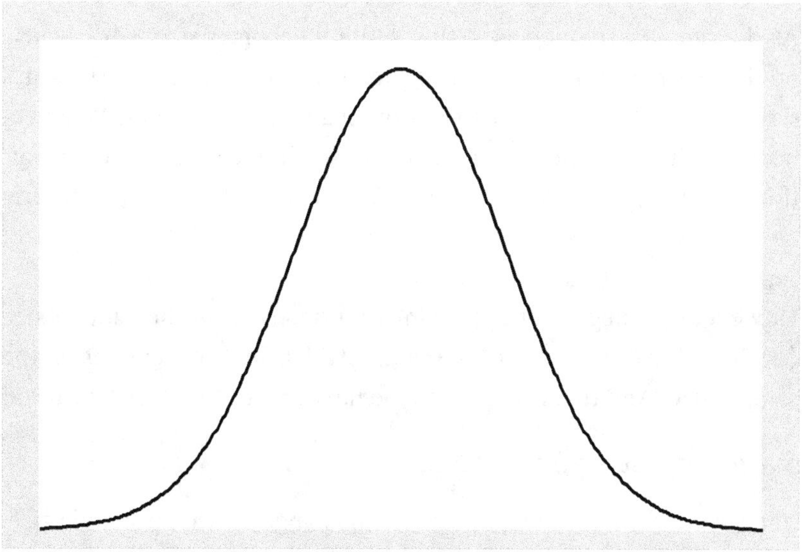

It's not surprising that a normally distributed variable produces a sampling distribution that also follows the normal distribution. But, surprisingly, nonnormal population distributions can also create normal sampling distributions.

Specifically, the sampling distribution of the mean will approximate a normal distribution with mean μ and standard deviation $\sigma / \sqrt{n}$ when the sample size is large enough, even when the parent distribution is nonnormal. The sampling distribution converges on a normal distribution as the sample size increases.

Why does this happen? Intuitively, when you take an average, you are combining many independent observations into a single number. Each observation contributes only a fraction to the final result. In a skewed population, individual observations might occasionally fall far out in a long tail. However, it is unlikely that many extreme tail values will appear together in a single random sample. Most samples contain a mix of typical values and perhaps an occasional extreme one. When averaged together, those extremes are diluted by the other observations.

As the sample size increases, this dilution becomes stronger because no single value has much influence on the mean. The irregularities and skewness of the original distribution begin to smooth out. What remains is the combined effect of many small, independent contributions. The combined pattern tends to form the familiar, symmetric bell-shaped curve.

However, the degree of approximation depends on the sample size (n). Recall that the shape of the sampling distribution depends on the sample size. And that brings us to the next part of the CLT definition!

Sufficiently Large Sample Size

The shape of the sampling distribution changes with the sample size. And, the definition of the central limit theorem states that when you have a sufficiently large sample size, the sampling distribution starts to approximate a normal distribution. How large does the sample size have to be for that approximation to occur?

It depends on the shape of the variable's distribution in the underlying population. The more the population distribution deviates from normality, the larger the sample size must be. Statisticians typically say a sample size of 30 is often sufficient. That's a conservative number, and frequently even 20 is adequate. However, strongly skewed distributions can require larger sample sizes. We'll see the sample size aspect in action during the following simulations.

Let's get more specific about the CLT and parameter values of the normal distribution. What values do these parameters converge on?

As the sample size increases, the sampling distribution converges on a normal distribution where the mean equals the population mean, and the standard deviation equals $\sigma/\sqrt{n}$. Those are the same values as for cases where the parent distribution is normal.

Simulations

Now the fun part! There is a mathematical proof for the central theorem, but that goes beyond the scope of this introductory book. However, I will show how it works empirically by using statistical simulation software. As I did before, I'll define population distributions and have the software draw many thousands of random samples from it. The software will calculate the mean for each sample and then graph the sample means in a histogram to display the sampling distribution of the mean.

For the following examples, I'll vary the sample size to show how that affects the sampling distribution. To produce the sampling distribution, I'll draw 500,000 random samples because that creates a smooth distribution in the histogram.

Keep this critical difference in mind. While I'll collect a consistent 500,000 samples per condition, the sample sizes will vary, which affects the shape of the sampling distributions.

I'll show you how the central limit theorem works with three different parent distributions: moderately skewed, severely skewed, and a uniform distribution. The first two distributions skew to the right and follow the lognormal distribution. The probability distribution plot displays the distribution of values in two populations. Notice how the dashed distribution is much more severely skewed. It actually extends quite a way off the graph! We'll see how this affects the sampling distributions.

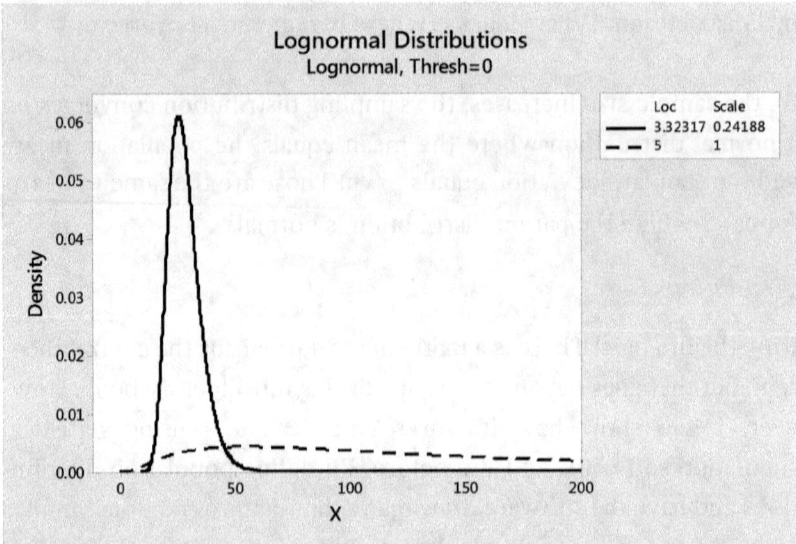

Let's see how the central limit theorem handles these two distributions and the uniform distribution.

Moderately Skewed Distribution

The following graph shows the moderately skewed lognormal distribution. This distribution describes the lognormal probability distribution of body fat percentages that I show near the end of Chapter 13. It reflects a real-world skewed distribution. I use the simulation software to draw random samples from this population 500,000 times for each sample size (5, 20, 40).

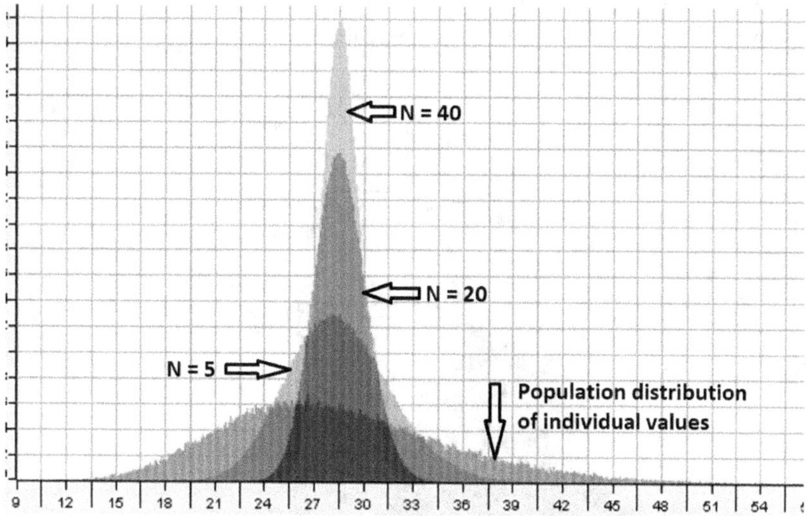

In the graph, the broadest distribution represents the skewed distribution of individual values in the population. The other distributions represent the sampling distributions of the means for different sample sizes. The distribution for a sample size of 5 is still skewed because each sample is so small. However, the sampling distributions for the larger sample sizes of 20 and 40 are not visibly skewed. They are approximately normal. The N = 40 distribution has the tightest range, indicating that sample means are unlikely to fall far from the population mean.

As the sample size increases, the sampling distributions more closely approximate the normal distribution and become more tightly clustered around the population mean, just as the central limit theorem states!

Very Skewed Distribution

Now, let's try this with the very skewed lognormal distribution. I follow the same process but use larger sample sizes of 40, 60, and 80. I do not include the population distribution in this one because it is so skewed that it messes up the X-axis scale!

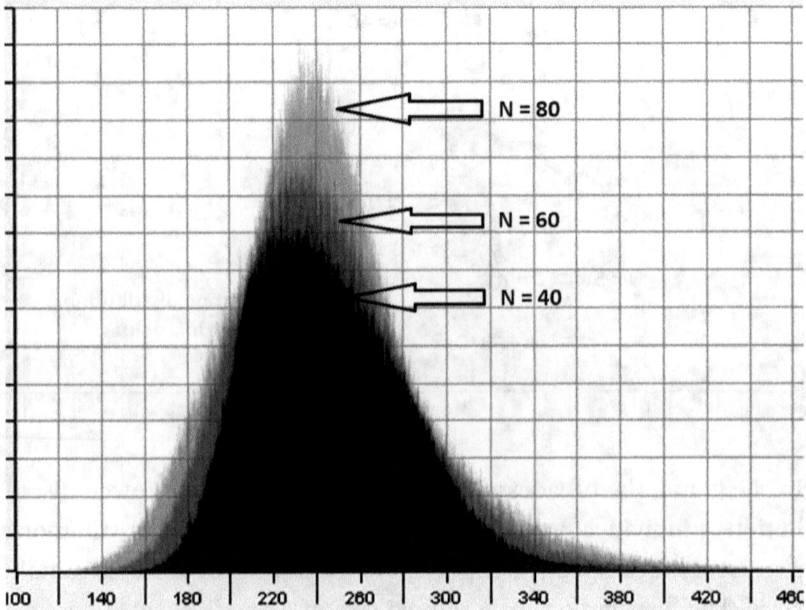

The population distribution is extremely skewed. It's probably more skewed than real data tend to be. As you can see, even with the largest sample size (N = 80), the sampling distribution of the mean remains skewed to the right. However, it is less skewed than the sampling distributions for the smaller sample sizes. Also, notice how the peaks of the sampling distribution shift to the right as the sample size increases.

Eventually, with a large enough sample size, the sampling distributions will become symmetric, and the peak will stop shifting and center on the actual population mean. Given the lognormal parameters, I can calculate that the population mean is 244.7. The peak of the N = 80 curve is near 240, so it is closing in on the correct value.

If your population distribution is extremely skewed, be aware that you might need a substantial sample size for the central limit theorem to kick in and produce sampling distributions that approximate a normal distribution! However, please note that the extremely skewed example is severe enough to surpass the skew present in most real datasets.

Uniform Distribution

Now, let's change gears and look at an entirely different type of distribution. Imagine that we roll a die and take the average value of the rolls. The probabilities of rolling the numbers on a die follow a discrete uniform distribution because they all have the same chance of occurring. Can the central limit theorem work with discrete numbers and uniform probabilities? Let's see!

In the graph, I use the same procedure as before. In this example, the sample size refers to the number of times we roll the die. The process calculates the mean for each sample.

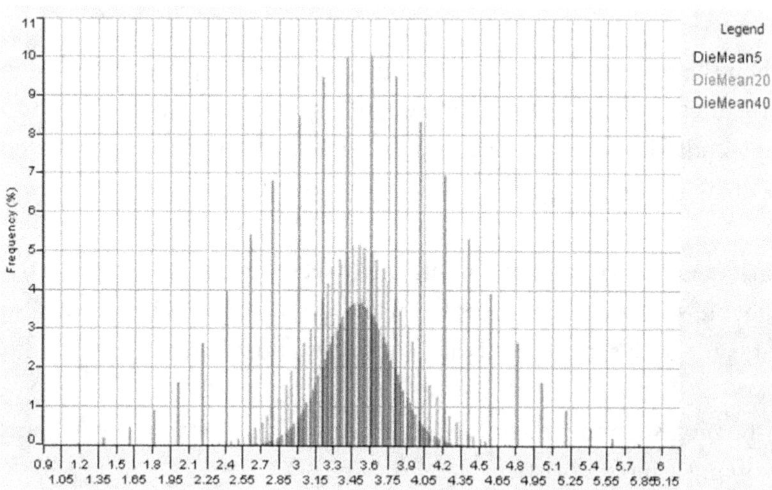

In the chart, I use sample sizes of 5, 20, and 40. We'd expect the average to be $(1 + 2 + 3 + 4 + 5 + 6) / 6 = 3.5)$. All the sampling distributions center on this value. As we increase the sample size, the sampling distributions more closely approximate a normal distribution and have a tighter range. Just as the central limit theorem predicts!

Also, notice that for a uniform parent distribution, even a small sample size of 5 produces a sampling distribution that closely approximates a normal distribution. Uniform distributions are symmetric (i.e., not skewed), so their sampling distributions converge on a normal distribution with very small sample sizes.

You could perform a similar experiment using the binomial distribution with coin flips and obtain the same types of results when it comes to, say, the probability of getting heads. All thanks to the central limit theorem!

Let's summarize the past several sections. When the parent distribution is normal, the sampling distribution of the mean is also normal with the following properties:

$\bar{X} \sim N(\mu, \sigma / \sqrt{n})$

When the parent distribution is nonnormal, the sampling distribution of the mean becomes approximately normal with the same center (μ) and standard error ($\sigma / \sqrt{n}$) for sufficiently large sample sizes (often $n \geq 30$, with some exceptions).

Consequently, we can rely on this single set of properties to model the probability behavior of sample means. And as you will see, it also works for both sample totals and sample proportions.

In the next section, you will learn how to solve probability problems involving sample statistics.

Solving Sample Statistics Problems

Now that we understand sampling distributions and the central limit theorem, we are ready to use them.

In this section, you will learn how to solve probability problems involving sample statistics. Instead of finding probabilities for single observations, we will now find probabilities for:

- Sample means
- Sample totals
- Sample proportions

The methods for these problems are remarkably similar to those in Chapter 13 because you will again use the normal distribution, z-scores, and cumulative z-tables (or calculators). Be sure that you are familiar with the methods in that chapter before proceeding.

The main difference is that while you are working with sampling distributions that follow a normal distribution, you must use the appropriate standard error instead of the population standard deviation.

To use these methods, the sampling distribution must be normal or approximately normal. That occurs under the following conditions:

1. If the parent population is normally distributed, the sampling distribution of the mean is normal for any sample size. In this case, you do not need to consider sample size when using the normal model.
2. If the parent population is not normally distributed, the central limit theorem states that the sampling distribution of the mean is approximately normal for sufficiently large sample sizes. Many statisticians use a sample size of $n \geq 30$ as a general guideline. See the caveats in the previous section.

One additional assumption underlies these methods: the observations in the sample must be independent. Independence ensures that individual values do not systematically influence one another. When sampling is random and the population is large relative to the sample size, this assumption is typically reasonable. If observations are strongly dependent, the formulas for standard errors might not apply.

In practice, the normal model works surprisingly well in many situations. The central limit theorem explains why.

In the problems that follow, I will work through the solutions step by step. Solutions for each sample statistics relies on the same underlying process. When the sampling distribution is normal or approximately normal, we standardize the statistic and use the normal distribution to calculate probabilities.

See Appendix A for the complete z-table that we use in the following sections.

Sample Means

We begin with probability problems involving the sample mean. These problems apply the sampling distribution of the mean and the central limit theorem using the z-score methods you learned.

The solution's structure will feel familiar. The key difference is that we are now working with the sampling distribution of $\bar{X}$ (i.e., sample means) rather than the distribution of individual observations.

Use this method when a probability problem focuses on the average of a sample rather than on a single observation. Look for wording such as "the mean of a random sample," "the average of n observations," or "a sample of size n has an average of..." These phrases signal that the random variable in the problem is the sample mean.

These problems typically provide the population mean and standard deviation, along with a sample size, and ask for the probability that the sample mean is greater than, less than, or between specific values.

Before proceeding, confirm that the normal model applies. As discussed earlier, this occurs when the parent population is normal or when the sample size is sufficiently large to justify a normal approximation under the central limit theorem.

If the question involves finding a probability for individual outcomes rather than sample averages, use the methods in Chapter 13 instead.

After determining that the sampling distribution of the mean is normal or approximately normal, follow these steps:

1. **Identify the parameters and sample size.**
   Record the population mean (μ), population standard deviation (σ), and sample size (n).
2. **Calculate the standard error of the mean.**
   Standard error = $\sigma / \sqrt{n}$
3. **State the sampling distribution.**
   $\bar{X} \sim N(\mu, \sigma / \sqrt{n})$
4. **Convert the sample mean value to a z-score.**
   $z = (\bar{x} - \mu) / (\sigma / \sqrt{n})$
5. **Use the standard normal distribution.**
   Use a z-table or calculator to find the required probability.

These steps mirror the procedure in Chapter 13. The key change is that you are standardizing a sample mean within its sampling distribution using the standard error rather than standardizing an individual observation within its parent population using the standard deviation.

After you obtain the z-score, the exact probability depends on the question's phrasing. If the problem asks for the probability that the sample mean is "at least" or "greater than" a value, you will find a right-tail probability. If it asks for "at most" or "less than," you will find a left-tail probability. If it asks for a probability "between" two values, you will calculate the area between two z-scores. After finding the z-score, use the same z-table techniques from the previous chapter to determine the appropriate area under the normal curve.

Note: If the problem requires you to find a sample mean associated with a probability, you'll need to convert the z-score to the sample mean ($\bar{x}$) in natural units. For the standard normal distribution for individual values (x), we used $x = \mu + z\sigma$. Here, we need to adapt that to the distribution of sample means by using the following instead:

$$\bar{x} = \mu + z \times \left(\sigma / \sqrt{n}\right)$$

This formula reflects the fact that we're still using μ for the population mean, but using the standard error of the mean for the variability.

Worked Example

A team of researchers is screening potential participants for a study that requires unusually strong performance on reasoning tasks. IQ scores in the population follow a normal distribution with a mean of 100 and a standard deviation of 15. The researchers randomly select 5 people and calculate the sample mean IQ.

What is the probability that the sample of 5 people will have an average IQ of 110 or higher?

Because the parent population is normal, the sampling distribution of the mean is normal even with a small sample size, such as n = 5.

Step 1: Identify the given values.
Population mean: $\mu = 100$
Population standard deviation: $\sigma = 15$
Sample size: $n = 5$
We want: $P(\bar{X} \geq 110)$

Step 2: Calculate the standard error of the mean.
Standard error = σ / √n = 15 / √5
√5 = 2.236
So, standard error = 15 / 2.236 = 6.708

Step 3: State the sampling distribution.
$\bar{X} \sim N(\mu, \sigma / \sqrt{n})$
So: $\bar{X} \sim N(100, 6.708)$

Step 4: Convert the sample-mean cutoff to a z-score.
$z = (\bar{x} - \mu) / (\sigma / \sqrt{n})$
z = (110 − 100) / 6.708
z = 10 / 6.708
z = 1.49

Step 5: Use the cumulative z-table to find the probability. Cumulative z-tables give the area to the left of a z-score. Look up z = 1.49 in the cumulative z-table:

Z	0.05	0.06	0.07	0.08	0.09
1.0	0.85314	0.85543	0.85769	0.85993	0.86214
1.1	0.87493	0.87698	0.87900	0.88100	0.88298
1.2	0.89435	0.89617	0.89796	0.89973	0.90147
1.3	0.91149	0.91309	0.91466	0.91621	0.91774
1.4	0.92647	0.92785	0.92922	0.93056	0.93189

For z = 1.49, the cumulative probability is 0.93189.

That means:
$P(\bar{X} \leq 110) = 0.93189$

But we want:
$P(\bar{X} \geq 110)$

So, we use the complement to obtain the right-tail probability:
$P(\bar{X} \geq 110) = 1 - 0.93189 = 0.06811$

The probability that a random sample of 5 people has an average IQ of 110 or higher is about 0.068, or 6.8%.

The following graph illustrates the solution to this problem. Unlike the similar PDF graphs in the previous chapter, this one displays the distribution of sample means for N = 5 rather than the distribution of individual observations.

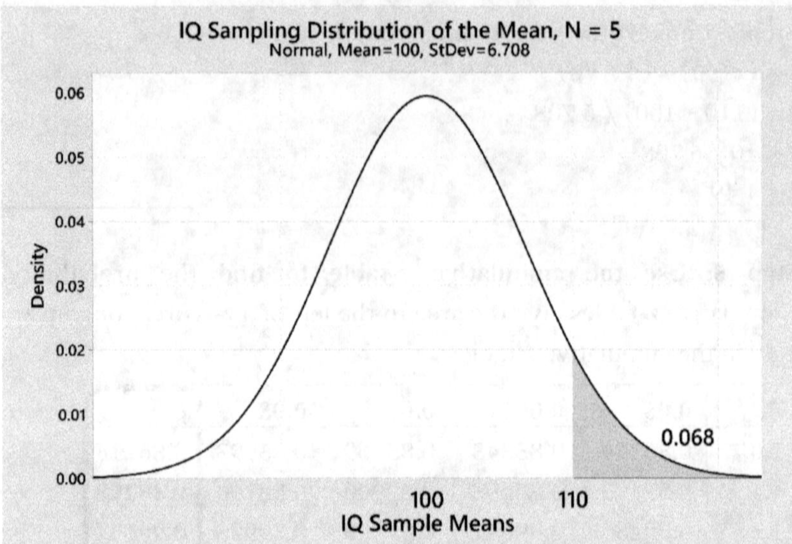

Sample Totals

In some probability problems, the quantity of interest is not the sample average but the sum of all observations in the sample. For example, you might be asked about the total weight of several items, the total time required to complete several tasks, or the total revenue from a group of customers.

Although the wording changes from “mean” to “total,” the underlying method is nearly identical to what we used for sample means.

The sample total is directly related to the sample mean. In fact, the total equals the mean multiplied by the sample size. Because of this relationship, the sampling distribution of the total has a center and variability that scale with n. The mean of the total equals n times the population mean ($n\mu$), and the standard deviation of the total equals the population standard deviation multiplied by the square root of the sample size ($\sigma\sqrt{n}$).

Note that when you are given the information necessary to solve a sample mean probability problem, you can also solve a sample total problem, and vice versa. A sample total is simply n times the sample mean. Because of this relationship, the sampling distribution of the total and the sampling distribution of the mean contain the same information expressed on different scales. If you know μ and σ, you can calculate either distribution and convert between them by multiplying or dividing by n. The underlying probability calculations use the same normal methods in both cases.

This scaling reflects the fact that when we add independent observations, their expected values add directly, while their variability increases at a slower rate, proportional to $\sqrt{n}$.

Use this method when a probability problem refers to the sum or total of values in a random sample. Look for wording such as "the total of n observations," "the combined weight," "the sum of the sample," or similar language indicating that all sample values are being added together.

As with sample means, the problem will provide the population mean and standard deviation, along with a sample size, and it will ask for the probability that the total is greater than, less than, or between certain values.

Before proceeding, confirm that the normal model applies. As discussed earlier, this occurs when the parent population is normal or when the sample size is sufficiently large to justify a normal approximation under the central limit theorem.

After determining that the sampling distribution of the total is normal (or approximately normal), follow these steps:

1. **Identify the given values and the sample size.**
   Record the population mean (μ), population standard deviation (σ), and the sample size (n).
2. **Calculate the mean and standard deviation of the sample total.**
   Mean of the total = $n\mu$
   Standard deviation of the total = $\sigma\sqrt{n}$
3. **State the sampling distribution of the total.**
   Let T represent the sample total. Then:
   $T \sim N(n\mu, \sigma\sqrt{n})$
4. **Convert the total value to a z-score.**
   $z = (T - n\mu) / (\sigma\sqrt{n})$
5. **Use the standard normal distribution.**
   Use a z-table or calculator to find the required probability.

These steps mirror the procedure used for sample means. The difference is that we are now standardizing a total rather than an average, so both the center and the spread are scaled by the sample size.

As with sample mean problems, after you obtain the z-score, the exact probability you calculate depends on how the problem is phrased.

Note: If the problem requires you to find a sample total associated with a probability, you'll need to convert the z-score to the sample total (T) in natural units. For the distribution of sample means, use the following:

$$T = n\mu + z \times \left(\sigma\sqrt{n}\right)$$

This formula reflects the fact that we're using $n\mu$ for the population mean and the standard deviation of the total for the variability.

Worked Example

A produce company packs apples into shipping boxes that hold 30 apples each. The boxes have a maximum safe load of 2.50 kg. Apples are randomly placed into boxes from a large supply.

Individual apple weights have a population mean of $\mu = 0.080$ kg (80 grams) and a standard deviation of $\sigma = 0.012$ kg (12 grams).

What is the probability that a box of 30 apples has a total weight less than 2.50 kg?

The problem doesn't specify the population distribution. However, because the sample size is $n = 30$, the sampling distribution of the total should be approximately normal.

Let T represent the total weight of the 30 apples.

Step 1: Determine the mean and standard deviation of the total.
Mean of the total: $n\mu = 30(0.080) = 2.40$ kg
Standard deviation of the total: $\sigma\sqrt{n}$
$= 0.012\sqrt{30}$
$= 5.477$
$= 0.012 \times 5.477 = 0.0657$ kg

Step 2: State the sampling distribution of the total.

By the central limit theorem, the sampling distribution of T is approximately normal:

$T \sim N(2.40, 0.0657)$

Step 3: Convert the cutoff value to a z-score.

$z = (T - n\mu) / (\sigma\sqrt{n})$
$z = (2.50 - 2.40) / 0.0657$
$z = 0.10 / 0.0657$
$z = 1.52$

Step 4: Use the cumulative z-table.

Look up z = 1.52:

Z	0.00	0.01	0.02	0.03	0.04
1.4	0.91924	0.92073	0.92220	0.92364	0.92507
1.5	0.93319	0.93448	0.93574	0.93699	0.93822
1.6	0.94520	0.94630	0.94738	0.94845	0.94950
1.7	0.95543	0.95637	0.95728	0.95818	0.95907
1.8	0.96407	0.96485	0.96562	0.96638	0.96712
1.9	0.97128	0.97193	0.97257	0.97320	0.97381

For z = 1.52, the area to the left is 0.93574.

So: $P(T < 2.50) = 0.93574$.

The probability that a box of 30 randomly selected apples weighs less than 2.50 kg is approximately 0.936 (93.6%).

That means about 6.4% of boxes would exceed the 2.50 kg limit.

Add-on: Finding a Box Capacity That Limits Overweight Boxes to 1%

In the previous solution, we found that a 2.50 kg limit results in about 6% of boxes being overweight. Suppose the company wants to redesign the box so that only 1% of the boxes are overweight.

Let X be the maximum safe box capacity. We want:

$P(T > X) = 0.01$

That is the same as:
$P(T \leq X) = 0.99$

From the earlier work, the sampling distribution of the total weight is:
$T \sim N(2.40, 0.0657)$

The key difference from the previous calculation is that we are not finding a probability from a given cutoff. Instead, we are given a probability and must find the cutoff value X. Hence, we must find the cumulative probability of 0.99 in the table, take its z-score, and convert that score to the raw value.

Step 1: Find the z-score for a cumulative probability of 0.99

Z	0.00	0.01	0.02	0.03	0.04
2.0	0.97725	0.97778	0.97831	0.97882	0.97932
2.1	0.98214	0.98257	0.98300	0.98341	0.98382
2.2	0.98610	0.98645	0.98679	0.98713	0.98745
2.3	0.98928	0.98956	0.98983	0.99010	0.99036
2.4	0.99180	0.99202	0.99224	0.99245	0.99266

$P(Z \leq z) = 0.99$ corresponds to $z = 2.33$

Step 2: Convert that z-score back to the original scale

Plug in the parameter values for the distribution of sample totals ($T \sim N(2.40, 0.0657)$) and the z-score (2.33) into the following equation:

$$
\begin{aligned}
x &= n\mu + z(\sigma\sqrt{n}) \\
&= (2.40) + (2.33 \times 0.0657) \\
&= 2.40 + 0.1531 \\
&= 2.5531 \text{ kg}
\end{aligned}
$$

To limit overweight boxes to about 1%, the company should use boxes with a maximum safe capacity of approximately 2.55 kg.

A very small increase in box strength (2.50 → 2.55kg) will markedly reduce the proportion of overweight boxes. Because tail probabilities decrease rapidly as we move farther from the mean in standard deviation units, relatively small increases in capacity can substantially reduce the proportion of overweight boxes.

This small change in box capacity is a simple, cost-effective solution to reduce overweight boxes!

Sample Proportions

Some probability problems involve proportions rather than measurements. Instead of working with averages or totals, we are interested in the proportion of observations in a sample that fall into a particular category.

For example, a probability question might ask you about the proportion of defective items in a shipment, the proportion of voters who support a candidate, or the proportion of patients who respond positively to a treatment.

A sample proportion is the average of a binary variable that takes the value 1 when an observation has the characteristic of interest and 0 otherwise. Because proportions are averages, the central limit theorem applies to them under specific conditions.

Let p represent the population proportion, and $\hat{p}$ is a sample proportion. The sampling distribution of the sample proportion has the following properties:

Expected value of the sampling distribution:

$$E(\hat{p}) = p$$

Standard error of the proportion:

$$\sqrt{p(1-p)/_n}$$

Thus, the sampling distribution of the sample proportion centers on the true population proportion, and its variability decreases as the sample size increases.

When the sample size is sufficiently large, the sampling distribution of $\hat{p}$ is approximately normal:

$$\hat{p} \sim N(\,p, \sqrt{p(1\ -\ p)\ /\ n}\,)$$

Before applying the normal model to a sample proportion problem, confirm that the conditions for the normal approximation are satisfied. The sampling distribution of $\hat{p}$ is approximately normal when the sample size is large enough that both np and n(1 – p) are at least 5 (some texts use 10 as a more conservative guideline).

These quantities (np and n(1 – p)) represent the expected number of successes and failures in a sample of size n. When both are sufficiently large, the underlying binomial distribution is not excessively skewed, and the normal approximation is reasonable.

As in the previous sections, independence of observations is also required, which is typically satisfied when sampling is random and the population is large relative to the sample size.

After determining that the sampling distribution of the sample proportion is approximately normal, follow these steps:

1. **Identify the given values and the sample size.**
   Record the population proportion (p) and the sample size (n).
2. **Calculate the standard error of the proportion.**
   Standard error = √p(1 – p) / n
3. **State the sampling distribution.**
   $\hat{p}$ ~ N(p, √p(1 – p) / n)
4. **Convert the sample proportion value to a z-score.**
   z = ($\hat{p}$ – p) / √p(1 – p) / n
5. **Use the standard normal distribution.**
   Use a z-table or calculator to find the required probability.

Note: If the problem requires you to find a proportion associated with a probability, you'll need to convert the z-score to a sample proportion ($\hat{p}$). For the distribution of sample proportions, use the following:

$$\hat{p} = p + z \times \left({}^{p(1-p)}/_{n} \right)$$

This formula reflects the fact that we're using p for the population proportion and the standard error of the proportion for the variability.

Worked Example

A lightbulb manufacturer tracks defects during production. Based on long-run quality data, the company estimates that 4% of all lightbulbs are defective. Lightbulbs are packaged in boxes containing 200 lightbulbs. An inspector randomly selects a box of 200 lightbulbs from a shipment to check that the proportion of defective bulbs is less than the mandated 5%.

What is the probability that fewer than 5% of the sampled lightbulbs are defective in a sample?

When $\hat{p}$ is the sample proportion defective, we want to find:

$P(\hat{p} < 0.05)$

Because this is a random sample and the sample size is large, the normal model for the sampling distribution of $\hat{p}$ should be reasonable. We can quickly check:

- $np = 200(0.04) = 8$ (expected defective bulbs)
- $n(1 - p) = 200(0.96) = 192$ (expected non-defective bulbs)

Both values are larger than 5, so the sampling distribution of $\hat{p}$ is approximately normal.

Step 1: Identify the given values.
Population proportion: $p = 0.04$
Sample size: $n = 200$
We want: $P(\hat{p} < 0.05)$

Step 2: Calculate the standard error of the proportion.
Standard error = $\sqrt{p(1 - p) / n}$
$= \sqrt{[\ 0.04(0.96) / 200\ ]}$
$= \sqrt{[\ 0.0384 / 200\ ]}$
$= \sqrt{[\ 0.000192\ ]}$
$= 0.01386$

Step 3: State the sampling distribution.
$\hat{p} \sim N(\ p, \sqrt{p(1 - p) / n}\ )$
So: $\hat{p} \sim N(0.04, 0.01386)$

Step 4: Convert the cutoff proportion to a z-score.
$z = (\hat{p} - p) / \sqrt{[\ p(1 - p) / n\ ]}$
$= (0.05 - 0.04) / 0.01386$
$= 0.01 / 0.01386$
$= 0.72$

Step 5: Use the cumulative z-table.

Look up z = 0.72:

Z	0.00	0.01	0.02	0.03	0.04
0.4	0.65542	0.65910	0.66276	0.66640	0.67003
0.5	0.69146	0.69497	0.69847	0.70194	0.70540
0.6	0.72575	0.72907	0.73237	0.73565	0.73891
0.7	0.75804	0.76115	0.76424	0.76730	0.77035
0.8	0.78814	0.79103	0.79389	0.79673	0.79955
0.9	0.81594	0.81859	0.82121	0.82381	0.82639

Therefore: $P(\hat{p} < 0.05) = 0.76424$

The probability that fewer than 5% of the 200 sampled lightbulbs are defective is about 0.764, or 76.4%.

The following graph illustrates this solution by showing the sampling distribution of the proportions with the shaded area of $\hat{p} < 0.05$.

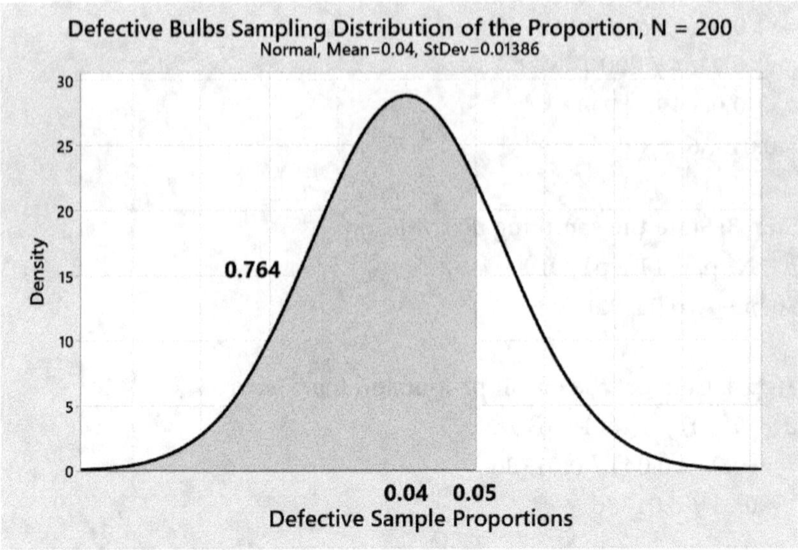

Conversely, that means the probability that 5% or more are defective in a sample is:

$1 - 0.764 = 0.236$

So about 23.6% of such samples will contain at least 5% defective lightbulbs, even when the true defect rate is only 4%.

Notice that in both the apple total weight and defective lightbulb examples, the expected value lies on the desirable side of the cutoff. Yet the probability of exceeding the cutoff is not zero. In the defective lightbulb example, even though the overall population percentage of defective bulbs is 4%, the probability of a sample of 200 exceeding the 5% cutoff is surprisingly large at 23.6%.

Why does this happen?

Think back to our discussions about expected values and the standard deviation in Chapters 10 and 11. The expected value is not a guarantee. It tells you the center of the distribution and the long-run average. Variability, reflected in the standard deviation or standard error, determines how widely outcomes spread around that center and, therefore, how much risk exists in the tails. Nothing "went wrong" in these scenarios. It is simply random variability behaving exactly as probability theory predicts.

Review and Next Steps

In this chapter, you shifted from modeling individual outcomes to modeling statistics calculated from random samples. Earlier chapters described the behavior of single observations or single realized values from a process. Here, you treated sample means, totals, and proportions as random variables with their own probability distributions.

You learned that when we repeatedly draw random samples from the same fixed population, the resulting statistics vary from sample to sample in a predictable way. Sampling distributions describe that variability. Further, variability determines the risk of extreme outcomes. The expected value marks the center, but it does not guarantee a specific result.

You saw that when the parent distribution is normal, the sampling distribution is normal. And when the parent distribution is not normal, the Central Limit Theorem tells us that the sampling distribution of the mean becomes approximately normal for sufficiently large samples.

Hence, you can frequently calculate probabilities for sample statistics using the standard normal distribution methods developed earlier.

This chapter assumes that the population parameters are known. We treated μ, σ, and p as given, fixed values that allowed us to model the behavior of sample statistics precisely. In practice, we rarely know population parameters with certainty, even though they still govern the behavior of sampling distributions.

The next chapter builds directly on this one. Instead of calculating probabilities for sample statistics when parameters are known, you will see how sampling distributions allow us to draw conclusions about unknown population parameters. That shift moves us from probability modeling into the foundations of statistical inference.

Formula Quick Reference

Sampling Distribution	**Center**	**Standard Error**	**Statistics → Z** **Z → statistics**
Means	μ	$\sigma / \sqrt{n}$	$z = (\bar{x} - \mu) / (\sigma / \sqrt{n})$ $\bar{x} = \mu + z \times (\sigma/\sqrt{n})$
Totals	$n\mu$	$\sigma\sqrt{n}$	$z = (T - n\mu) / (\sigma\sqrt{n})$ $T = n\mu + z \times (\sigma\sqrt{n})$
Proportions	p	$\sqrt{p(1-p)/n}$	$z = (\hat{p} - p) / \sqrt{p(1-p)/n}$ $\hat{p} = p + z \times \sqrt{(p(1-p)/n)}$

Practice Problems Chapter 14

Note: Appendix A contains the cumulative z-table.

Question 1: A food scientist studies sugar content in strawberries.

She collects 1,000 random samples of 20 strawberries each and calculates the average sugar content. She then creates a graph showing the distribution of these 1,000 averages.

Does this graph represent a distribution of individual strawberry sugar content or a sampling distribution? Explain what values are being plotted.

Question 2: A population has a mean of 75 and a standard deviation of 10. Samples of size 25 are repeatedly drawn, and the sample mean is calculated for each sample.

What does the standard error of the mean describe in this context? Be specific about what is varying and what is not.

Question 3: A population is normally distributed with mean $\mu = 60$ and standard deviation $\sigma = 12$. Two sampling distributions of the mean are created: one with $n = 16$ and one with $n = 64$. Everything else remains the same.

Compare the two sampling distributions in terms of:

- **A.** Their centers
- **B.** Their variability

Explain your reasoning.

Question 4: The average test score in a large population is 70. Samples of 40 students are taken, and the sampling distribution of the mean is approximately normal.

Suppose:
$P(\bar{x} > 75) = 0.04$

Interpret this probability in context.

Question 5: A company knows that 20% of its products are returned by customers. Samples of 150 orders are taken, and the sampling distribution of p^ is approximately normal.

Suppose:
$P(\hat{p} \geq 0.25) = 0.03$

Interpret this probability in context.

Question 6: For each of the following scenarios, determine whether the sampling distribution described is approximately normal. Assume independence in all cases.

Indicate cases where the normal approximation is reasonable. Consider the following populations:

- **A.** Strongly right-skewed. Sample size n = 10.
- **B.** Moderately skewed. Sample size n = 35.
- **C.** Normally distributed. Sample size n = 5.
- **D.** A proportion problem with p = 0.02 and n = 100.
- **E.** A proportion problem with p = 0.40 and n = 50.
- **F.** A population is uniform. Sample size n = 6.

Question 7: A population has standard deviation $\sigma = 20$. Samples of size n are drawn, and the standard error of the mean is calculated. For what value of n is the standard error of the mean equal to the population standard deviation? Explain your reasoning.

Question 8: A population has mean $\mu = 5$ and standard deviation $\sigma = 2$. Consider samples of size n.

As the sample size increases:

- What happens to the mean of the sample total?
- What happens to the standard deviation of the sample total?

Explain how each changes and why.

Question 9: A study finds that: $P(\bar{x} < 48) = 0.02$

Which of the following interpretations is correct?

A. 2% of individual observations have values below 48.
B. 2% of samples have an average below 48.
C. 2% of the population means are below 48.
D. 2% of values in each sample are below 48.

Question 10: A company fills 40 small bottles with a liquid supplement. The amount in each bottle varies randomly, with a population mean of 250 mL and a population standard deviation of 6 mL. Assume the relevant sampling distribution is approximately normal.

What is the probability that the combined amount in a randomly selected set of 40 bottles is less than 9,960 mL?

Question 11: A pharmaceutical company produces a medication where, on average, 3% of doses fail a quality test. Regulations require that no more than 5% of doses in a batch fail inspection.

An inspector randomly selects 400 doses from a large batch.

What is the probability that the sample will meet the regulatory requirement?

Question 12: A fitness app tracks daily step counts for users in a city. The population mean is 7,800 steps with a population standard deviation of 2,100 steps. A researcher randomly selects 49 users and records their average step count.

What is the probability that the sample mean is greater than 8,300 steps?

Question 13: A customer service center finds that 18% of incoming calls result in a complaint. A manager monitors random samples of 250 calls at a time.

What sample proportion of complaints marks the lowest 10% of the sampling distribution?

Then, briefly comment on the manager's idea of treating samples below this cutoff as evidence of improved performance.

CHAPTER 15

Statistical Inference

In the previous chapter, you learned how sampling distributions describe the probability behavior of sample statistics. Now you will learn how those same distributions allow us to draw conclusions about unknown population parameters as we transition to inferential statistics.

Inferential statistics uses sample data to draw conclusions about entire populations. More specifically, you use sample statistics to estimate population parameters.

Up to this point, we have spent several chapters developing tools that describe how randomness behaves. You learned how probability distributions model long-run patterns. You saw how sample means and sample proportions vary from sample to sample. You examined how the Central Limit Theorem explains why the normal distribution shape appears repeatedly.

Sampling distributions give you a way to measure how much a statistic is expected to fluctuate from sample to sample simply because of random sampling. That fluctuation is the key to inference. Without a model for that variability, you have no statistical way to assess how close the sample result is likely to be to the population value. In other words, the model evaluates the precision of the sample statistic.

Precision in this context refers to how tightly sample estimates are expected to cluster around the true population parameter. When the sampling distribution is tightly concentrated, estimates are more precise because they vary less across repeated samples and tend to fall close to the population parameter.

Sampling distributions convert the random variability of sample statistics into quantifiable probabilities, making statistical inference possible. Let's start by understanding inferential statistics in more detail and then move on to learning how sampling distributions help us.

Learning About Populations from Samples

Statistics probably surround you. You can find them everywhere, such as in the news media, surveys, commercials, and so on. Often, these statistics aren't meant to describe only the specific group of subjects that were measured. Instead, the goal is to infer properties about a larger population. This practice is called inferential statistics.

For example, when we read survey results, we are not learning about just the opinions of those who responded to the survey, but about an entire population. Or, when we see averages, such as health measures and salaries, we're learning about them on a population scale, not just the few subjects in the study. Consequently, inferential statistics can provide extremely meaningful information.

Inferential statistics is a powerful tool because it allows you to use a relatively small sample to learn about an entire population. Typically, it is impractical to measure every member of a population. Instead, we collect a random sample from a small portion of the population, measure it, and use the sample data to estimate population properties. Using appropriate inferential statistical procedures, you can draw reasonable conclusions from samples about the populations they represent.

Researchers tend to focus on understanding population parameters because science is usually uninterested in the properties of only one small sample. For example, scientists don't care about a new medicine's mean effect on a small group, a sample statistic. Instead, they want to generalize the results and understand the impact on a whole population, a parameter.

You gain tremendous benefits by working with samples because it is usually impossible to measure the entire population. However, there are tradeoffs. Most samples are tiny compared to the whole population. Consequently, when you estimate a population's characteristics from a sample, the estimates are unlikely to match the population value exactly.

For example, a medicine's mean effect in a sample is unlikely to equal its mean population effect exactly. Inferential statistics uses various analytical tools to account for this error in the results.

At a broad level, inferential statistics requires the following three steps:

1. Define the population we are studying.
2. Draw a representative sample from that population.
3. Use analyses that incorporate sampling distributions.

The first step is identifying the population you want to study. That's relatively simple. Let's dig into the latter two steps!

Representative Samples

Inferential statistics involves generalizing the sample results to the population. To make that generalization, you must be confident that the sample accurately represents the population. This requirement affects how you should obtain that sample. If you use an incorrect method, the sample might not represent the population, leading you to erroneous conclusions.

In inferential statistics, simple random sampling is the most well-known method to obtain an unbiased sample. Using this method, researchers have an equal probability of selecting all items in the population. However, there are other forms of random sampling.

This book doesn't go into the mechanics of obtaining representative samples because that can be complicated. For the purposes of this book, know that statistical inference assumes that randomness enters the process through the way the sample is obtained. Random sampling justifies using probability models to describe how sample statistics fluctuate from one sample to another. Without random sampling, the sampling distribution framework no longer provides a reliable basis for drawing conclusions about the population.

Using Sampling Distributions in Estimation

In the previous chapter, you learned that a sample statistic is not a fixed quantity. If we repeatedly draw samples of the same size from a population, the statistic's value changes from sample to sample. The statistic's sampling distribution describes its pattern of variation over many samples.

Now we'll use sampling distributions to look at it from the opposite direction. Given a statistic's long-run pattern, what can we learn about a population parameter from a single sample statistic?

Start with a simple idea. Suppose we collect one random sample and calculate a sample proportion or a sample mean. That number is our best estimate of the population value. It is called a *point estimate* because it is a single number chosen to estimate an unknown parameter.

A point estimate is a single value that best estimates a population parameter. For instance, the sample mean estimates the population mean. And the sample proportion estimates the population proportion. When the sample is representative and selected randomly, these estimates tend to center around the true population values rather than consistently being too high or too low.

Imagine a study of adult women in the U.S. finds that the random sample has a mean height of 63.5 inches (1.61 m). We can use that sample mean as the estimate of the population mean. That value is our best estimate of the parameter and it's equally likely to be too high or too low.

However, by understanding sampling distributions, we know the statistic varies from sample to sample and that our particular estimate is incorrect by some amount. In fact, it's virtually guaranteed that sample statistics from a study will be at least somewhat wrong compared to their corresponding population parameters.

The difference between a sample statistic and a population parameter is sampling error. However, because the parameters are unknown, the degree of error is not apparent from the statistic alone.

For example, our sample's mean height (63.5 inches) won't exactly equal the population mean. How far off is it likely to be?

The sampling distribution tells us how much variation to expect solely due to random sampling. More specifically, the standard error (SE) for the distribution summarizes the expected variation. It measures how far a statistic typically falls from the true population parameter when samples are drawn at random.

Hence, the SE allows us to evaluate the precision of a sample estimate. A small standard error means the sampling distribution is tightly clustered around the parameter, so our estimate is likely to be close. A larger standard error indicates a more spread-out sampling distribution, so greater fluctuation is expected. In this case, we're more likely to obtain a sample estimate that is far from the correct parameter value.

The precision of an estimate relates to the magnitude of the standard error as follows:

- **High-precision estimates**: The standard error is low, and the sample estimate tends to be close to the population parameter.
- **Low-precision estimates**: SE is high, and the estimates tend to be farther from the parameter.

Some error is unavoidable in inferential statistics because samples are typically minuscule relative to the population. Samples introduce fuzziness because there's a margin of error around all point estimates of population parameters. We need to account for that error when inferring population properties from a sample by incorporating sampling distributions.

Estimation in statistical inference begins with a statistic from a random sample. We recognize that it varies across samples because of sampling error. We use the sampling distribution, and the standard error in particular, to quantify that variability and construct a range of plausible values for the population parameter. Without the sampling distribution, we would have only a point estimate and no way to judge its precision.

Sampling distributions connect what we observe in a single sample to probabilities associated with the population. That probability information is what allows us to move from description to inference. Because we know how the statistic behaves across many possible samples, we can infer the properties of the population from which the sample was drawn.

In the rest of this chapter, we'll cover two of the main statistical inference tools, confidence intervals and hypothesis tests. In short, these tools do the following:

- **Confidence intervals**: Construct a range of plausible values for population parameters.
- **Hypothesis tests**: Assess competing claims about population parameters.

For each of these methods, I'll start by covering the conceptual details and interpretations, and then move on to calculating them for several examples.

Note: I have various free calculators on my website for the margin of error, confidence intervals, and various hypothesis tests for you. Find the link in the book's Introduction for all the free resources.

Confidence Intervals

A confidence interval (CI) is a range of values that is likely to contain the value of an unknown population parameter. These intervals represent a plausible domain for the parameter given the characteristics of your sample data, specifically the standard error of the sampling distribution.

CIs have an advantage over point estimates. We know that any given point estimate almost certainly does not equal the parameter exactly. Conversely, CIs generally do contain the parameter. The trade-off is that CIs give us a range of values that is likely correct rather than a single point that is likely wrong.

For example, our sample mean of adult women is 63.5 inches. That's our point estimate. Suppose the confidence interval for height is [62, 65]. We can be confident that the average population height is between 62 and 65 inches.

That's the nature of working with samples! In most cases, you won't know the exact value of a population parameter.

Ok, we're confident that the population parameter lies within the confidence interval. How confident are we exactly? That's where the confidence level comes in.

Confidence levels are the long-run percentage of times that the population value falls within the CI. Standard confidence levels are 90, 95, and 99%. 95% is the most common. For studies that use a 95% confidence level, you'd expect 95% of their results to have CIs that contain the population value.

For example, a 95% confidence level suggests that if you draw 20 random samples from the same population, you'd expect 19 of the confidence intervals to include the population value, as the following graph shows. The highlighted interval does not contain the population value represented by the dashed line.

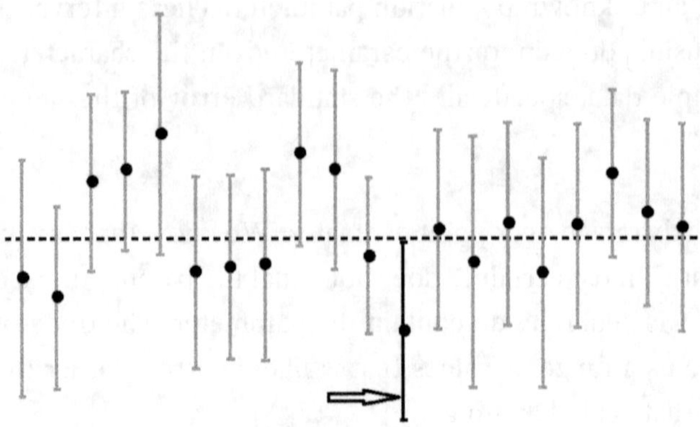

The horizontal dashed line is the population parameter, which is usually unknown. Each dot is the sample's point estimate for the parameter. Vertical lines represent the CIs. Notice that 19 of 20 (95%) CIs contain the parameter. The graph illustrates how confidence intervals are not perfect but usually correct.

The CI procedure provides meaningful estimates because it produces ranges that usually contain the parameter. Hence, they present plausible values for the parameter and help you navigate the uncertainty of how well a sample estimates a value for an entire population.

The width of the confidence interval around the point estimate reveals the precision. If the range is narrow, the margin of error is small, and there is only a tiny range of plausible values. That's a precise estimate. However, if the interval is wide, the margin of error is large, and the actual parameter value is likely to fall somewhere within that more extensive range. That's an imprecise estimate.

For example, imagine we are collecting random samples and measure IQ scores. Suppose our sample has a mean IQ of 101. If the confidence interval is [99 103], you can be confident that the actual population parameter falls within that range. It's a fairly tight range, indicating high precision. Our point estimate of 101 is likely wrong by just an IQ point or two. However, a broader CI, such as [90 110], indicates lower precision and our point estimate can be off by a larger amount. It could be wrong by 10 points!

Unlike a point estimate by itself, a confidence interval provides information about precision, allowing you to understand how close the estimate is likely to be to the parameter value. The narrower the interval, the more precise the estimate.

Now, it's time to learn how to calculate confidence intervals. We'll start with CIs for proportions and then move on to means.

Margin of Error for Proportions

Surveys generally report the uncertainty around proportion and percentage estimates using the margin of error (MOE) rather than CIs, but they're the same thing, just presented differently. The margin of error is the distance from the sample estimate to either end of the interval. In statistical analysis more broadly, this same idea is usually expressed as a confidence interval, which states the entire range of plausible values rather than just the distance from the estimate.

In other words, the margin of error describes the half-width of the interval, while a confidence interval describes the full range. However, both terms refer to the same underlying concept.

For instance, a survey reports that 52% of respondents support a policy with a margin of error of ±3%. That MOE is equivalent to a confidence interval of 49% to 55%.

MOEs and CIs both indicate how close you can expect the survey results to be to the correct population value using a confidence level and a range. Both methods use the same math under the hood.

Similarly, margin of error assesses the precision of a survey's estimates. A smaller MOE suggests that the survey's results tend to be close to the correct population values. Conversely, larger MOEs indicate that the survey's estimates can be further away from the parameter values.

Maximum Margin of Error for a Survey

In a survey, the size of the margin of error varies depending not only on the sample size but also on the proportion itself. Surveys frequently cite its maximum MOE. That's the value you'll see in the media when they report a survey's margin of error. Knowing this value is beneficial because the margins of error for all percentages in a survey will not exceed this value.

Survey proportions and percentages are based on binary outcomes. Each respondent either selects an option or does not. The sample proportion represents the fraction of respondents who chose that option.

The MOE depends on the standard error of the sample proportion. That SE comes from the standard deviation of the underlying binomial process generating the data, which determines how much the proportion varies from sample to sample. Consequently, the MOE's behavior depends on that distribution.

Recall from Chapter 11 that the standard deviation of a binomial distribution is maximized when the probability equals 0.5. Hence, the largest MOE for a given sample size occurs when the proportion is 50%.

For example, 50% of the respondents prefer Candidate Smith. As the percentage for an item moves away from 50%, either higher or lower, the MOE decreases if you hold sample size constant.

Let's see this phenomenon in action! The following graph displays the 95% margin of error for a sample size of 1,000. The vertical axis represents the MOE, while the horizontal axis indicates percentages in the survey. The highest MOE occurs at the vertical line where the survey result is 50%, and it diminishes in both directions from that peak.

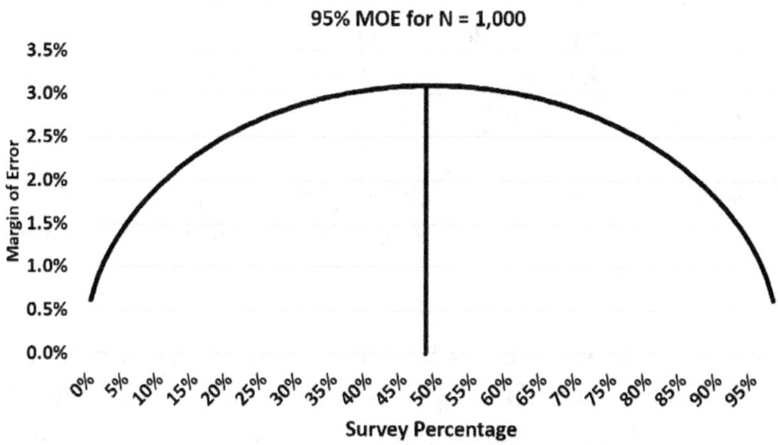

In the formula section, I'll show you how to find the maximum and other MOEs.

On this book's resource page on my website, you can download the Excel file I made to create this graph. In this file, you can change the sample size to see how it affects the MOE for all percentages.

Margin of Error Formula

The margin of error formula for surveys relates to proportions and the binomial distribution. When you add and subtract the MOE from the survey proportion, you obtain a confidence interval of the proportion.

In statistics, most margin of error and confidence interval calculations involve multiplying a standard error by a critical value.

A critical value is a cutoff value in a probability distribution that determines the boundary of a specified probability region. In this book, we use critical z-scores, which are critical values taken from the standard normal distribution (See Appendix A).

The critical z-score marks the cutoff points that leave a chosen proportion of probability in the middle of the distribution. For example, selecting a 95% confidence level means choosing the z-score that captures the central 95% of the standard normal distribution, leaving 2.5% in each tail.

You set the confidence level by choosing the corresponding critical z-score. This value comes from the cumulative z-table we've been using the past several chapters.

In practice, you will rarely need to use the z-table to find critical z-scores because most applications use the standard confidence levels shown in the following table. Among these, the 95% confidence level is by far the most common.

Confidence Level	Z-score
90%	1.64
95%	1.96
99%	2.58

In the following formula, the square root term is the standard error of the proportion, and you multiply it by the critical z-score:

$$MOE = z^* \times \sqrt{\frac{\hat{p} \times (1-\hat{p})}{n}}$$

Where:

- z* is the critical value that corresponds to your confidence level.
- $\hat{p}$ is the sample proportion.
- n is the sample size.

As mentioned earlier, the standard error term comes from the formula for the standard deviation of a binomial distribution. The binomial standard deviation describes variability in counts, while the standard error describes variability in proportions. The standard error of a sample proportion comes from rescaling the binomial standard deviation from counts to proportions, which results in dividing by the sample size within the square root.

Maximum Margin of Error Formula

To find the maximum MOE that surveys typically report, we need to set p to 0.5 and use the z-score for our confidence level. 95% is the standard confidence level. Consequently, I'll use the corresponding z-score of 1.96. Because we're setting the values of these two inputs (z and p), the maximum margin of error formula becomes a direct transformation of the survey's sample size, as shown in the following formula.

$$Max\ 95\%\ MOE = 1.96 \times \sqrt{\frac{0.5(1-0.5)}{n}} = \frac{0.98}{\sqrt{n}}$$

For any survey, all we need to enter into the formula is the sample size to find its maximum margin of error. Each sample size has its own maximum value. For example, if our sample size is 1000, our maximum MOE is the following:

$$\frac{0.98}{\sqrt{1000}} = 0.03099 = 3.099\%$$

This maximum MOE corresponds to the peak in the previous chart.

Margin of Error for Other Percentages

It's easy to use the margin-of-error formula for other percentages. Simply enter the percentage as a proportion (i.e., 75% = 0.75), sample size, and z-score into the formula.

For example, a survey item finds that 75% of the respondents agree with the governor's policy. 1000 individuals responded to this survey. To find the 95% margin of error for this percentage, we do the following:

$$95\%\ MOE = 1.96 \times \sqrt{\frac{0.75(1-0.75)}{1000}} = 0.02684$$

We can be 95% confident that the population percentage for this survey item is within 75% ± 2.684%, or 72.316 – 77.684%.

Additionally, notice that the MOE for this specific percentage (75%) is 2.684%, compared to the maximum MOE of 3.099% for 50% at a sample size of 1,000. You can find both points on the earlier chart.

Caveats and Cautions for MOEs

The margin of error reflects only random sampling variability. It does not account for other potential sources of error, such as biased wording, non-representative samples, or systematic exclusions of certain groups. Even a small margin of error cannot correct for those issues.

In addition, reported margins of error typically apply to individual percentages, not to differences between percentages. Comparing two responses requires analyzing the sampling variability of their difference, which involves additional probability calculations beyond the single-proportion case discussed here.

Confidence Intervals of the Mean

Confidence intervals (CIs) are similarly helpful for understanding sample means ($\bar{x}$). These CIs give you a range of plausible values for the population mean (μ) and help you assess the precision of the sample estimate. For example, if you take a random sample of adult women in the U.S. and measure their heights, the confidence interval of the mean gives you a range of plausible values for the parameter.

Formula

As with confidence intervals for proportions, CIs for means account for sampling randomness by using critical values, standard errors, and sample sizes. Again, you'll use critical z-scores to calculate your confidence interval of the mean.

The following formula finds the confidence interval of a mean. Like the proportion interval, it uses the same approach of taking the critical z-score and multiplying it by the standard error. However, this time you're using the standard error of the mean (SEM).

$$\text{CI} = \bar{x} \pm z^* \times \frac{\sigma}{\sqrt{n}}$$

Where:

- $\bar{x}$ = sample mean
- z* = critical z-score
- σ = population standard deviation
- n = sample size
- $\sigma / \sqrt{n}$ = the standard error of the mean

In practice, the population standard deviation σ is usually unknown and must be estimated using the sample standard deviation s. In this example, we use the sample estimate of the standard deviation (s) instead of σ when calculating the confidence interval. When σ is unknown, analysts often use a t-critical value rather than a z-score. In this chapter, I use z-scores for simplicity so we can focus on the core ideas of confidence intervals and hypothesis testing without introducing additional complexity.

Worked Example

Let's move on to using this formula to find a confidence interval! Suppose a researcher randomly selects a random sample of 25 households and obtains their fuel costs. We want to calculate the 95% confidence interval of the mean.

Imagine we have only the following summary information and know that the population follows a normal distribution:

- Sample mean: 330.6
- Standard deviation: 154.2
- N = 25

Fortunately, that's all we need to calculate our 95% confidence interval of the mean.

Our next step is to look up the critical z-score. It's the same value that we used for the 95% MOE calculations, 1.96.

Let's enter all this information into the formula.

First, I'll calculate the margin of error:

$$z^* \times \frac{s}{\sqrt{n}} = 1.96 \times \frac{154.2}{\sqrt{25}} = 60.44$$

Next, I'll take the sample mean and add and subtract the margin of error from it:

- 330.6 + 60.44 = 391.04
- 330.6 – 60.44 = 270.16

The 95% confidence interval for the mean fuel cost is [270.16, 391.04]. We can be 95% confident that the population mean falls within this range.

What Affects CI Widths?

Ok, so you want narrower CIs for their greater precision. What conditions produce tighter ranges?

The width of a confidence interval reflects how much a statistic is expected to fluctuate across random samples. That fluctuation is determined by the sampling distribution, so anything that affects sampling variability also affects interval width.

Think back to the terms in the CI formulas because they directly impact the CI. Hence, sample size, variability, and the confidence level affect the widths.

Sample size is a key factor in determining the width of the interval. Larger samples reduce the standard error, which makes the sampling distribution more tightly clustered around the true population parameter. Recall the apple weight simulations in Chapter 14. As a result, confidence intervals become narrower and estimates become more precise.

Increasing the sample size is also the primary way researchers can intentionally improve precision. While many aspects of a study are constrained by practical limits, researchers often have direct control over how many observations they collect, making sample size a key design choice for reducing uncertainty.

Variability also matters. Greater variability in the population leads to greater variability in sample statistics, thereby increasing the standard error and widening confidence intervals. Unlike sample size, however, population variability is usually an inherent characteristic of the population itself. Researchers typically must work with the natural variation present in the population rather than control it directly.

Confidence level affects interval width as well, but it is a methodology choice separate from your sample's characteristics. A higher confidence level means capturing a larger portion of the sampling distribution's probability in the center. Doing so requires a larger critical value, which expands the margin of error and produces a wider interval. Lower confidence levels use smaller critical values and therefore produce narrower intervals, but with less coverage probability.

For example, if you increase the confidence level (e.g., 95% to 99%) while holding the sample size and variability constant, the confidence interval widens. Conversely, decreasing the confidence level (e.g., 95% to 90%) narrows the range.

I've found that many students find the effect of changing the confidence level on the width to be counterintuitive.

Imagine you take your knowledge of a subject area and indicate you're 95% confident that the correct answer lies between 15 and 20. Then I ask you to give me your confidence for it falling between the tighter range of 17 and 18. The correct answer is less likely to fall within the narrower interval, so your confidence naturally decreases.

Conversely, I ask you to indicate your confidence that the answer is between 10 and 30. That's a much wider range, and the correct value is more likely to be in it. Consequently, your confidence grows.

Confidence levels involve a tradeoff between confidence and the interval's spread. To have more confidence that the parameter falls within the interval, you must widen the interval. Conversely, your confidence necessarily decreases if you use a narrower range.

The general formula of MOE = Critical Value × Standard Error helps you understand how all those factors work mathematically.

Smaller margins of error produce narrower confidence intervals. By considering this equation, you can see that the following conditions create smaller MOEs:

- Larger sample sizes, because n is in the denominator of the SE.
- Smaller standard deviations, because they're in the numerator of the SE.
- Lower confidence levels, because they correspond to smaller critical values by which you multiply the SE.

In summary, confidence intervals rely directly on the sampling distributions you studied earlier. When we draw random samples, the value of a statistic, such as a sample mean or sample proportion varies, from sample to sample. The sampling distribution describes how those statistics are expected to fluctuate around the population parameter.

Using the standard error and the normal distribution, we can determine how far a statistic typically falls from the parameter purely due to random sampling. A confidence interval uses that information to place bounds around the sample estimate, creating a range of values consistent with the expected variability of the sampling distribution.

Hypothesis Testing

From Estimation to Testing Claims

Confidence intervals focus on estimating unknown population parameters. They provide a range of plausible values consistent with data from a random sample. In many situations, however, the goal is slightly different. Instead of estimating a parameter, we want to evaluate a specific claim about that parameter. Enter hypothesis testing.

Hypothesis tests are a form of inferential statistics that assess mutually exclusive hypotheses about population parameters and determine which one the sample data support. In short, these tests allow you to use sample data to make claims about population parameters.

For example, a hypothesis test can determine whether a medication is effective in a population rather than only in a specific sample. Making that leap from conclusions about a specific sample to one about an entire population requires using probabilities derived from sampling distributions.

Suppose researchers develop a medication to affect cognitive performance. IQ scores in the general population are defined to have an average of 100. To determine whether the medication affects IQ, researchers give it to a random sample of 36 participants and then measure their IQ scores.

The question then becomes whether the average IQ score in this sample differs from the population benchmark of 100. Questions like this shift the focus from estimation to testing.

Imagine that researchers collected a random sample of participants who took the medication and found that the sample mean IQ score is

105.5. Because the general population's average IQ is 100, the sample mean is 5.5 points higher.

Why perform a hypothesis test at all? We can clearly see that the sample average is higher. Isn't that enough to conclude that the medication increases IQ?

The situation is not quite that simple because we are analyzing a sample rather than the entire population. Working with samples has major advantages, because it is usually impossible to measure every individual in a population. However, the tradeoff for working with a manageable sample is that we must account for sampling error, the gap between a sample statistic and the population parameter it estimates.

In this example, the sample statistic is the sample mean IQ of 105.5. The population parameter is μ (mu), the true average IQ score for everyone who would take the medication. That value is unknown.

If the medication has no effect on IQ, the population mean IQ of those who take it is the same as that of those who don't (100). In this scenario, even though the researchers obtained a sample mean of 105.5 for those taking the medicine, the overall mean for that population might still be only 100.

In other words, this observed difference might be due solely to sampling variability. If the researchers drew another random sample and administered the drug, the next sample mean might be closer to the population benchmark of 100. Looking at the sample mean alone can't tell us the likelihood of that possibility.

Hypothesis tests allow us to answer that probability question by quantifying whether the observed result is more consistent with sampling error alone or with a true population effect. They do that by finding the likelihood of the observed sample outcome given hypothesized parameter values and sampling distributions.

For the medication example, a hypothesis test will tell us how likely it is to observe a sample mean of 105.5 if the true population mean is actually 100.

Hypothesis testing approaches this problem by starting with a proposed value for the population parameter. Using the sampling distribution, we then ask a probability question:

Under the assumed parameter value, what is the probability of observing a sample statistic at least as extreme as the one we obtained?

If the observed statistic were very unlikely under that assumption, the data provide evidence against the hypothesized parameter claim.

This reasoning relies on the same ideas you have already seen. The sampling distribution describes how much a statistic is expected to vary from sample to sample. By comparing our observed statistic to that distribution, we can determine whether the result falls within the range of typical outcomes or in a region that would rarely occur if the claim were correct.

In this way, hypothesis testing and confidence intervals are closely related. Both rely on the sampling distribution and the standard error to quantify sampling variability.

Confidence intervals use that information to estimate a range of plausible parameter values. CIs start with the sample statistic at the center and add a margin of error around it.

Conversely, hypothesis tests create a sampling distribution that centers on the hypothesized parameter value and determines how likely the observed statistic is in that distribution. In other words, it assesses how compatible the observed sample data are with the hypothesized parameter value.

Testing the Hypotheses

Hypothesis testing uses a standardized framework for evaluating claims about population parameters. It begins by translating a research question into two competing statements about a population parameter. These statements are called hypotheses. They do not describe sample results. Instead, they make claims about the unknown population parameter that the sample data evaluate.

Inferential testing uses the following two hypotheses:

- **Null hypothesis (H_0):** A statement that the population parameter equals a specific value. It represents the status quo, no effect, or a benchmark assumption that we initially treat as true.
- **Alternative hypothesis (H_1 or H_a):** A statement that contradicts the null hypothesis. It represents the possibility that the population parameter differs from the value stated in the null. It usually represents a relationship or treatment effect.

In every study or experiment, researchers assess an effect or relationship. This effect can be the effectiveness of a new drug, building material, or other intervention that has benefits. There is a benefit or relationship that the researchers hope to identify. Unfortunately, no effect can exist. In statistics, we call this lack of an effect the null hypothesis. Researchers assume this notion of no effect is correct until they have sufficient evidence to suggest otherwise, similar to how a trial presumes innocence.

In this context, the analysts don't necessarily believe the null hypothesis is correct. In fact, they typically want to reject it because that leads to more exciting finds about an effect or relationship. The new IQ medication works!

You can think of it as the default theory that requires sufficiently strong evidence to reject. Like a prosecutor, researchers must collect sufficient evidence to overturn the presumption of no effect. Investigators must work hard to set up a study and a data collection system to obtain evidence that can reject the null hypothesis.

Hypothesis testing works by asking whether the observed sample result would be unusually unlikely if the null hypothesis were true. If the result would be very unlikely under that assumption, the data provide enough evidence against the null hypothesis that you can reject it. However, if the result is reasonably likely, we do not have enough evidence to reject it.

In this manner, hypothesis testing seeks to find evidence against the null hypothesis rather than to prove it true.

IQ Medication Example

Returning to the IQ medication example, IQ scores are designed to have a population average of 100. For our null hypothesis, we'll assume the medication has no effect on IQ score. Hence, the population mean for the IQ medication population under the null hypothesis is also $\mu = 100$.

The hypotheses for this test are the following:

- **H_0:** $\mu = 100$
- **H_1:** $\mu \neq 100$

The null hypothesis states that the medication does not change the average IQ level from the population benchmark. The alternative hypothesis states that the true mean IQ for people taking the medication differs from 100.

Assume that we drew a random sample of 36, administered the medication, and measured the IQ scores. The following is a summary of our findings:

- Sample size: n = 36
- Sample IQ Mean: $\bar{x}$ = 105.5
- Population SD: σ = 15 IQ Points

I'll first show a graphical depiction of the hypothesis test using the sampling distribution in IQ points because that's more intuitive. After we establish the process visually, I'll cover the calculations using z-scores and a table.

We need to set up the sampling distribution under the null hypothesis to determine the likelihood of obtaining our sample mean of 105.5 if the null is true. This distribution uses the null hypothesis mean (μ = 100) and the standard error of the mean.

The null hypothesized mean is 100, indicating that the medication has no effect. Let's calculate the SEM.

$$SEM = \frac{\sigma}{\sqrt{n}} = \frac{15}{6} = 2.5$$

Therefore, by using the null value for μ and the SEM for σ, the sampling distribution of the mean that assumes the null hypothesis is true is $\bar{x} \sim N(100, 2.5)$.

The following graph depicts this sampling distribution and shows which sample means are more and less likely under the null hypothesis. We can place our sample mean of 105.5 in this distribution. This larger context helps us see how unlikely our sample mean is if the null hypothesis is true (μ = 100).

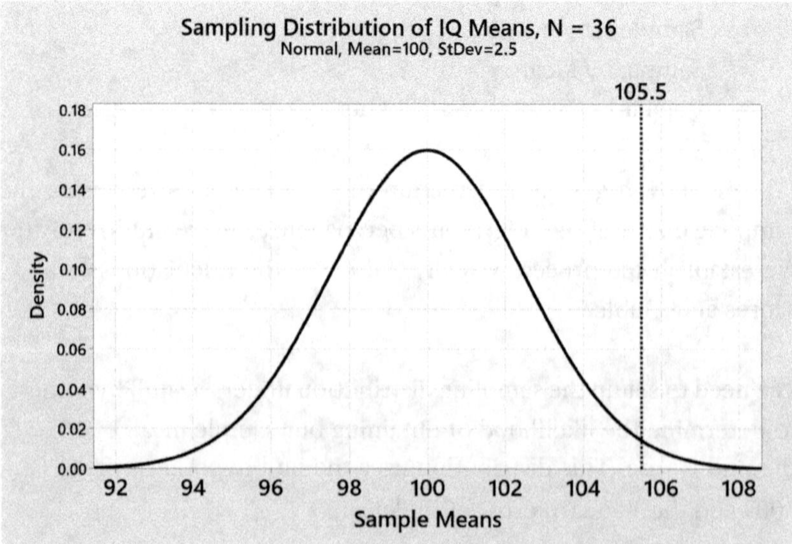

The graph displays the distribution of sample means under the null hypothesis. The most likely values are near 100 because this distribution assumes the true population mean is 100. However, given random sampling error, it would not be surprising to observe sample means ranging approximately from 94 to 106, with values closer to 100 occurring more frequently.

In other words, if the population mean is still 100 and given the observed variability, our study's sample mean (105.5) isn't the most likely value, but it's not completely implausible either.

Is our sample mean *so* unlikely that we can reject the notion that the population mean is 100?

The next step is to measure how far our sample mean is from the null value and determine the likelihood of that distance. This distance is known as the observed effect, and it is the following for our study:

105.5 – 100 = 5.5

Next, I'll shade the regions on both sides of the distribution that are at least as far away as 5.5 from the null (100 ± 5.5), equivalent to $\bar{x} < 94.5$ and $\bar{x} > 105.5$. This process graphs the two-tailed probability of observing a sample mean at least as extreme as the one we observed.

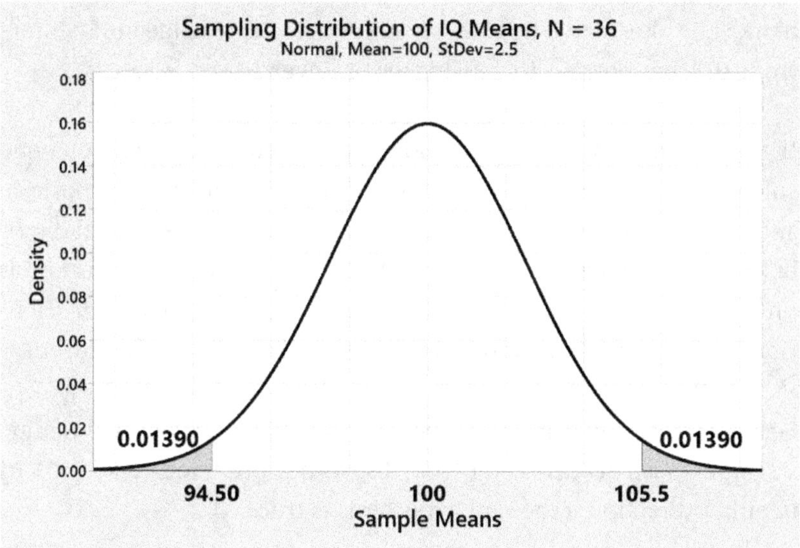

The total probability of the two shaded regions is 2 × 0.01390 = 0.02780.

If the null hypothesis mean (100) is true and you drew many random samples, you would expect to observe sample means at least 5.5 points away from 100 about 2.78% of the time. In other words, results at least as far from 100 as our observed effect would occur with this probability if the null is true.

You probably already know the name of this probability. It's the p-value! This probability quantifies our earlier visual assessment of the sampling distribution.

P-value Definition: The probability of observing our sample data, or more extreme, if the null hypothesis is true.

Why include both sides for a two-tailed probability?

Although our sample mean is a single value, 105.5, the hypothesis we are testing concerns deviations in either direction from the benchmark. The alternative hypothesis states that the true mean IQ differs from 100, but not specifically higher or lower.

If the null hypothesis is true and any difference we observe is due solely to random sampling variability, then unusually large positive and negative differences are equally likely. Because the normal distribution is symmetric, a result that is far above the benchmark is just as unlikely as one that is equally far below it. In the sampling distribution, those possibilities appear in the two tails of the normal curve.

So, the p-value indicates that there is only a 2.78% chance of obtaining a sample mean at least as far from 100 as our observed value (105.5) in either direction if the null hypothesis is true.

Is that small enough to conclude that the null hypothesis is false?

Look at the sampling distribution again. There is no special location on the curve where you can definitively draw this conclusion. Instead, the likelihood decreases as we move away from the null value. So, where do we draw the line and decide that a result is far enough away to be able to reject the null?

Significance Levels

A significance level, also known as alpha (α), is an evidentiary standard that a researcher sets before the study. It defines how strongly the sample evidence must contradict the null hypothesis before you can reject it. The strength of the evidence is defined by the probability of rejecting a true null hypothesis. In practical terms, it is the risk of concluding that an effect exists when there is no real effect.

For instance, a significance level of 0.05 indicates a 5% risk of concluding that an effect exists when it does not, solely due to random sampling variability.

Smaller significance levels require stronger evidence to reject the null hypothesis and, therefore, make statistically significant findings harder to obtain. Larger significance levels make rejection easier but increase the risk of false positives. The choice depends on how costly it would be to falsely conclude that an effect exists, as well as the conventions of the research field.

In practice, researchers generally use a few standard significance levels. A level of 0.05 is the most common and represents a moderate balance between caution and leniency. A level of 0.01 is more conservative and requires stronger evidence to reject the null hypothesis. A level of 0.10 is less common and makes it easier to reject the null, but it increases the risk of false positives.

Statistical Significance

If your p-value is less than or equal to your alpha level, reject the null hypothesis. Your results are statistically significant. This condition indicates your sample data support the alternative hypothesis.

When the p-value is low, the null must go!

For our medication example, the sample effect is statistically significant because the p-value (0.0278) is less than the significance level of 0.05. Our data support the alternative hypothesis, which states that the population mean IQ for those taking the medication doesn't equal 100. Because the sample mean is greater than the null value, we conclude that the population of people who take the medication has a mean IQ greater than 100, or $\mu > 100$.

To estimate the effect size, you can calculate a 95% confidence interval of the mean. Using the method shown in the previous section, the

CI for the IQ medication study is [100.6, 110.4]. The mean IQ of the population taking the medicine is likely within that range.

Confidence intervals and hypothesis tests are closely connected. If a confidence interval does not include the null value, then that value lies in an unlikely region in the sampling distribution. In that case, the hypothesis test will produce a statistically significant p-value. Both methods rely on the same probability model and lead to consistent conclusions.

This agreement occurs when the confidence level matches the significance level. A 95% confidence interval corresponds to a hypothesis test with $\alpha = 0.05$ because both use the same cutoff points in the sampling distribution. In general, a confidence interval with level $(1 - \alpha)$ will agree with a hypothesis test conducted at significance level α.

For the medication example, notice that the 95% CI [100.6, 110.4] excludes the null value of 100, indicating the null is not a plausible value. This CI corresponds to statistically significant results where the p-value indicates you can reject the null hypothesis.

Caution! P-values are very frequently misinterpreted as the probability of rejecting a null hypothesis that is actually true. However, notice that the sampling distribution used to calculate p-values assumes the null hypothesis is true. Consequently, the p-value represents the probability of observing your sample data, or more extreme, given that the null is true.

That was a graphical representation of a 1-sample Z-test. That's a type of test that compares a single sample mean to a hypothesized population mean. This test is particularly easy to calculate by hand. Let's work through the calculations you need to perform to solve this type of problem.

Formulas and Calculations

This type of problem should be familiar to you because we looked at using z-scores and tables in Chapter 13 and assessed sample means in Chapter 14. You'll use methods from both chapters to solve this problem. The difference here is that you need to center the sampling distribution on the null hypothesis value rather than the sample mean.

Appendix A contains the complete z-table that we use in this section.

For starters, standardize the sample mean relative to the null hypothesis value. The formula for that is:

$z = (\bar{x} - \mu_0) / \text{SEM}$

Where:

- $\bar{x}$ is the sample mean.
- μ_0 is the null hypothesis value for the population mean.
- SEM is the standard error of the mean (standard deviation / sqrt(n)).

This process involves subtracting the null hypothesis value from the sample mean. That finds the observed effect. Then divide that by the standard error of the mean to obtain the z-score (i.e., the standardized effect size).

Let's plug our values into the formula. Here's the relevant information from our study.

- Sample size: n = 36
- Sample IQ Mean: $\bar{x}$ = 105.5
- Population SD: σ = 15 IQ Points

From the previous section, we found that the SEM for this example is 2.5.

And z = $(\bar{x} - \mu_0)$ / SEM = (105.5 – 100) / 2.5 = 5.5 / 2.5 = 2.20

Our z-test statistic is 2.20. Now what?

We need to find the two-tailed probability of obtaining that value or more extreme. Look back at Chapter 13 for finding the probability of z-scores being outside of two values. We can use the symmetry shortcut: find the cumulative z-score of -2.20 in the table for the less-than tail, then double it to include the greater-than tail.

Z	0.00	0.01	0.02	0.03	0.04
-2.5	0.00621	0.00604	0.00587	0.00570	0.00554
-2.4	0.00820	0.00798	0.00776	0.00755	0.00734
-2.3	0.01072	0.01044	0.01017	0.00990	0.00964
-2.2	0.01390	0.01355	0.01321	0.01287	0.01255
-2.1	0.01786	0.01743	0.01700	0.01659	0.01618
-2.0	0.02275	0.02222	0.02169	0.02118	0.02068

Recall that for more extreme than probabilities, $p = 2 \times P(Z \leq -|z|)$.

$p = 2 \times P(Z \leq -2.20) = 2 \times 0.0139 = 0.0278$

Using a few simple calculations, our answer matches the statistical software solution provided in the graphical approach.

Beware of Common Misinterpretations

Hypothesis testing provides a structured way to evaluate evidence, but its results are often misunderstood. Keep the following points in mind when interpreting results.

A small p-value does not mean the null hypothesis is probably false. Instead, it indicates that the observed result would be unlikely if the null hypothesis were true. The p-value measures the probability of the data given the null hypothesis, not the probability that the null hypothesis is true.

Statistical significance does not imply that an effect is large or practically important. With sufficiently large samples, even small and unimportant differences can produce statistically significant results. Always consider the size and practical relevance of an effect in addition to its statistical significance.

Failing to reject the null hypothesis does not prove that no effect exists. It simply means that the sample data do not provide strong enough evidence to detect one. The true effect might be small, the sample size might be too limited to detect it, or high variability might obscure it.

Bringing It All Together

Hypothesis testing is a direct application of sampling distributions. By assuming a population parameter value and examining how sample statistics behave under that assumption, we use probability to judge whether our observed data are typical or unusually extreme.

The p-value quantifies this idea. It measures how compatible the observed sample result is with the null hypothesis. Small p-values indicate that the observed result would rarely occur if the null hypothesis were true, providing evidence against it. Larger p-values indicate that the observed result is consistent with typical sampling variability when the null is true.

Confidence intervals and hypothesis tests provide two complementary ways to use the same probability model. Confidence intervals start with the sample statistic and estimate a range of plausible population values. Hypothesis tests start with a proposed population parameter and evaluate how plausible the sample result is under that assumption. Both approaches rely on the sampling distribution and lead to consistent conclusions.

In practice, statistical inference always involves uncertainty because we work with samples rather than entire populations. Probability gives us the tools to measure that uncertainty and make informed conclusions about population parameters.

These conclusions depend on satisfying the assumptions behind the method. Random sampling, independent observations, appropriate probability models, and reliable measurements all matter. Statistical inference accounts for uncertainty due to random sampling, but it can't correct for violations of these assumptions.

More advanced tests extend these ideas to multiple samples, different data types, and more complex research designs. Those methods build on the same probability foundation introduced here.

Review and Next Steps

In this chapter, you completed the transition from probability theory to statistical inference. You saw how sampling distributions allow us to use sample data to make conclusions about population parameters. Confidence intervals use probability to estimate ranges of plausible values, while hypothesis tests use the same probability models to evaluate specific claims. Both approaches rely on the variability inherent in random sampling and the structure of sampling distributions.

You also learned how margins of error and confidence intervals express precision, how sample size and variability affect uncertainty, and how p-values measure the compatibility of observed data with an assumed population value. Throughout, probability provided the framework for distinguishing typical sampling variation from evidence of real population effects.

This book has focused on the foundations that make statistical reasoning possible. Many important areas build on these ideas. Further study in probability includes advanced probability distributions, simulation methods, and Bayesian probability. These topics expand the tools for modeling randomness in increasingly complex settings.

Statistical methods also extend far beyond the introductory techniques presented here. My book *Hypothesis Testing: An Intuitive Guide* develops formal testing procedures in greater depth, including additional test types, error rates, and study design considerations. For modeling relationships among multiple variables, my book *Regression Analysis: An Intuitive Guide* explores methods for explaining and predicting outcomes using advanced statistical models.

If you enjoyed learning how to make decisions using data and probability, consider my book *Thinking Analytically*. In it, developing a probability mindset forms the second level of my five-level analytical thinking pyramid. The other levels explore additional ways to reason more effectively and make better decisions using data.

The probability principles you've learned here form the foundation for all these areas. Whether you continue into theoretical probability, applied statistics, or data science, the core ideas of sampling variability and probabilistic reasoning remain essential.

Formula Quick Reference

Standard Error of the Mean (SEM)	σ / sqrt(n)
Standard Error of a Proportion	SE($\hat{p}$) = sqrt($\hat{p}$ (1 − $\hat{p}$) / n)
Margin of Error (General Form)	MOE = Critical Value × Standard Error
Margin of Error for a Proportion	MOE = z* × sqrt($\hat{p}$ (1 − $\hat{p}$) / n)
Confidence Interval for a Mean	$\bar{x}$ ± z* × SEM
z-Test Statistic (1-Sample Z)	z = ($\bar{x}$ − μ_0) / SEM
Two-Tailed p-Value (z-Test)	p = 2 × P(Z ≤ −|z|)

Practice Problems Chapter 15

Question 1: A researcher tests whether a new study method affects exam scores. The p-value is 0.032. Which interpretation is correct?

A. There is a 3.2% probability that the null hypothesis is true.
B. There is a 3.2% probability of rejecting a true null hypothesis.
C. There is a 3.2% chance of observing results at least this extreme if the null is true.
D. There is a 96.8% probability that the alternative hypothesis is true.

Question 2: A study finds that a new fertilizer increases crop yield by 0.5%, and the result is statistically significant (p = 0.01). What is the correct interpretation?

A. The fertilizer produces a meaningful increase in yield.
B. The fertilizer has no effect because the increase is small.
C. The increase is unlikely to be due to random sampling variability alone.
D. The fertilizer will increase yield by 0.5% in all future samples.

Question 3: A researcher tests whether a training program improves reaction time. The p-value is 0.12, using $\alpha = 0.05$. What is the correct conclusion?

A. Accept the null hypothesis.
B. Reject the null hypothesis.
C. Fail to reject the null hypothesis.
D. The training program has no effect.

Question 4: A researcher studies whether a new diet changes average cholesterol levels. The null hypothesis is $\mu = 200$. A 99% confidence interval for the mean is: [201.5, 215.0]. Using $\alpha = 0.01$, are the results statistically significant? Explain why or why not.

Question 5: A researcher increases the sample size while keeping all other factors the same. What happens?

- **A.** The standard error increases.
- **B.** The margin of error increases.
- **C.** The standard error decreases.
- **D.** The population mean changes.

Queston 6: What does the margin of error represent?

- **A.** The probability that the estimate is incorrect.
- **B.** The typical amount the sample estimate differs from the population parameter due to random sampling variability.
- **C.** The total error in the study.
- **D.** The range of values in the population.

Question 7: Why is probability necessary for statistical inference?

- **A.** Because population parameters are random.
- **B.** Because sample statistics are random.
- **C.** Because all data are normally distributed.
- **D.** Because larger samples eliminate uncertainty.

Question 8: What does a significance level of 0.01 represent?

- **A.** A 1% probability that the null hypothesis is true.
- **B.** A 1% risk of rejecting a true null hypothesis due to random sampling variability.
- **C.** A 99% probability that the alternative hypothesis is true.
- **D.** The probability of obtaining the observed sample mean.

Queston 9: A researcher changes the significance level from 0.05 to 0.01. Which statement is correct?

- **A.** Less evidence is required to reject the null hypothesis, and the risk of false positives increases.
- **B.** More evidence is required to reject the null hypothesis, and the risk of false positives decreases.
- **C.** More evidence is required to reject the null hypothesis, and the risk of false positives increases.
- **D.** The amount of required evidence does not change.

Question 10: Consider a normal distribution with a mean of 50 and a standard deviation of 10.

What is the relative probability of observing a value less than 38 compared to a value greater than 62?

- **A.** Greater than.
- **B.** Less than.
- **C.** Equal to.
- **D.** Cannot be determined.

Question 11: A city surveys 1,200 randomly selected residents about whether they support adding more protected bike lanes downtown. The survey finds that 62% support the proposal. Use a 95% confidence level.

- **A.** Calculate the margin of error for this survey percentage.
- **B.** Interpret the margin of error in context.

Question 12: A researcher studies whether a new alert system improves driver reaction scores. The standard benchmark score is 50. After using the new system, a random sample of 25 drivers has a mean reaction score of 57, with a standard deviation of 20. Assume the sampling distribution of the sample mean follows an approximately normal distribution.

The sample mean appears higher than the benchmark, but the researcher wants to know whether that apparent improvement is strong enough to rule out random sampling variability. Use a 95% confidence level and a two-sided hypothesis test with $\alpha = 0.05$.

- **A.** Calculate the 95% confidence interval for the population mean.
- **B.** Calculate the two-sided p-value.
- **C.** Interpret both results.

Do the confidence interval and hypothesis test agree?

Appendix A: Cumulative Standard Normal (z) Table

A cumulative z-table, also known as the standard normal table, provides the area under the curve to the left of a z-score, or $P(Z \leq z)$. This area represents the cumulative probability associated with a z-score in the standard normal distribution. Use the z-table to find probabilities corresponding to ranges of z-scores, find z-scores associated with probabilities, and calculate confidence intervals and p-values for z-tests.

The z-table is divided into two sections: negative and positive z-scores. Negative z-scores are below the mean and positive z-scores are above the mean. The row and column headers define the z-score, while the table cells contain the corresponding cumulative probabilities.

The table begins on the next page. The first two pages contain negative z-scores, and the second two pages contain positive z-scores.

Z	0.00	0.01	0.02	0.03	0.04	0.05	0.06	0.07	0.08	0.09
-3.9	0.00005	0.00005	0.00004	0.00004	0.00004	0.00004	0.00004	0.00004	0.00003	0.00003
-3.8	0.00007	0.00007	0.00007	0.00006	0.00006	0.00006	0.00006	0.00005	0.00005	0.00005
-3.7	0.00011	0.00010	0.00010	0.00010	0.00009	0.00009	0.00008	0.00008	0.00008	0.00008
-3.6	0.00016	0.00015	0.00015	0.00014	0.00014	0.00013	0.00013	0.00012	0.00012	0.00011
-3.5	0.00023	0.00022	0.00022	0.00021	0.00020	0.00019	0.00019	0.00018	0.00017	0.00017
-3.4	0.00034	0.00032	0.00031	0.00030	0.00029	0.00028	0.00027	0.00026	0.00025	0.00024
-3.3	0.00048	0.00047	0.00045	0.00043	0.00042	0.00040	0.00039	0.00038	0.00036	0.00035
-3.2	0.00069	0.00066	0.00064	0.00062	0.00060	0.00058	0.00056	0.00054	0.00052	0.00050
-3.1	0.00097	0.00094	0.00090	0.00087	0.00084	0.00082	0.00079	0.00076	0.00074	0.00071
-3.0	0.00135	0.00131	0.00126	0.00122	0.00118	0.00114	0.00111	0.00107	0.00104	0.00100
-2.9	0.00187	0.00181	0.00175	0.00169	0.00164	0.00159	0.00154	0.00149	0.00144	0.00139
-2.8	0.00256	0.00248	0.00240	0.00233	0.00226	0.00219	0.00212	0.00205	0.00199	0.00193
-2.7	0.00347	0.00336	0.00326	0.00317	0.00307	0.00298	0.00289	0.00280	0.00272	0.00264
-2.6	0.00466	0.00453	0.00440	0.00427	0.00415	0.00402	0.00391	0.00379	0.00368	0.00357
-2.5	0.00621	0.00604	0.00587	0.00570	0.00554	0.00539	0.00523	0.00508	0.00494	0.00480
-2.4	0.00820	0.00798	0.00776	0.00755	0.00734	0.00714	0.00695	0.00676	0.00657	0.00639
-2.3	0.01072	0.01044	0.01017	0.00990	0.00964	0.00939	0.00914	0.00889	0.00866	0.00842
-2.2	0.01390	0.01355	0.01321	0.01287	0.01255	0.01222	0.01191	0.01160	0.01130	0.01101
-2.1	0.01786	0.01743	0.01700	0.01659	0.01618	0.01578	0.01539	0.01500	0.01463	0.01426
-2.0	0.02275	0.02222	0.02169	0.02118	0.02068	0.02018	0.01970	0.01923	0.01876	0.01831

Z	0.00	0.01	0.02	0.03	0.04	0.05	0.06	0.07	0.08	0.09
-1.9	0.02872	0.02807	0.02743	0.02680	0.02619	0.02559	0.02500	0.02442	0.02385	0.02330
-1.8	0.03593	0.03515	0.03438	0.03362	0.03288	0.03216	0.03144	0.03074	0.03005	0.02938
-1.7	0.04457	0.04363	0.04272	0.04182	0.04093	0.04006	0.03920	0.03836	0.03754	0.03673
-1.6	0.05480	0.05370	0.05262	0.05155	0.05050	0.04947	0.04846	0.04746	0.04648	0.04551
-1.5	0.06681	0.06552	0.06426	0.06301	0.06178	0.06057	0.05938	0.05821	0.05705	0.05592
-1.4	0.08076	0.07927	0.07780	0.07636	0.07493	0.07353	0.07215	0.07078	0.06944	0.06811
-1.3	0.09680	0.09510	0.09342	0.09176	0.09012	0.08851	0.08691	0.08534	0.08379	0.08226
-1.2	0.11507	0.11314	0.11123	0.10935	0.10749	0.10565	0.10383	0.10204	0.10027	0.09853
-1.1	0.13567	0.13350	0.13136	0.12924	0.12714	0.12507	0.12302	0.12100	0.11900	0.11702
-1.0	0.15866	0.15625	0.15386	0.15151	0.14917	0.14686	0.14457	0.14231	0.14007	0.13786
-0.9	0.18406	0.18141	0.17879	0.17619	0.17361	0.17106	0.16853	0.16602	0.16354	0.16109
-0.8	0.21186	0.20897	0.20611	0.20327	0.20045	0.19766	0.19489	0.19215	0.18943	0.18673
-0.7	0.24196	0.23885	0.23576	0.23270	0.22965	0.22663	0.22363	0.22065	0.21770	0.21476
-0.6	0.27425	0.27093	0.26763	0.26435	0.26109	0.25785	0.25463	0.25143	0.24825	0.24510
-0.5	0.30854	0.30503	0.30153	0.29806	0.29460	0.29116	0.28774	0.28434	0.28096	0.27760
-0.4	0.34458	0.34090	0.33724	0.33360	0.32997	0.32636	0.32276	0.31918	0.31561	0.31207
-0.3	0.38209	0.37828	0.37448	0.37070	0.36693	0.36317	0.35942	0.35569	0.35197	0.34827
-0.2	0.42074	0.41683	0.41294	0.40905	0.40517	0.40129	0.39743	0.39358	0.38974	0.38591
-0.1	0.46017	0.45620	0.45224	0.44828	0.44433	0.44038	0.43644	0.43251	0.42858	0.42465
0.0	0.50000	0.49601	0.49202	0.48803	0.48405	0.48006	0.47608	0.47210	0.46812	0.46414

Z	0.00	0.01	0.02	0.03	0.04	0.05	0.06	0.07	0.08	0.09
0.0	0.50000	0.50399	0.50798	0.51197	0.51595	0.51994	0.52392	0.52790	0.53188	0.53586
0.1	0.53983	0.54380	0.54776	0.55172	0.55567	0.55962	0.56356	0.56749	0.57142	0.57535
0.2	0.57926	0.58317	0.58706	0.59095	0.59483	0.59871	0.60257	0.60642	0.61026	0.61409
0.3	0.61791	0.62172	0.62552	0.62930	0.63307	0.63683	0.64058	0.64431	0.64803	0.65173
0.4	0.65542	0.65910	0.66276	0.66640	0.67003	0.67364	0.67724	0.68082	0.68439	0.68793
0.5	0.69146	0.69497	0.69847	0.70194	0.70540	0.70884	0.71226	0.71566	0.71904	0.72240
0.6	0.72575	0.72907	0.73237	0.73565	0.73891	0.74215	0.74537	0.74857	0.75175	0.75490
0.7	0.75804	0.76115	0.76424	0.76730	0.77035	0.77337	0.77637	0.77935	0.78230	0.78524
0.8	0.78814	0.79103	0.79389	0.79673	0.79955	0.80234	0.80511	0.80785	0.81057	0.81327
0.9	0.81594	0.81859	0.82121	0.82381	0.82639	0.82894	0.83147	0.83398	0.83646	0.83891
1.0	0.84134	0.84375	0.84614	0.84849	0.85083	0.85314	0.85543	0.85769	0.85993	0.86214
1.1	0.86433	0.86650	0.86864	0.87076	0.87286	0.87493	0.87698	0.87900	0.88100	0.88298
1.2	0.88493	0.88686	0.88877	0.89065	0.89251	0.89435	0.89617	0.89796	0.89973	0.90147
1.3	0.90320	0.90490	0.90658	0.90824	0.90988	0.91149	0.91309	0.91466	0.91621	0.91774
1.4	0.91924	0.92073	0.92220	0.92364	0.92507	0.92647	0.92785	0.92922	0.93056	0.93189
1.5	0.93319	0.93448	0.93574	0.93699	0.93822	0.93943	0.94062	0.94179	0.94295	0.94408
1.6	0.94520	0.94630	0.94738	0.94845	0.94950	0.95053	0.95154	0.95254	0.95352	0.95449
1.7	0.95543	0.95637	0.95728	0.95818	0.95907	0.95994	0.96080	0.96164	0.96246	0.96327
1.8	0.96407	0.96485	0.96562	0.96638	0.96712	0.96784	0.96856	0.96926	0.96995	0.97062
1.9	0.97128	0.97193	0.97257	0.97320	0.97381	0.97441	0.97500	0.97558	0.97615	0.97670

Z	0.00	0.01	0.02	0.03	0.04	0.05	0.06	0.07	0.08	0.09
2.0	0.97725	0.97778	0.97831	0.97882	0.97932	0.97982	0.98030	0.98077	0.98124	0.98169
2.1	0.98214	0.98257	0.98300	0.98341	0.98382	0.98422	0.98461	0.98500	0.98537	0.98574
2.2	0.98610	0.98645	0.98679	0.98713	0.98745	0.98778	0.98809	0.98840	0.98870	0.98899
2.3	0.98928	0.98956	0.98983	0.99010	0.99036	0.99061	0.99086	0.99111	0.99134	0.99158
2.4	0.99180	0.99202	0.99224	0.99245	0.99266	0.99286	0.99305	0.99324	0.99343	0.99361
2.5	0.99379	0.99396	0.99413	0.99430	0.99446	0.99461	0.99477	0.99492	0.99506	0.99520
2.6	0.99534	0.99547	0.99560	0.99573	0.99585	0.99598	0.99609	0.99621	0.99632	0.99643
2.7	0.99653	0.99664	0.99674	0.99683	0.99693	0.99702	0.99711	0.99720	0.99728	0.99736
2.8	0.99744	0.99752	0.99760	0.99767	0.99774	0.99781	0.99788	0.99795	0.99801	0.99807
2.9	0.99813	0.99819	0.99825	0.99831	0.99836	0.99841	0.99846	0.99851	0.99856	0.99861
3.0	0.99865	0.99869	0.99874	0.99878	0.99882	0.99886	0.99889	0.99893	0.99896	0.99900
3.1	0.99903	0.99906	0.99910	0.99913	0.99916	0.99918	0.99921	0.99924	0.99926	0.99929
3.2	0.99931	0.99934	0.99936	0.99938	0.99940	0.99942	0.99944	0.99946	0.99948	0.99950
3.3	0.99952	0.99953	0.99955	0.99957	0.99958	0.99960	0.99961	0.99962	0.99964	0.99965
3.4	0.99966	0.99968	0.99969	0.99970	0.99971	0.99972	0.99973	0.99974	0.99975	0.99976
3.5	0.99977	0.99978	0.99978	0.99979	0.99980	0.99981	0.99981	0.99982	0.99983	0.99983
3.6	0.99984	0.99985	0.99985	0.99986	0.99986	0.99987	0.99987	0.99988	0.99988	0.99989
3.7	0.99989	0.99990	0.99990	0.99990	0.99991	0.99991	0.99992	0.99992	0.99992	0.99992
3.8	0.99993	0.99993	0.99993	0.99994	0.99994	0.99994	0.99994	0.99995	0.99995	0.99995
3.9	0.99995	0.99995	0.99996	0.99996	0.99996	0.99996	0.99996	0.99996	0.99997	0.99997

My Other Books

Introduction to Statistics: An Intuitive Guide

Learn statistics without fear! Build a solid foundation in data analysis. Be confident that you understand what your data are telling you and that you can explain the results to others! I'll help you intuitively understand statistics by using simple language and deemphasizing formulas.

This guide starts with an overview of statistics and why it is so important. We proceed to essential statistical skills and knowledge about different types of data, relationships, and distributions. Then we move to using inferential statistics to expand human knowledge, how it fits into the scientific method, and how to design and critique experiments—whether it's your own or another researcher's.

Learn the fundamentals of statistics in this 255-page book:

- Why is the field of statistics vital in our data-driven society?
- Interpret graphs and summary statistics.
- Find relationships between different types of variables.
- Understand the properties of data distributions.
- Use measures of central tendency and variability.
- Interpret correlations and percentiles.
- Use probability distributions to calculate probabilities.
- Learn about the normal and binomial distributions in depth.
- Grasp the differences between descriptive and inferential statistics.
- Use data collection methodologies properly and understand sample size considerations.
- Access free downloadable datasets so you can try it yourself.

Currently available in print and as an ebook at most retailers!
Learn more on my website: statisticsbyjim.com/store

Hypothesis Testing: An Intuitive Guide

Build a solid foundation for understanding how hypothesis tests work and become confident that you know when to use each type of test, how to use them properly to obtain reliable results, and interpret the results correctly. Chances are high that you'll need a working knowledge of hypothesis testing to produce new findings yourself and to understand the work of others. I present a wide variety of tests that assess characteristics of different data types. I focus on helping you grasp key concepts, methodologies, and procedures while deemphasizing equations. Learn how to use these tests painlessly!

In today's data-driven world, we hear about making decisions based on the data all the time. Hypothesis testing plays a crucial role in that process, whether you're in academia, making business decisions, or in quality improvement. Without hypothesis tests, you risk drawing the wrong conclusions and making bad decisions. The world today produces more data and more analyses designed to influence you than ever before. Are you ready for it?

In this 367-page book, build the skills and knowledge you'll need for effective hypothesis testing, including the following:

- Why you need hypothesis tests and how they work.
- Using significance levels, p-values, confidence intervals.
- Interpreting the results.
- Select the correct type of hypothesis test to answer your question.
- Learn how to test means, medians, variances, proportions, distributions, counts, correlations for continuous and categorical data, and outliers.
- One-Way ANOVA, Two-Way ANOVA and interaction effects.
- Estimate a good sample size for your study.
- Checking assumptions and obtaining reliable results.

- Manage the error rates for false positives and false negatives.
- Understand sampling distributions, central limit theorem, and statistical power.
- Know how t-tests, F-tests, chi-squared tests, and post hoc tests work.
- Learn about the differences between parametric, nonparametric, and bootstrapping methods.
- Examples of different types of hypothesis tests.
- Downloadable datasets so you can try it yourself.

Currently available in print and as an ebook at most retailers!

Free samples & learn more on my website: statisticsbyjim.com/store

Regression Analysis: An Intuitive Guide

Over the course of this full-length book, you'll progress from a beginner to a skilled practitioner. I'll help you intuitively understand regression analysis by focusing on concepts and graphs rather than equations and formulas. I use everyday language so you can grasp regression at a deeper level.

Learn practical tips for performing your analysis and interpreting the results. Feel confident that you're analyzing your data properly and able to trust your results. Know that you can detect and correct problems that arise.

This 336-page book covers the following:

- How regression works and when to use it.
- Selecting the correct type of regression analysis.
- Specifying the best model.
- Understanding main effects, interaction effects, and modeling curvature.
- Interpreting the results.
- Assessing the fit of the model.
- Generating predictions and evaluating their precision.
- Checking the assumptions and resolving issues.
- Examples of different types of regression analyses.
- Downloadable datasets for the examples.

Currently available in print and as an ebook at most retailers.
See all my books on my Amazon Author Page!
https://www.amazon.com/author/jim-frost

Thinking Analytically: A Guide for Making Data-Driven Decisions

Transform how you think about problems, prepare to outsmart misinformation, and avoid common pitfalls by developing an analytical mindset that cuts through the noise.

In an age where data and statistical claims surround us, the ability to critique them is essential. *Thinking Analytically* guides you through the complexities of real-world data. Offering more than just interpretation, it focuses on how to think about and solve problems effectively.

Thinking Analytically is for anyone who wants to be an informed consumer of information. Decision-makers, data analysts, scientists, and interested readers will find its insights valuable. The book offers practical techniques and examples for critiquing findings, understanding the processes behind data, and unraveling problems.

This book introduces the *Thinking Analytically* pyramid, covering:

1. **Cognitive Foundation**: Understand how cognitive limitations impact data interpretation.
2. **Probability Mindset**: Improve probabilistic thinking and avoid common errors.
3. **Data Quality**: Learn how subtle problems can distort the results.
4. **Experimental Design**: Explore how context affects the conclusions you can draw.
5. **Analytics**: Produce sound results by integrating all pyramid levels.

You'll develop a critical eye for evaluating data and the skills to ask the right questions. Whether you're making business decisions, interpreting studies, or navigating everyday information, this book improves your analytical understanding.

References

Beran, J., Vesikari, T., Wertzova, V., Karvonen, A., Honegr, K., Lindblad, N., . . . Devaster, J. (2009). Efficacy of inactivated split-virus influenza vaccine against culture-confirmed influenza in healthy adults: a prospective, randomized, placebo controlled trial. *J Infect Dis*, 1861-9.

Bridges, C., Thompson, W., Meltzer, M., Reeve, G., Talamonti, W., Cox, N., . . . Fukuda, K. (2000). Effectiveness and cost-benefit of influenza vaccination of healthy working adults: A randomized controlled trial. *JAMA*, 1655-63.

Croson, R., & Sundali, J. (2005). The gambler's fallacy and the hot hand: Empirical data from casinos. *Journal of Risk and Uncertainty*, 195-209.

Goode, B., O'Reilly, C., & al, e. (2009). Outbreak of Escherichia coli O157:H7 Infections After Petting Zoo Visits, North Carolina State Fair. *Arch Pediatr Adolesc Med*, 42-48.

Halladay, J., Yankaskas, B., & Bowling, J. (2010). Positive Predictive Value of Mammography: Comparison of Interpretations of Screening and Diagnostic Images by the Same Radiologist and by Different Radiologists. *American Journal of Roentgenology*, 782-785.

Kahneman, D. (2011). *Thinking, Fast and Slow.* Straus and Girouz.

Mello, J., Bittelbrunn, F., Rockenbach, M., & al., e. (2017). Breast cancer mammographic diagnosis performance in a public health institution: a retrospective cohort study. *Insights into Imaging*, 581-588.

Monto, A., Ohmit, S., Petrie, J., Johnson, E., Truscon, R., Teich, E., . . . Victor, J. (2009). Comparative efficacy of inactivated and live-attenuated influenza vaccines. *N Engl J Med*, 1260-7.

Tversky, A., & Kahneman, D. (1973). Availability: A heuristic for judging frequency and probability. *Cognitive Psychology*, 207-232.

Tversky, A., & Kahneman, D. (1973). On the psychology of prediction. *Psychological Review*, 237-257.

Tversky, A., & Kahneman, D. (1981). The framing of decisions and the psychology of choice. *Science*, 453-458.

Tversky, A., & Kahneman, D. (1983). Extension versus intuitive reasoning: The conjunction fallacy in probability judgment. *Psychological Review*, 293-315.

Recommended Citation for This Book

Frost, J. (2026). Probability Fundamentals: An Intuitive Guide to Understanding Uncertainty. Statistics By Jim Publishing.

Index

About the Author

I'm Jim Frost, and I have extensive experience in academic research and consulting projects. In addition to my statistics website, I am a regular columnist for the American Society of Quality's *Statistics Digest*. Additionally, my most recent journal publication as a coauthor is *The Neutral Gas Properties of Extremely Isolated Early-Type Galaxies III* (2019) for the American Astronomical Society.

I've been the "data/stat guy" for research projects that range from osteoporosis prevention to analysis of online user behavior. My role has been to design the proper research settings, collect a large amount of valid measurements, and figure out what it all means. Typically, I'm the first person on the project to learn about new findings while interpreting the results of the statistical analysis. Even if the findings are not newsworthy, that thrill of discovery is an awesome job perk!

I love statistics and analyzing data! I've been performing statistical analysis on-the-job for 20 years and helping people learn statistics for over ten years at a statistical software company. I love talking and writing about statistics. While working at the statistical software company, I learned how to present statistics in a manner that makes statistics more intuitive.

I want to help you learn statistics. But I'm not talking about learning all the equations. Don't get me wrong. Equations are necessary. Equations are the framework that makes the magic, but the truly fascinating aspects are what it all means. I want you to learn the true essence of statistics. I'll help you intuitively understand statistics by focusing on concepts and graphs. After all, you use statistical software so you don't have to worry about the formulas and instead focus on understanding the results.

I've spent over a decade working at a major statistical software company. When you work on research projects, you generally use a regular group of statistical analyses. However, when you work at a statistical software company, you need to know of all the analyses that are in the software! I helped people use our software to gain insights and maximize the value of their own data regardless of their field.

Statistics is the field of learning from data. That's amazing. It gets to the very essence of discovery. Statistics facilitates the creation of new knowledge. Bit by bit, we push back the frontier of what is known. That is what I want to teach you! My goal is to help you to see statistics through my eyes–as a key that can unlock discoveries that are in your data.

The best thing about being a statistician is that you get to play in everyone's backyard. —John Tukey

I enthusiastically agree! If you have an inquisitive mind, statistical knowledge, and data, the potential is boundless. You can play in a broad range of intriguing backyards!

That interface between a muddled reality and obtaining orderly, valid data is an exciting place. This place ties together the lofty goals of scientists to the nitty-gritty nature of the real world. It's an interaction that I've written about extensively in this book and on my blog, and I plan to continue to do so. It's where the rubber meets the road.

One of the coolest things about the statistical analysis is that it provides you with a toolkit for exploring the unknown. Christopher Columbus needed many tools to navigate to the New World and make his discoveries. Statistics are the equivalent tools for the scientific explorer because they help you navigate the sea of data that you collect.

You'll be increasingly thankful for these tools when you see a worksheet filled with numbers and you're responsible for telling everyone what it all means.

Read more on my website: StatisticsByJim.com

www.ingramcontent.com/pod-product-compliance
Lightning Source LLC
LaVergne TN
LVHW052210200926
848481LV00001B/72
9798991193597